Acidic Proteins of the Nucleus

This is a volume in
CELL BIOLOGY
A series of monographs
Editors: *D. E. Buetow, I. L. Cameron, and G. M. Padilla*
A complete list of the books in this series appears at the end of the volume.

Acidic Proteins of the Nucleus

Edited by

IVAN L. CAMERON

Department of Anatomy
University of Texas Health Science Center
 at San Antonio
San Antonio, Texas

JAMES R. JETER, Jr.

McArdle Laboratory for Cancer Research
University of Wisconsin Medical Center
Madison, Wisconsin

ACADEMIC PRESS New York San Francisco London 1974
A Subsidiary of Harcourt Brace Jovanovich, Publishers

COPYRIGHT © 1974, BY ACADEMIC PRESS, INC.
ALL RIGHTS RESERVED.
NO PART OF THIS PUBLICATION MAY BE REPRODUCED OR
TRANSMITTED IN ANY FORM OR BY ANY MEANS, ELECTRONIC
OR MECHANICAL, INCLUDING PHOTOCOPY, RECORDING, OR ANY
INFORMATION STORAGE AND RETRIEVAL SYSTEM, WITHOUT
PERMISSION IN WRITING FROM THE PUBLISHER.

ACADEMIC PRESS, INC.
111 Fifth Avenue, New York, New York 10003

United Kingdom Edition published by
ACADEMIC PRESS, INC. (LONDON) LTD.
24/28 Oval Road, London NW1

Library of Congress Cataloging in Publication Data

Cameron, Ivan L
        Acidic proteins of the nucleus.
        (Cell biology series)
        Includes bibliographies.
        1. Nucleoproteins.   2.   Cell nuclei.   I.   Jeter,
James R., joint author.   II.   Title.   [DNLM: 1.   Cell
nucleus.   2.   Cytology.   3.   Proteins.   QH595 A181]
QH595.C32   574.8'732   74-5691
ISBN 0–12–156930–6

PRINTED IN THE UNITED STATES OF AMERICA

Contents

List of Contributors ix
Preface xi

1 DNA-BINDING PROTEINS AND TRANSCRIPTIONAL CONTROL IN PROKARYOTIC AND EUKARYOTIC SYSTEMS
Vincent G. Allfrey

    I. Introduction 2
   II. Transcriptional Control in Prokaryotes—A Model for Higher Organisms 3
  III. Transcriptional Control by Modification of RNA Polymerases .. 12
  IV. Nuclear Nonhistone Proteins and Transcriptional Control in Eukaryotes 13
      References 21

2 ISOLATION OF THE NUCLEAR ACIDIC PROTEINS, THEIR FRACTIONATION, AND SOME GENERAL CHARACTERISTICS
Gordhan L. Patel

    I. Introduction 30
   II. Isolation of Cell Nuclei and Subnuclear Components 32
  III. Acidic Proteins of the Chromatin 35
  IV. Acidic Proteins of the Nucleolus 47
   V. Acidic Proteins of the Nucleoplasm 50
  VI. Concluding Remarks 52
      References 53

3 EXTRACTION AND CHARACTERIZATION OF THE PHENOL-SOLUBLE ACIDIC NUCLEAR PROTEINS
Wallace M. LeStourgeon and Wayne Wray

    I. Introduction 60
   II. Development of Procedure Using Aqueous Phenol to Dissociate and Solubilize the Nuclear Acidic Proteins 61

v

III. Chemical Characteristics and Biochemical Activity of the Phenol-Soluble Residual Acidic Proteins 66
IV. Fractionation of the Nuclear Proteins 91
V. Preparation of Protein for Electrophoretic Separation 96
VI. Electrophoretic Separation of the Phenol-Soluble Proteins 98
VII. Concluding Remarks 99
References 100

4 ACIDIC NUCLEAR PHOSPHOPROTEINS
Lewis J. Kleinsmith

I. Introduction 103
II. Isolation and Fractionation of Acidic Nuclear Phosphoproteins 104
III. Phosphate Metabolism of Acidic Nuclear Phosphoproteins 111
IV. Functional Properties of Acidic Nuclear Phosphoproteins 119
V. Role of Acidic Phosphoproteins in Nuclear Function 129
VI. Concluding Remarks 132
References 133

5 CHARACTERIZATION OF NUCLEAR PHOSPHOPROTEINS IN *Physarum polycephalum*
Bruce E. Magun

I. Introduction 137
II. Phosphate Content of Phosphoproteins 139
III. Pulse Labeling in $^{32}P_i$ 145
IV. Labeling of Nuclei *in Vitro* with [γ-^{32}P]ATP 147
V. Kinetics of Phosphate Turnover 151
VI. Phosphorylation during Starvation 154
VII. Summary 157
References 158

6 THE NUCLEAR ACIDIC PROTEINS IN CELL PROLIFERATION AND DIFFERENTIATION
Wallace M. LeStourgeon, Roger Totten, and Arthur Forer

I. Introduction 159
II. The Heterogeneity of the Nuclear Acidic Proteins and Cell Proliferation 161
III. The Major Acidic Nuclear Proteins and Specific Gene Regulation 171
IV. The Contractile Proteins of Isolated Chromatin and Considerations of Their Possible Role in the Regulation of Cell Proliferation 174
References 187

7 NONHISTONE PROTEINS OF DIPTERAN POLYTENE NUCLEI
H. D. Berendes and P. J. Helmsing

I. Introduction	191
II. Cytology and Cytochemistry of Genome Activity	192
III. Chemical Modification of Polytene Chromosome Proteins	199
IV. Qualitative and Quantitative Changes in Protein during Gene Activation	201
V. The Possible Role of Proteins in Gene Activation	206
References	209

8 ACIDIC NUCLEAR PROTEINS AND THE CELL CYCLE
James R. Jeter, Jr., and Ivan L. Cameron

I. Introduction	213
II. Synchronous Cell Cycle Systems, a Comparative Analysis	215
III. Acidic Proteins of the Nucleus in Relation to the Cell Cycle	221
IV. Summary and Conclusions	240
References	242

9 THE ROLE OF NUCLEAR ACIDIC PROTEINS IN BINDING STEROID HORMONES
Thomas C. Spelsberg

I. Introduction	248
II. Primary Site of Action of Steroid Hormones	249
III. Cytoplasmic Acidic Proteins as Receptors for Steroid Hormones	250
IV. Intracellular Distribution of Steroid Hormones	252
V. Intranuclear Localization of Steroid Hormones: Chromatin Binding	262
VI. Chromatin-Binding Sites for Hormone–Receptor Complexes: "The Acceptor Molecules"	263
VII. Identification and Characterization of the "Acceptor" Molecule in Chick Oviduct Chromatin Which Binds the Progesterone–Receptor Complex	268
References	290

10 THE ROLE OF ACIDIC PROTEINS IN GENE REGULATION
R. Stewart Gilmour

I. Introduction	297
II. Isolation and Characterization of Acidic Proteins	298
III. Distribution and Specificity of the Acidic Proteins	300

IV.	Some Metabolic Aspects of Acidic Proteins	301
V.	The Biological Assessment of the Acidic Proteins	303
VI.	Gene Regulation in Eukaryotes	307
	References	313

Author Index 319

Subject Index 334

List of Contributors

Numbers in parentheses indicate the pages on which the authors' contributions begin.

Vincent G. Allfrey (1), The Rockefeller University, New York, New York

H. D. Berendes (191), Department of Genetics University of Nijmegen, The Netherlands

Ivan L. Cameron (213), Department of Anatomy, University of Texas Health Science Center at San Antonio, San Antonio, Texas

Arthur Forer (159), Biology Department, York University, Downsview, Ontario, Canada

R. Stewart Gilmour (297), The Beatson Institute for Cancer Research, Glasgow, Scotland

P. J. Helmsing (191), Department of Genetics, University of Nijmegen, The Netherlands

James R. Jeter, Jr. (213), McArdle Laboratory for Cancer Research, University of Wisconsin Medical Center, Madison, Wisconsin

Lewis J. Kleinsmith (103), Department of Zoology, The University of Michigan, Ann Arbor, Michigan

Wallace M. LeStourgeon (59, 159), Department of Molecular Biology, Vanderbilt University, Nashville, Tennessee

Bruce E. Magun (137), Department of Anatomy, University of Tennessee Medical Unit, Memphis, Tennessee

Gordhan L. Patel (29), Department of Zoology, University of Georgia, Athens, Georgia

Thomas C. Spelsberg (247), Department of Endocrine Research, Mayo Clinic, Rochester, Minnesota

Roger Totten (159), Department of Plant Pathology, Montana State College, Bozeman, Montana

Wayne Wray (59), Department of Cell Biology, Baylor College of Medicine, Texas Medical Center, Houston, Texas

Preface

The nucleus is the central repository of genetic information within the cell, and this genetic information is encoded in a sequence of deoxyribonucleotides. Each somatic cell nucleus of an organism contains the same complete compliment of genetic material except in a few clearly defined cases where some specific chromosomes are discarded during somatic cell differentiation, as occurs in the gall midge and in ascaris. It has become apparent that only a small part of this total genetic information is expressed in any given cell at any given time. How is the transcription of this genetic information controlled? Many feel this to be the most exciting question in modern biology. This question has led many workers to suggest that the nuclear proteins are the regulators of genetic expression in cells.

During the early 1940's Edgar and Ellen Stedman at Edinburgh revived interest in the nucleoproteins, particularly the histones, as possible repressors of specific chromosomal genes. Information concerning the histones is nicely summarized in a series of monographs (A. Kossal, "The Portamines and Histones," Longmans, Green and Co., London, 1928; J. Bonner and P.O.P. Ts'o, eds., "The Nucleohistones," Holden-Day, San Francisco, 1964; H. Busch, "Histones and Other Nuclear Proteins," Academic Press, New York, 1965; D. M. P. Phillips, ed., "Histones and Nucleohistones," Plenum, London, 1971; and L. S. Hnilica, "The Structure and Functions of Histones," The Chemical Rubber Co., Cleveland, 1972). Despite a wealth of information on the chemistry and structure of histones it has been disappointing to find that these proteins have only a limited and nonspecific role in the regulation of genetic expression. It also seems clear that this group of proteins has limited heterogeneity and has remained virtually unchanged during the evolution of eukaryotic cells. Consequently, interest in the histones as specific regulators of genetic expression has waned. More recently interest has turned to the heterogeneous group of nucleoproteins which the

Stedmans called "chromosomin." Recent workers refer to these heterogeneous proteins as the nonhistone or the acidic nuclear proteins, and it is felt by many that some of the proteins in this group act as regulators of the expression of specific genes. It is now clear that many of these proteins also play structural, enzymatic, mobility, receptor, and probably other roles in the nucleus.

To strongly distinguish this group of nuclear proteins from the histones we have chosen to refer to them as acidic nuclear proteins rather than nonhistone nuclear proteins even though we feel, as do most of the workers in the field, that these terms are equitable. Historically these proteins were referred to as acidic in nature because they were insoluble in dilute mineral acids and their amino acid composition showed a preponderance of acidic over basic amino acid residues. However, isoelectric foscusing data have shown that a few of these proteins are actually basic rather than acidic in nature, therefore it is probably more correct to refer to the whole group as nonhistone rather than acidic proteins.

Because of the large and growing interest in the acidic nuclear proteins we decided that it would be worthwhile to bring together into one monograph the diverse and scattered information on this topic. Although it seems of considerable value to collate information in a rapidly growing field, we are only too aware that this monograph is doomed to obsolescence. Nevertheless, we are hopeful that the book will stand as a notable historical landmark in the field.

This work gives a broad account of much of what is presently known of the acidic nuclear proteins. Although some of the chapters deal with various approaches for isolating, separating, and characterizing these proteins, much of the book is directed toward elucidating the functional role that these nuclear proteins may play in differential gene expression. At present a variety of techniques is used by many of the laboratories investigating these proteins. Taking into account the way the histone story developed we feel that firm conclusions about the role of the acidic nuclear proteins will not be possible until there are accepted standard high-resolution techniques for the study of these proteins.

The contributors to this book are all active and eminent researchers in their field. Each chapter contains some previously unpublished work and points the way to new and promising areas

Preface   xiii

for future research. We feel that this monograph will serve as a good reference background to the acidic nuclear proteins and will stimulate researchers to unravel the exciting questions concerning the role of these proteins.

Ivan L. Cameron
James R. Jeter, Jr.

Acidic Proteins
of the Nucleus

DNA-Binding Proteins and Transcriptional Control in Prokaryotic and Eukaryotic Systems

VINCENT G. ALLFREY

I. Introduction	2
A. The Problem of Differential Transcription	2
B. Involvement of DNA-Associated Proteins	2
C. Histones and Chromatin Structure	2
II. Transcriptional Control in Prokaryotes—A Model for Higher Organisms	3
A. The *lac* Repressor	4
B. Repression of the *hut* Operons in *Salmonella*	6
C. The Lambda Repressor	7
D. Rho Factor—Geometry of the Functional Oligomers	8
E. Positive Transcriptional Control by DNA-Binding Proteins	9
F. Low Molecular Weight RNA Initiation Proteins in *E. coli*	11
G. DNA-Unwinding Proteins	11
III. Transcriptional Control by Modification of RNA Polymerases	12
IV. Nuclear Nonhistone Proteins and Transcriptional Control in Eukaryotes	13
A. Correlations with Gene Activity	13
B. Specificity in DNA-Binding by Nuclear Nonhistone Proteins	15
C. Species-Selective DNA Binding	16
D. Sequence-Specific DNA Binding	17
E. Transcription of Unique and Repetitive DNA Sequences	20
F. Protein Phosphorylation and Gene Activation	21
References	21

I. INTRODUCTION

A. The Problem of Differential Transcription

The process of cell differentiation, leading to the diversity of specialized cells in higher organisms, involves a selective transcription of different regions of the total chromosomal DNA to produce the RNA populations characteristic of different cell types. This selective, but partial utilization of the genome occurs despite the presence in most diploid somatic cells of a complete array of chromosomes and a DNA complement sufficient to specify the formation of an entire organism (1).

It is now clear that individual cells suppress the transcription of most of their DNA (2) while they selectively activate a relatively small number of genes necessary for the synthesis, assembly, processing, and postsynthetic modification of enzymes and structural proteins characteristic of the cell type and the species. This is evident in the limited nature of their RNA transcripts (2–4). In the course of embryonic development and aging, different sets of genes are activated or repressed in response to programmed signals from the nucleus, the cytoplasm, the cell membrane, and the environment.

B. Involvement of DNA-Associated Proteins

The two aspects of transcriptional control—suppression of the template activity of most of the DNA and activation of RNA synthesis at particular genetic loci—require the participation of proteins associated with the chromatin complex. These proteins influence the structure of the genetic material, strengthen or weaken its interactions with RNA polymerases, and transmit physiological control signals for gene activation or repression in response to hormones, cyclic nucleotides, and other types of stimuli.

The main protein fractions concerned with structure and function of chromatin include the histones—the basic suppressive components of chromatin—and the more acidic nuclear proteins involved in positive and selective aspects of genetic control.

C. Histones and Chromatin Structure

Speculations on the regulatory role of histones date back to Edgar and Ellen Stedman who, in 1950, proposed that "the basic proteins of cell nuclei are gene inhibitors, each histone or protamine being capable of suppressing the activities of certain groups of genes" (5). This surmise, years ahead of its time, was the first suggestion that gene function could be regulated by DNA-associated proteins. Its first experimental verifications came twelve years later in studies of histone inhibition of RNA synthesis in isolated cell nuclei (6) and in chromatin fractions (7). Further developments, however, made it clear that the number and complexity of histone fractions were far too limited for the multitudinous, selective, and highly specific aspects of transcriptional control which must take place at all genetic loci in the chromosomes of differentiated cells. (For histone reviews, see Refs. 8–11.) Histones with few exceptions (such as F_1 and F_{2c}) are not tissue specific; they are remarkably uniform in distribution, relative proportions, and composition in widely different cell types, and they show an impressive evolutionary stability (12). It is now considered that their role is largely structural; their synthesis is tightly coupled to DNA synthesis (13–15); their metabolism in interphase cells is largely concerned with postsynthetic modifications of structure and charge that modulate their interactions with DNA (16). While some modifications (such as acetylation) of histones correlate with differences in the morphology and functional state of the chromatin (16–20), others, such as histone serine phosphorylation, occur at particular stages of the cell cycle (21–24). Such postsynthetic modifications of histone fractions appear to be involved in the control of chromatin structure. In a general way, they may influence patterns of transcription by altering the physical state of the DNA template (16), but, in themselves, histone modification reactions such as acetylation are not a sufficient cause for the initiation of RNA synthesis at particular gene loci (16, 25). Moreover, the results of many chromatin reconstitution experiments (e.g., 26–29) have shown that the source of the histone does not determine the nature of the RNA transcribed. It follows that selectivity in transcription involves chromosomal components other than histones. This view is supported by research on transcriptional control in both eukaryotic and prokaryotic systems.

II. TRANSCRIPTIONAL CONTROL IN PROKARYOTES— A MODEL FOR HIGHER ORGANISMS

Enormous contributions to the understanding of genetic control mechanisms have come from studies of bacteria and viruses, beginning with the classic observations on enzyme induction and repression in *Escherichia coli* (30, 31) and progressing to recent studies of DNA-binding proteins that activate transcription from specific genetic loci (32, 33).

The importance of these studies as a model for regulatory mechanisms in eukaryotes deserves new emphasis. The following summary presents some features of prokaryotic transcriptional control that have particular relevance to the study of the DNA-binding properties and functions of nuclear proteins in higher organisms.

The binding of proteins to DNA has functional consequences ranging from repression to activation of RNA synthesis. An important point in the following discussion is that the recognition of specific DNA sequences is achieved by appropriate structure and conformational states of polypeptide chains *alone*. In the examples to be considered, there is no need to invoke supplementary mechanisms (such as base-pairing of attached complementary RNA sequences) to assist the protein in the search for the proper control sequences in the DNA molecule.

A. The *lac* Repressor

The *Escherichia coli lac* repressor is one of the first DNA-binding proteins that it has been possible to study in full genetic and chemical detail. The protein (a product of the *i* gene) regulates the expression of the lactose operon at a site on the bacterial chromosome called the *lac* operator (30). The repressor binds to the operator with great specificity (31) and this interaction prevents transcription of the *lac* messenger RNA. Derepression occurs as a result of a conformational change in the repressor due to a binding of a low molecular weight "inducer" molecule (a galactoside). Upon derepression, the *lac* operon mRNA is transcribed and the enzymes necessary for the utilization of lactose are synthesized (30).

The two activities of the repressor protein, operator-binding (31, 34) and inducer-binding (35–37), are separable. Mutations (i^s) are known in which the repressor protein retains its affinity for DNA and loses much of its affinity for the inducer (37). Conversely, there are mutations (i^{-d}) which do not affect the inducer-binding properties of the repressor, but eliminate its binding to the *lac* operator. Recent studies have provided the chemical basis for these genetic differences.

The nucleotide sequence of the *lac* operator region is now known to consist of a segment, 24 base pairs long (38).

```
5' . . . TGGAATTGTGAGCGGATAACAATT . . . 3'
3' . . . ACCTTAACACTCGCCTATTGTTAA . . . 5'
```

The only molecular species of *lac* repressor known to interact with the operator is a stable tetramer of identical subunits of molecular weight 38,000. The binding reaction is specific for double-stranded DNA (34).

1. DNA-Binding Proteins and Transcriptional Control

The binding reaction involves the amino-terminal portion of the repressor protein molecule. A mutant repressor, lacking the first 42 amino acids is unable to repress, although it retains its affinity for the inducer (isopropyl-β-D-thiogalactoside) and can maintain a stable tetrameric conformation (39). Other mutations in the *i* gene affecting the ability of the protein to repress *in vivo* transcription also involve alterations in the amino-terminal region of the repressor molecule (e.g., serine for proline at residue 16, threonine to alanine at residue 19, and alanine to valine at residue 53) (40). The amino-terminal sequence of the first 59 residues has been determined (41) as:

Met-Lys-Pro-Val-Thr-Leu-Tyr-Asp-Val-Ala-Glu-Tyr-Ala-Gly-Val-
-Ser-Tyr-Gln-Thr-Val-Ser-Arg-Val-Val-Asn-Gln-Ala-Ser-His-Val-
-Ser-Ala-Lys-Thr-Arg-Glu-Lys-Val-Glu-Ala-Ala-Met-Ala-Glu-Leu-
-Asn-Tyr-Ile-Pro-Asn-Arg-Val-Ala-Gln-Gln-Leu-Ala-Gly-Lys-

The basic amino acids (Arg, Lys, His) outnumber the acidic (Asp, Glu) 8 to 5 in this region of the molecule. The extremely fast rate of association between the *lac* repressor and operator DNA (42) almost certainly is due to an electrostatic attraction between the positively charged site on the protein and the negatively charged phosphate groups in the operator. This type of binding would also account for the decreased stability of the complex at high ionic strength. The situation is reminiscent of histone–DNA interactions, especially for those histones [e.g., histone F2A1 (IV)] which have an asymmetric distribution of basic amino acids and bind DNA through their highly positive amino-terminal regions (see Ref. 43).

The COOH-terminal region of the *lac* repressor is essential to its stability and tetrameric organization (44), while more than 80 residues can be removed from the NH_2-terminal region without loss of tetrameric structure or inducer-binding properties (41). Thus, there is a clear separation of function in different regions of the polypeptide chain. (This naturally raises questions about the function of histone regions that do not participate in DNA binding.)

It has been proposed that the region of the *lac* repressor which interacts with DNA does so by protrusion into the deep groove of the double helix (44). The role of the tetramer in such binding is still obscure, but recent studies of rho factor (to be discussed below) may be pertinent to the general function of oligomers in DNA binding.

Careful kinetic studies of the *lac* repressor–operator interaction (42) show the binding equilibrium constant (at 0.05 M ionic strength) to be 1×10^{-13} M. The binding is specific for double-stranded DNA (34, 45). Competition experiments have been used to measure the interaction of the *lac* repressor with many natural and synthetic DNA's that do *not* contain the *lac* operator. The results show that the repressor also binds to natural DNA's of high A + T content and that poly[d(A–T)] is a good competitor (46), as would now be expected from the high A + T content of the operator sequence (38). However, some indication of the specificity of the repressor–DNA interaction is provided by a comparison of the equilibrium-binding constants: 10^{-8} M for poly-[d(A–T)] and 10^{-13} M for operator DNA.

Binding of repressors to nonoperator DNA sequences must be regarded as an important aspect of the overall binding mechanism, because the search for operator sequences *in vivo* must involve transient associations with nonoperator DNA. Depending on the nucleotide composition of the DNA and the avidity of the repressor, a fraction of the repressor molecules may not be in a position to block transcription at the appropriate operator loci. This suggests that repressors with relatively low specificity for particular operators may occur in relatively high concentration. That repressor proteins may vary greatly in their affinity for operator regions is indicated by the occurrence of the X86 repressor, resulting from a mutation in the *i* gene to give a *lac* repressor which binds to the operator 40 times more tightly than does the wild-type repressor (47).

Conversely, alterations in the composition of the operator DNA can influence the binding of the wild-type repressor. The *lac* repressor binds 10 times tighter to 5-bromodeoxyuridine (BrdU)-substituted *lac* operator than it does to normal *lac* operator (48). This has prompted the interesting speculation that, since BrdU selectively blocks the expression of many differentiated cell functions in eukaryotic cells, operator regions of eukaryotic DNA's are modified to produce tighter binding of their regulatory proteins (48).

B. Repression of the *hut* Operons in *Salmonella*

The genes coding for the enzymes of histidine utilization (*hut*) in *Salmonella typhimurium* are clustered in two adjacent operons. A single repressor, a product of the *hut* C gene, regulates both operons. As in the case of the *lac* repressor in *E. coli*, a low molecular weight "inducer" (urocanate) inactivates the repressor and weakens its binding to the operator DNA sequences (49).

An important feature of this system is that the *hut* repressor regulates its own synthesis, because the *hut* C gene is a member of one of the *hut* operons and is inducible. Thus, as induction proceeds, further induction becomes more difficult.

A second basic feature is that the repressor has more than one site of attachment to the chromosome. This feature, which also occurs in the *lambda* (λ) phage repressor, is important in considering the binding of regulatory proteins to eukaryotic DNA's in which some of the "repetitive" sequences are believed to play a role in transcriptional control.

C. The Lambda Repressor

The genetic analysis of bacteriophage λ has revealed an intricate set of interlocking controls for expression of the phage genome in both the lytic and integrated states (50).

The λ phage repressor, a product of the *cI* gene, is the regulatory element which maintains repression of the viral DNA in the lysogenic state (51). Two operators are recognized by the same protein. These operators occur within a small region of the DNA (the immunity or *imm* region) about 3000 base pairs apart. Each operator independently regulates transcription of a separate operon on either side of the *cI* gene. Repressor bound to the left-hand operator (O_L) blocks leftward transcription of gene N, and repressor bound to the right-hand operator (O_R) blocks rightward transcription of gene *tof*.

The repressor recognizes these operators highly specifically and binds to them with much greater affinity than for other sequences in λ or *E. coli* DNA. Repressor oligomers (probably dimers) bind to the operators in a complex way. The nature of the binding has been studied by analyzing the short DNA duplexes protected from nuclease digestion when λ repressor combines with λ DNA (52). As the ratio of repressor to operator is increased, the length of the protected DNA fragment increases in 6 steps ranging from 35 to 100 base pairs. Analysis of the fragments indicates that, at each of the operators (O_L and O_R), the repressor first binds to a unique site and that 5 additional sites are filled in, in linear rightward or leftward progression. The nucleotide sequences of O_L and O_R are not identical nor is their affinity for the repressor identical.

The repressor interacts with double-stranded DNA, and the operator duplex does not unwind when the repressor binds to it (52). One molecular mechanism consistent with these results is that the repressor dimer binds to a preferred site in each operator region. Monomers are then added to adjacent sites each 15 base pairs long.

It is known that 100–300 nucleotide pairs intervene between the promotor locus of the left-hand operator (O_L) and the actual starting point of RNA transcription (53). The O_L region may constitute all or part of this intervening segment. Thus, the RNA polymerase would have to traverse up to six repressor-binding sites before beginning transcription. This could explain why the λ repressor is able to block transcription by RNA polymerase molecules already attached to the DNA template (54).

There are several important aspects of this system that may have particular relevance to studies of transcriptional control in higher organisms: (a) The repressor can combine with and influence the function of DNA regions that are not identical in nucleotide sequence. (b) The repressor can block transcription from opposite DNA strands (O_L and O_R regulate transcription in opposite directions). (c) The repressor can block RNA chain initiation if added before RNA polymerase, and it can prevent chain elongation if the polymerase is bound to DNA before the repressor (54). (d) The inhibition of transcription at O_L prevents expression of other genes such as the N gene, the products of which are involved in *positive* aspects of genetic control. The synthesis of active N protein allows transcription to occur in other regions of the genome. (e) The expression of the N gene appears to be subject to multiple negative controls, not only by the λ repressor but also by a repressor product of the *tof* gene (55).

These are wonderfully instructive examples of interlocking controls which act to regulate the production of proteins needed for gene activation or repression in a relatively simple life form. It can be assumed that interlocking control systems in higher organisms, although far more complex, will utilize some of the same general principles.

D. Rho Factor—Geometry of the Functional Oligomers

A protein factor, rho, which terminates transcription at defined sites on bacteriophage λ (56) and on the replicating form of bacteriophage *fd* DNA (57) has been purified (58, 59). The monomeric form of the protein has a molecular weight of 50,000 daltons. Under the electron microscope the purified rho preparation appears as a unique particle consisting of six subunits of identical size arranged in a plane around an empty core. The approximate diameters of the hexagonal particle and each subunit were estimated at 115 and 40 Å, respectively (59).

Electron microscope studies of the binding of rho to DNA of the replicating form of the *fd* phage show that the hexameric unit is attached so that the DNA strand fits in the empty core of the rho complex. It is postulated that the rho factor binds to the specific termination sites on DNA by slipping the DNA strand into the core through a built-in gap between the subunit particles (59).

1. DNA-Binding Proteins and Transcriptional Control

The result of the attachment is to terminate RNA transcription at specific, not random, points along the template. In studies of transcription of the galactose operon in *E. coli* (60), it was shown that the rho factor terminates the growth of nascent mRNA chains at stop signals within the operon. At high rho concentrations, the stop occurs at the end of the first structural gene (*galE*). Similarly, rho affects transcription of the *lac* operon in *E. coli* to reduce the size of the RNA transcript from a length corresponding to the entire operon to the approximate size of the *lacZ* gene at the operator-proximal end of the operon (60).

Such specific RNA termination by DNA-binding proteins is likely to be a general phenomenon applicable to eukaryotes as well as prokaryotes. It will be interesting to see whether the attachment of eukaryotic termination factors to DNA also involves encirclement of the DNA double helix.

E. Positive Transcriptional Control by DNA-Binding Proteins

The preceding examples have dealt almost exclusively with negative aspects of genetic control, stressing the ability of DNA-binding proteins to form oligomers which restrict RNA chain initiation or elongation, or which promote chain termination. Prokaryotic organisms also employ DNA-binding proteins to facilitate transcription. The N protein of λ phage has already been mentioned briefly. Another example is provided by the study of the mechanism of activation of the *gal* and *lac* operons in *E. coli*.

The galactose operon consists of three linked structural genes (*galE*, *galT*, and *galK*) that direct the synthesis of the enzymes, UDP-galactose-4-epimerase, galactose-1-*P*-uridyl transferase, and galactokinase (61, 62). Each of the genes is estimated to contain about 1100 base pairs. The operator–promotor region lies at the *galE* end of the operon (62) and is controlled by a specific repressor in conjunction with the inducer (D-galactose). Transcription initiates at the *gal* promotor sequence.

The induction of transcription of the *gal* operon requires its interaction with a specific protein referred to as the cyclic adenosine-3′,5′-monophosphate receptor protein (CRP) (63, 64) or the catabolite gene activator protein (CAP) (65). The protein has a molecular weight of about 45,000 and is composed of two apparently identical subunits of molecular weight 22,500. The protein is basic with an isoelectric point of 9.12 (64).

CRP requires the presence of cyclic AMP in order to stimulate transcription (63, 65). [This dependency accounts for the phenomenon of catabolite repression. Bacterial operons such as *gal* are subject to glucose repression because the catabolism of glucose leads to reduced cyclic AMP levels in the cell (66, 67). This repression can be reversed by the addition of cyclic AMP (68, 69).]

The binding of cyclic AMP to the receptor involves only one binding site per protein molecule of 45,000 molecular weight (64). The complex is essential for the accurate initiation of transcription of both the *gal* and *lac* operons. A mutant strain lacking the CRP protein does not make *lac* mRNA despite the addition of exogenous cyclic AMP (70).

The fact that the CRP protein stimulates RNA chain initiation at the *lac* (71) and *gal* (72) operons has been attributed to DNA binding at the promoter regions of both operons (73–76). CRP does not act by replacing the sigma factor of the bacterial RNA polymerase (75) but the sigma factor is essential for proper initiation and asymmetric transcription by the enzyme.

Genetic analysis of the system supports the idea that the *lac* promoter region is composed of at least two sites (73). A deletion, L1, of the region interacting with the CRP protein permits a residual low level of *lac* expression. This could be explained if there exists a weak transcription-initiation site between the region defined by the deletion L1 and the *lac* operator (77). In the absence of CRP and cyclic AMP, RNA polymerase holoenzyme interacts only weakly with this site to give 2% of the usual rate of RNA initiation. The role of CRP and cyclic AMP is to attach to the adjacent site (defined by L1) and to increase the interaction between RNA polymerase and the first promotor site.

Several features of this system warrant emphasis in models of transcriptional control in eukaryotic systems:

a. A protein factor interacts with DNA to *stimulate* transcription—presumably by facilitating the binding of the RNA polymerase to the promoter region. This is in accord with observations that certain nuclear proteins stimulate transcription *in vitro* from *free* DNA (78–82).

b. The protein which stimulates transcription requires a cofactor (cyclic AMP), the binding of which is essential to its function. This is an important point in considering the significance of protein modification reactions in higher organisms, such as the early phosphorylation which often precedes increased rates of RNA synthesis (78, 79, Chapter 4 by L. J. Kleinsmith). The model may be particularly relevant in cases where the phosphorylation of nuclear proteins is selectively mediated by cyclic AMP (83).

c. More than one gene is affected by the binding of a single regulatory protein. This obviates the need to invoke a "one gene:one positive control protein" model which, in the final analysis, would require unending amplifications of control elements.

d. Removal of a repressor is not, in itself, a sufficient cause to initiate transcription at a blocked genetic locus.

F. Low Molecular Weight RNA Initiation Proteins in E. coli

Cukier-Kahn et al. (84) have described two heat-stable protein factors, H_1 and H_2, which stimulate DNA-directed RNA synthesis. The first is a neutral polypeptide of molecular weight 8000–11,000; H_2 is a basic polypeptide of molecular weight 6000–7000 (84). Both factors strongly stimulate transcription from double-stranded viral DNA templates using the E. coli RNA polymerase holoenzyme. No effect is observed with single-stranded templates. The action of the factors is synergistic. Both H_1 and H_2 bind readily to DNA templates. Since their enhancing effect on RNA synthesis occurs at the level of chain initiation, rather than chain elongation, it has been proposed that they operate by changing the conformation of DNA at promoter regions (84).

G. DNA-Unwinding Proteins

The organization of the genetic material in the form of a stable double helix poses a number of obstacles in DNA replication and genetic recombination. Recent work suggests that the physiological mechanism for DNA unwinding involves a lowering of helix stability by binding of special proteins to single strands.

The first protein of this type to be characterized was detected in E. coli infected with bacteriophage T4. It is the product of phage gene-32 (85–87). This DNA-unwinding protein stimulates the rate of DNA synthesis on single-stranded DNA templates as much as 5- to 10-fold. It is required for genetic recombination and replication of T4 DNA.

A similar protein has recently been purified from the host organism. The E. coli DNA-unwinding protein has a molecular weight of about 22,000 and it binds tightly and cooperatively to single-stranded DNA (87). Like the T4 gene-32 protein, the E. coli DNA-unwinding protein lowers the stability of double-stranded DNA and strongly stimulates in vitro DNA synthesis on appropriate templates (87).

The existence of similar DNA-unwinding proteins in higher organisms (e.g., in calf thymus) has been noted (88). It follows that fractionations of nuclear proteins by affinity chromatography on single-stranded DNA's may select for proteins of this type (see below).

III. TRANSCRIPTIONAL CONTROL BY MODIFICATION OF RNA POLYMERASES

The examples thus far considered have dealt with mechanisms for transcriptional or replicative control by proteins which interact directly with the DNA template. An entirely different mechanism of transcriptional control also exists in both prokaryotic and eukaryotic organisms. It involves modification or substitution of RNA-polymerizing enzymes. For example, (a) In the infection of *E. coli* by virulent phages such as T7, "early" viral functions are directed by phage genes transcribed by the host RNA polymerase; "late" genes are transcribed by a T7-induced RNA polymerase, the product of the *1* gene (89, 90). The host RNA polymerase transcribes only the leftmost 20% of the phage genome (91–93). The remaining 80% requires the T7 RNA polymerase for efficient transcription. (b) In the case of phage T4, many modifications of RNA polymerase occur in the course of infection. The alpha subunits of the enzyme are adenylated (94) and the beta subunits show an altered electrophoretic mobility in urea gels (95). Changes occur in the tryptic peptides of all four RNA polymerase subunits (96) and the association of subunits and other factors to form an active rapidly sedimenting complex is impaired (97). (c) RNA polymerase in *Bacillus subtilis* undergoes at least two major changes in subunit structure during the course of sporulation. During the first hour the vegetative sigma factor is lost, with an accompanying change in template specificity of the polymerase. Later in sporulation one of the beta subunits disappears and is replaced by a smaller polypeptide of molecular weight 110,000 (98, 99). A new polymerase-binding protein of molecular weight 70,000 is also synthesized during the transition (100).

In eukaryotic cells, the situation is complicated by the presence of multiple RNA polymerases (101–105) which differ in intranuclear localization and show a division of function (101, 102). RNA polymerase I(A) is primarily involved in the synthesis of ribosomal RNA in the nucleolus. It is insensitive to α-amanitin. Polymerase II(B) is a nucleoplasmic enzyme which is completely inhibited by α-amanitin. The various RNA polymerases can be resolved by chromatographic procedures. Changes in polymerase elution profiles have been noted during cellular differentiation, e.g., during oogenesis in the sea urchin (106).

The activity of eukaryotic RNA polymerases is also modifiable by protein factors. Yeast cells contain a factor, pi, which stimulates transcription from a variety of DNA templates by rat liver RNA polymerases I and II (107). The pi factor is a heat-stable polypeptide with a molecular weight of 12,000. Its interaction with the liver polymerases makes them insensitive to the inhibitor rifamycin AF-013.

Proteins that stimulate RNA polymerase activity have been purified from Novikoff ascites cells (108). One type of activity is heat stable, the other is heat labile. Both factors stimulate the activity of the ascites cell RNA polymerase II by severalfold. They appear to function independently. Neither factor stimulates the activity of *E. coli* RNA polymerase.

Protein factors have been prepared from calf thymus and from rat liver that stimulate the activity of RNA polymerase II, but do not enhance transcription by bacterial RNA polymerase (109, 110). A factor stimulating transcription by RNA polymerase I has been prepared from isolated rat liver nucleoli (111).

Thus, it is clear that there are multiple mechanisms for influencing the activity and specificity of the RNA polymerases. In the analysis and testing of proteins that modify transcription, a clear distinction will have to be made between changes in the enzymes and alterations in the DNA template.

IV. NUCLEAR NONHISTONE PROTEINS AND TRANSCRIPTIONAL CONTROL IN EUKARYOTES

A. Correlations with Gene Activity

The cell nuclei of higher organisms contain large numbers of nonhistone proteins associated with the chromatin. Clearly, not all of these proteins are involved in the control of transcription. Many are concerned with chromosome structure and mobility, with enzymatic modifications of nucleic acids and histones, and with the processing and transport of gene products. However, there is good evidence that some of the nonhistone proteins influence the rate and specificity of RNA synthesis and transmit signals from the cytoplasm, the cell membrane, and other regions of the nucleus to specific genetic loci.

It has been known for many years that the nonhistone proteins of the nucleus include components with high rates of synthesis and turnover (112–115). Different chromatin types were found to vary in their contents of nonhistone proteins (116), and strong correlations were soon noted between the "residual" protein content of the chromatin and the RNA synthetic capacity of the tissue (117).

The main implication of these early observations—that proteins associated with DNA in chromatin influence its activity in RNA synthesis—drew further support from chromatin fractionation studies which showed preferential localization of the nonhistone proteins in the active or euchromatic fractions of the nucleus (118–122). Subsequent developments in this area have, on the whole, tended to confirm the view of the diversity of the nonhistone proteins and the involvement of some of them in transcriptional control. Some of the evidence can be summarized briefly:

a. The distribution of the nonhistone proteins varies in different somatic tissues (80, 123–130) as would be expected if some were involved in differential gene activity. Immunological evidence for tissue-specificity of the nonhistone proteins (126–128) should help dispel lingering doubts about differential extractability being the major reason for the observed differences in electrophoretic patterns of proteins from different nuclear types.

b. The nature and amount of the nonhistone proteins change during embryogenesis (131–134).

c. The complement of nuclear nonhistone proteins is dramatically altered during the differentiation of particular cell types (135–138).

d. The synthesis of *specific* nonhistone proteins is augmented at times of gene activation by cortisol (139), estradiol (140), and glucagon (141).

e. The activated "puffing" regions of insect chromosomes accumulate specific nonhistone proteins after treatment with ecdysone or other stimuli (142).

f. Increased synthesis of nonhistone proteins is observed at early stages of cell proliferation in salivary gland cells stimulated by isoproterenol (143) and in lymphocytes stimulated by phytohemagglutinin (144) and concanavalin A (145).

g. When resting WI-38 fibroblasts are stimulated to proliferate by changing the medium, the incorporation of amino acids into nuclear proteins is promptly increased (146).

h. The "transformation" of WI-38 cells by SV40 virus results in an immunologically detectable alteration in the nuclear nonhistone proteins (128).

i. Changes in the metabolism of the nuclear acidic proteins occur during gene activation by insulin (147), aldosterone (148), and phenobarbital (149).

j. Many of the nuclear nonhistone proteins are phosphorylated (80, 81, 125, 138, 150–165). (See Chapter 4 by L. J. Kleinsmith.) The phosphorylation of certain proteins is an early event in embryonic development (150) and is augmented in gene activations induced by mitogenic agents (145, 151) or by hormones such as cortisol (78, 79), testosterone (152, 153), and gonadotropins (165).

k. Cyclic AMP modifies the phosphorylation of some, but not all proteins of liver nuclei (83). Changes in cyclic AMP levels during the cell cycle of synchronized HeLa cell cultures correlate well with changing rates of nuclear protein phosphorylation (24, 79, 154). There are clear differences in rates of phosphorylation of the nuclear proteins at different stages of the cell cycle. Phosphorylation is most active at periods when RNA synthesis is high (G_1 and S) and it is minimal in the G_2–M period when RNA synthesis is suppressed (24, 72, 154, 155).

l. Protein phosphorylation is most pronounced over the "puffing" regions of insect chromosomes (156).

m. Direct tests for the influence of nuclear nonhistone proteins on the rate of RNA synthesis or the nature of the RNA transcript show strong indications of their involvement in positive control. Some of these experiments are discussed in more detail in the context of DNA-binding properties of nuclear polypeptides.

B. Specificity in DNA-Binding by Nuclear Nonhistone Proteins

It is already known that the nuclear nonhistone protein fraction includes components with a high degree of selectivity in their interactions with DNA. The specificity of the interaction is indicated by two lines of evidence: (a) discrimination in the formation of DNA–protein complexes, and (b) selective activation of transcription in reconstituted chromatin fractions.

DNA binding by nuclear proteins other than histones has been observed repeatedly (24, 78, 80, 125, 166–172). The problem of specificity has been approached experimentally by the use of DNA affinity chromatography (24, 166, 168, 169, 172) and by the centrifugal isolation of the DNA–protein complex (80, 125, 170).

An important consideration in both techniques is the problem of protein denaturation during the isolation procedure. Nuclear proteins extracted in phenol (80, 125) or in 6 M urea containing 0.4 M guanidine hydrochloride (24, 172) must be reassociated with DNA under effective renaturing conditions (167, 173, 174). That proteins treated with urea and guanidine hydrochloride can be effectively renatured is illustrated by recent studies showing 85–95% reactivation of human carbonic anhydrases B and C (175). Similarly, proteins such as chymotrypsinogen A and bovine serum albumin exposed to phenol will recover their enzymatic activities and immunological properties during appropriate dialysis procedures (176).

C. Species-Selective DNA Binding

Nonhistone proteins prepared from rat liver and kidney nuclei by the phenol procedure have been "annealed" to DNA under renaturing conditions and then centrifuged in sucrose density gradients to separate the DNA–protein complexes. Many, but not all of the proteins form complexes with rat liver DNA under these conditions. Little or no binding was observed when rat liver or kidney nonhistone proteins were added under the same conditions to DNA's from calf thymus, human placenta, dog liver, or bacterial cells (80, 125). In the case of rat DNA, only a small fraction (less than 13%) of the total rat liver or kidney nuclear proteins could be recovered as soluble DNA–protein complexes. However, the nature of the protein present on the DNA was characteristic of the tissue of origin (80).

It was further established that the complex of nuclear nonhistone protein and homologous DNA was much more active as a template for RNA synthesis (using the *E. coli* RNA polymerase) than was the equivalent amount of free DNA. This stimulation of transcription did not occur for all DNA's; e.g., mixtures of rat liver nuclear proteins and calf thymus DNA were not more active than thymus DNA alone (80). Thus, the selectivity of the interactions between nonhistone proteins and DNA sequences is reflected both in DNA-binding properties and in the enhancement of transcription *in vitro*.

Similar results were reported for rat liver nonhistone proteins isolated in high salt concentrations and tested against a variety of DNA's, using the mammalian RNA polymerase II (B) (81). Stimulation was observed for rat DNA but not for salmon, calf, *E. coli*, or *C. perfringens* DNA's. Conversely, nuclear nonhistone proteins prepared from calf thymus were found to stimulate transcription from calf DNA, but not from rat DNA (82). An important point is that enhancement of RNA synthesis by nuclear nonhistone proteins occurs with both mammalian and bacterial RNA polymerases and depends on the source of the DNA, not of the enzyme.

The stability of the DNA–protein complex depends on the ionic strength of the medium. Many proteins dissociate even at low salt concentrations (80), while others show maximum binding at 0.14 M NaCl (170). The dissociation of nonhistone proteins from DNA–cellulose or DNA–aminoethyl Sepharose columns reveals a wide spectrum of DNA-binding affinities (24, 172).

Affinity chromatography on DNA cellulose (85, 86) has been employed to investigate the species specificity of the interactions between different DNA's and the nonhistone proteins of rat liver nuclei (166, 168). A

1. DNA-Binding Proteins and Transcriptional Control

small fraction of the nuclear proteins was found to combine selectively with rat DNA and not to combine with salmon DNA or *E. coli* DNA (166). Subsequent studies of ^{32}P-labeled nuclear phosphoproteins showed that binding to calf DNA, salmon DNA, and bacterial DNA's was much less effective than to DNA from rat liver or testes (169). The phosphorylated proteins that specifically bind to rat DNA–cellulose columns are highly heterogeneous, as judged by SDS–polyacrylamide gel electrophoresis, but the majority fall in the molecular weight range 30,000–70,000 (169). Similar conclusions about the size and heterogeneity of species-specific DNA-binding proteins were derived from studies of the elution of rat liver nuclear nonhistone proteins from tandem columns of *E. coli* DNA–cellulose and rat liver DNA–cellulose (168).

D. Sequence-Specific DNA Binding

Recent developments in DNA affinity chromatography have permitted a more incisive analysis of the interactions between nuclear proteins and DNA (24, 172).

The nonhistone proteins of calf thymus nuclei were extracted in 6 M urea containing 0.4 M guanidine hydrochloride (177). Histone contamination was removed by ion-exchange chromatography on Bio-Rex 70. The major peak emerging from the column (representing about 80% of the protein remaining in the dehistonized nuclei) was used for DNA affinity chromatography. The fraction comprises a complex mixture of nonbasic proteins which differ in molecular weight, amino acid composition, isoelectric point, degree of phosphorylation, and DNA-binding affinity. Banding patterns in unidirectional SDS–polyacrylamide gels show the presence of at least 30 components varying in molecular weight from 17,000 to 180,000 daltons.

Much greater complexity is revealed by electrophoresis in two dimensions (178), using isoelectric focusing in the first dimension to separate proteins on the basis of their isoelectric points, followed by electrophoretic separation according to size in the second dimension. The method reveals that most of the proteins have isoelectric points between pH 4.2 and 6.9 and may thus be regarded as "acidic" nuclear proteins. Arrays of spots appear in the electropherogram at the same molecular weight range but differing in isoelectric point. This is due to varying degrees of phosphorylation of the parent polypeptide chain (172).

In searching for sequence-specific interactions between nuclear proteins and DNA, recourse was made to methods which resolve DNA subfractions differing in C_0t value (179–181). The recognition of DNA sequences in higher organisms is complicated by the occurrence of large numbers of "repeated" or closely similar DNA sequences (179, 180, 182, 183). Fractions enriched in such sequences can be separated from the remainder of the DNA by shearing and heat denaturation of the DNA, followed by rapid reannealing of the multicopy strands and separation of the resulting duplexes by chromatography on hydroxyapatite. Other DNA subfractions with less common sequences can be separated following more prolonged reannealing and chromatography on hydroxyapatite columns (181). Using these techniques, calf thymus DNA subfractions of widely differing C_0t values were prepared and each fraction was covalently linked to aminoethyl Sepharose 4B (172). This made available a series of columns in which the DNA originated from the same species but differed in C_0t value or in strandedness.

Aliquots of the nuclear nonhistone protein fraction were complexed with each DNA subfraction under renaturing conditions (167, 172), and the columns were subsequently eluted by stepwise increments in the salt concentration of the eluting buffer. The type of separation achieved is shown in Fig. 1. Proteins eluted at each salt concentration were analyzed by electrophoresis in SDS–polyacrylamide gels. The separations achieved are highly reproducible, and proteins eluting at a given salt concentration emerge at the same salt concentration on rechromatography (172). The method resolves the nuclear nonhistone proteins into subsets of differing DNA affinities. Some proteins bind so strongly to DNA that their displacement is possible only at high salt concentrations or by the use of denaturing agents such as urea and guanidine hydrochloride. The fractions eluted at different salt concentrations contain different sets of proteins of varying complexity and size distribution (Fig. 1).

In comparisons of the elution diagrams and electrophoretic patterns of proteins emerging from parallel columns bearing DNA subfractions of different C_0t value, it became evident that certain proteins differed in their affinities for high C_0t (C_0t 225–4 $\times$ 10^4), intermediate C_0t (C_0t 6–225), and low C_0t (C_0t less than 6) DNA sequences. Obvious differences appeared in the side-by-side comparisons of the electrophoretic banding patterns of the protein sets eluted at a given salt concentration from each of the DNA columns. Apart from revealing differences in band concentrations or distribution, the patterns could be analyzed to determine whether a given protein is eluted more (or less) readily from high C_0t DNA or from low C_0t DNA (172). For example, bands representing proteins of molecular weight 30,000, 48,500, and 78,000 have greater affinity for single-stranded low C_0t DNA than for high C_0t DNA. Other proteins of molecular weight 37,500, 47,500, and 50,000 bind more strongly to the high C_0t sequences. A band of molecular weight 63,000 binds more strongly to intermediate C_0t DNA than to low C_0t DNA.

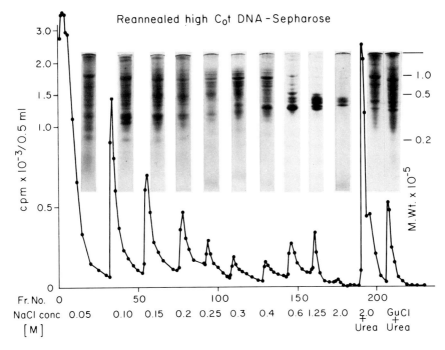

Fig. 1. Fractionation of calf thymus nuclear nonhistone proteins on double-stranded (reassociated) calf thymus DNA sequences of C_0t value 225–4 × 10⁴. The ³H-labeled nuclear proteins were combined with DNA covalently linked to aminoethyl Sepharose 4B under protein renaturing conditions (172). Elution was carried out in a stepwise salt gradient, monitoring each fraction, as indicated for radioactivity. The nature of the proteins eluted at each salt concentration was determined by electrophoresis in 10% polyacrylamide, 0.1% SDS gels at pH 7.4. The patterns are shown for each set at the appropriate peak of the elution diagram. The corresponding molecular weight scale is given at the right of the figure.

Similar studies were carried out using double-stranded (reassociated) DNA subfractions differing in C_0t value. Elution of the nonhistone proteins in stepwise salt gradients again revealed characteristic differences in DNA affinity (172). For example, bands representing proteins of molecular weight 34,000 and 61,000 have greater affinity for double-stranded low C_0t DNA, and thus require higher salt concentrations for their displacement from the "repeated" DNA sequences than from the "unique" sequences of the high C_0t DNA column. Conversely, a protein of molecular weight 50,000 binds more strongly to the high C_0t DNA sequences.

DNA-affinity chromatography of the nuclear nonhistone proteins depends not only on the C_0t value of the DNA but on its strandedness. This distinction becomes evident in comparisons of the elution patterns of the proteins released from aminoethyl Sepharose columns bearing DNA subfractions of the same C_0t value but differing in strandedness. In low C_0t DNA chromatography, proteins of molecular weight 50,000 and 61,000 have higher affinity for the double-stranded DNA form, while bands of molecular weight 40,500, 44,500, and 80,000 combine preferentially with single-stranded DNA. Similar strand-dependent associations are also evident in protein binding to high C_0t DNA sequences (172).

Thus, the chromatographic behavior of some of the nuclear nonhistone proteins is consistent with the recognition of the physical state of the DNA as well as its nucleotide sequence.

E. Transcription of Unique and Repetitive DNA Sequences

The indications of species' and sequence-specific interactions between nuclear nonhistone proteins and DNA have obvious relevance to the problem of RNA chain initiation at specific loci of the chromosomes. Since the original observations of Paul and Gilmour (28, 174, 184), there have been repeated indications that the nuclear nonhistone protein fraction includes components of polypeptide nature that alter the rate of transcription and influence the nature of the RNA transcript in reconstituted chromatin fractions (27–29, 185–188). It has been uniformly observed that the source of the histones does not modify the nature of the RNA synthesized in isolated or reconstituted chromatin (27–29), but varying the origin of the acidic nuclear proteins leads to altered patterns of transcription (27–29, 185–188).

In most of the experiments cited, the newly synthesized RNA was identified in terms of its DNA-hybridization properties. The conditions employed in such hybridization studies rarely permit detection of the transcripts from "unique" (single-copy) sequences of the DNA. Instead, the assay emphasizes the contribution from transcripts of the "repetitive" or nearly identical DNA sequences (182, 189–192). Nevertheless, there are ample indications that the high molecular weight nuclear precursors of cytoplasmic messenger RNA's contain transcripts from repetitive DNA sequences (193–195). Messenger RNA's from *Xenopus* embryos appear to be internally heterogeneous, containing a main part transcribed from unique DNA sequences and a smaller part transcribed from a family of homogeneously repeated DNA sequences (196). It follows that transcription of some reiterated sequences in isolated chromatin is an essential and normal aspect of the mechanism of RNA synthesis.

Most investigations of transcription from eukaryotic chromatin have employed RNA polymerases of bacterial origin. Recent studies indicate that *E. coli* RNA polymerase transcribes the gene for *Xenopus* ribosomal RNA and 5 S RNA in an aberrant manner, and moreover, the bacterial enzyme fails to recognize a mechanism of repression of ribosomal RNA synthesis in hybrid frogs (190). If one considers the evidence (cited previously) that modification of bacterial RNA polymerases greatly affects the patterns of transcription in prokaryotes, the objection to their use in studies of eukaryotic mechanisms of transcription is amplified. Fortunately, progress in the separation, purification, and assay of eukaryotic RNA polymerases has been rapid, and the availability of the proper RNA polymerizing enzymes should improve the accuracy and resolution of the *in vitro* transcription systems.

F. Protein Phosphorylation and Gene Activation

Studies of the effects of nuclear nonhistone proteins on RNA synthesis in isolated chromatin have not often considered the problem that such proteins are subject to postsynthetic modifications which may determine whether or not they can function. The binding of some nonhistone proteins to DNA, or their ability to facilitate the attachment of RNA polymerases to promoter regions, may well depend on their degree of phosphorylation. It is not yet clear whether the conformational changes induced in nuclear nonhistone proteins by the phosphorylation of serine and threonine residues result in altered binding affinities for *particular* DNA sequences. However, it is known that the phosphorylated forms of the proteins can bind to DNA in a specific manner (80, 125, 169). The relevance of this observation to the mechanism of gene activation is suggested by many reports of increased phosphorylation of nuclear proteins at times of gene activation (78, 79, 83, 145, 150–153, 165) and by *in vitro* studies relating the degree of phosphorylation of the associated nuclear proteins to the RNA synthetic capacity of a DNA or chomatin template (81, 186).

Phosphorylation of the nuclear nonhistone proteins may, like the binding of cyclic AMP to the CAP receptor protein of *E. coli* (discussed earlier), represent the critical variable in the interaction among regulatory proteins, DNA, and RNA polymerases. Further developments in this area promise to reveal that postsynthetic modifications of structure of DNA-binding proteins are a general mechanism for the control of chromosomal activity in prokaryotic and eukaryotic cells.

REFERENCES

1. Gurdon, J. (1962). *Develop. Biol.* **4**, 256.
2. Allfrey, V. G., and Mirsky, A. E. (1962). *Proc. Nat. Acad. Sci. U.S.* **48**, 1590.

3. Paul, J., and Gilmour, R. S. (1966). *J. Mol. Biol.* **16**, 242.
4. McCarthy, B. J., and Hoyer, B. H. (1964). *Proc. Nat. Acad. Sci. U.S.* **52**, 915.
5. Stedman, E., and Stedman, E. (1951). *Phil. Trans. Roy. Soc. London, Ser. B* **234**, 565.
6. Allfrey, V. G., Littau, V. C., and Mirsky, A. E. (1963). *Proc. Nat. Acad. Sci. U.S.* **49**, 1313.
7. Huang, R. C., and Bonner, J. (1962). *Proc. Nat. Acad. Sci. U.S.* **48**, 1216.
8. Stellwagen, R. H., and Cole, R. D. (1969). *Annu. Rev. Biochem.* **38**, 951.
9. Elgin, S. R., Froehner, S. C., Smart, J. E., and Bonner, J. (1971). *Advan. Cell Mol. Biol.* **1**, 2.
10. Wilhelm, J. A., Spelsberg, T. C., and Hnilica, L. S. (1971). *Sub-Cell. Biochem.* **1**, 39.
11. Phillips, D. M. P., ed. (1971). "Histones and Nucleohistones." Plenum, New York.
12. DeLange, R., and Smith, E. L. (1972). *Accounts Chem. Res.* **5**, 368.
13. Spalding, J., Kajiwara, K., and Mueller, G. C. (1966). *Proc. Nat. Acad. Sci. U.S.* **56**, 1535.
14. Robbins, E., and Borun, T. W. (1967). *Proc. Nat. Acad. Sci. U.S.* **57**, 409.
15. Hancock, R. (1969). *J. Mol. Biol.* **40**, 457.
16. Allfrey, V. G. (1971). *In* "Histones and Nucleohistones" (D. M. P. Phillips, ed.), p. 241. Plenum, New York.
17. Berlowitz, L., and Pallotta, D. (1972). *Exp. Cell Res.* **71**, 45.
18. Gorovsky, M. A., Pleger, G. L., Keevert, J. B., and Johmann, C. A. (1973). *J. Cell Biol.* **57**, 773.
19. Wangh, L. J., Ruiz-Carrillo, A., and Allfrey, V. G. (1972). *Arch. Biochem. Biophys.* **150**, 44.
20. Ord., M. G., and Stocken, L. A. (1968). *Biochem. J.* **107**, 403.
21. Oliver, D., Balhorn, R., Granner, D., and Chalkley, R. (1972). *Biochemistry* **11**, 3921.
22. Gurley, L. R., Walters, R. A., and Tobey, R. A. (1973). *Arch. Biochem. Biophys.* **154**, 212.
23. Balhorn, R., Bordwell, L., Sellers, L., Granner, D., and Chalkley, R. (1972). *Biochem. Biophys. Res. Commun.* **46**, 1326.
24. Allfrey, V. G., Inoue, A., Karn, J., Johnson, E. M., and Vidali, G. (1973). *Cold Spring Harbor Symp. Quant. Biol.* **38**, 785.
25. Ono, T., Terayama, H., Takaku, F., and Nakao, K. (1969). *Biochim. Biophys. Acta* **179**, 214.
26. Spelsberg, T. C., and Hnilica, L. S. (1970). *Biochem. J.* **120**, 435.
27. Spelsberg, T. C., Hnilica, L. S., and Ansevin, A. T. (1971). *Biochim. Biophys. Acta* **228**, 550.
28. Gilmour, R. S., and Paul, J. (1970). *FEBS Lett.* **9**, 242.
29. Stein, G. S., Chaudhuri, S., and Baserga, R. (1972). *J. Biol. Chem.* **247**, 3918.
30. Jacob, F., and Monod, J. (1961). *J. Mol. Biol.* **3**, 318.
31. Gilbert, W., and Müller-Hill, B. (1967). *Proc. Nat. Acad. Sci. U.S.* **58**, 245.
32. Zubay, G., Schwartz, D., and Beckwith, J. (1970). *Proc. Nat. Acad. Sci. U.S.* **66**, 104.
33. Anderson, W. B., Schneider, A. B., Emmer, M., Perlman, R. L., and Pastan, I. (1971). *J. Biol. Chem.* **246**, 5929.

34. Riggs, A. D., Bourgeois, S., Newby, R. F., and Cohn, M. (1968). *J. Mol. Biol.* **34**, 365.
35. Gilbert, W., and Müller-Hill, B. (1966). *Proc. Nat. Acad. Sci. U.S.* **56**, 1891.
36. Riggs, A. D., and Bourgeois, S. (1968). *J. Mol. Biol.* **34**, 361.
37. Jobe, A., Riggs, A. D., and Bourgeois, S. (1972). *J. Mol. Biol.* **64**, 181.
38. Gilbert, W., Maizels, N., and Maxam, A. (1973). *Cold Spring Harbor Symp. Quant. Biol.* **38**, 845.
39. Platt, T., Weber, K., Ganem, D., and Miller, J. H. (1972). *Proc. Nat. Acad. Sci. U.S.* **69**, 897.
40. Weber, K., Platt, T., Ganem, D., and Miller, J. H. (1972). *Proc. Nat. Acad. Sci. U.S.* **69**, 3624.
41. Platt, T., Files, J. G., and Weber, K. (1973). *J. Biol. Chem.* **248**, 110.
42. Riggs, A. D., Bourgeois, S., and Cohn, M. (1970). *J. Mol. Biol.* **53**, 401.
43. Sung, M. T., and Dixon, G. H. (1970). *Proc. Nat. Acad. Sci. U.S.* **67**, 1616.
44. Adler, K., Beyreuther, K., Fannig, E., Geisler, N., Gronenborn, B., Klemm, A., Müller-Hill, B., Pfahl, M., and Schmitz, A. (1972). *Nature (London)* **237**, 322.
45. Riggs, A. D., Suzuki, H., and Bourgeois, S. (1970). *J. Mol. Biol.* **48**, 67.
46. Lin, S. Y., and Riggs, A. D. (1972). *J. Mol. Biol.* **72**, 671.
47. Jobe, A., and Bourgeois, S. (1972). *J. Mol. Biol.* **72**, 139.
48. Lin, S. Y., and Riggs, A. D. (1972). *Proc. Nat. Acad. Sci. U.S.* **69**, 2574.
49. Hagen, D. C., and Magasanik, B. (1973). *Proc. Nat. Acad. Sci. U.S.* **70**, 808.
50. Hershey, J. (1971). "The Bacteriophage Lambda." Cold Spring Harbor Laboratory, Cold Spring Harbor, New York.
51. Kaiser, A. D., and Jacob, F. (1957). *Virology* **4**, 509.
52. Maniatis, T., and Ptashne, M. (1973). *Nature (London)* **246**, 133.
53. Blattner, F. R., Dahlberg, J. E., Boettiger, J. K., Fiandt, M., and Szybalski, W. (1972). *Nature (London) New Biol.* **237**, 232.
54. Wu, A. M., Ghosh, S., and Echols, H. (1972). *J. Mol. Biol.* **67**, 423.
55. Greenblatt, J. (1973). *Proc. Nat. Acad. Sci. U.S.* **70**, 421.
56. Roberts, J. (1970). *Cold Spring Harbor Symp. Quant. Biol.* **35**, 121.
57. Takanami, M., Okamoto, T., and Sugiura, M. (1972). *J. Mol. Biol.* **62**, 81.
58. Roberts, J. (1969). *Nature (London)* **224**, 1168.
59. Oda, T., and Takanami, M. (1972). *J. Mol. Biol.* **71**, 799.
60. DeCrombrugge, B., Adhya, S., Gottesman, M., and Pastan, I. (1973). *Nature (London), New Biol.* **241**, 260.
61. Echols, H., Remichek, J., and Adhya, S. (1963). *Proc. Nat. Acad. Sci. U.S.* **50**, 286.
62. Buttin, G. (1963). *J. Mol. Biol.* **7**, 183.
63. Emmer, M., DeCrombrugge, B., Pastan, I., and Perlman, R. (1970). *Proc. Nat. Acad. Sci. U.S.* **66**, 480.
64. Anderson, W. B., Schneider, B., Emmer, M., Perlman, R. L., and Pastan, I. (1971). *J. Biol. Chem.* **246**, 5929.
65. Zubay, G., Schwartz, D., and Beckwith, J. (1970). *Proc. Nat. Acad. Sci. U.S.* **66**, 104.
66. Makman, R. S., and Sutherland, E. W. (1965). *J. Biol. Chem.* **240**, 1309.
67. Pastan, I., and Perlman, R. L. (1970). *Science* **169**, 339.
68. Perlman, R. L., and Pastan, I. (1968). *J. Biol. Chem.* **243**, 5420.
69. Ullman, A., and Monod, J. (1968). *FEBS Lett.* **2**, 57.

70. Varmus, H. E., Perlman, R. L., and Pastan, I. (1970). *J. Biol. Chem.* **245**, 6366.
71. DeCrombrugge, B., Chen, B., Gottesman, M., Pastan, I., Varmus, H. E., Emmer, M., and Perlman, R. L. (1971). *Nature (London), New Biol.* **230**, 37.
72. Parks, J., Gottesman, M., Perlman, R. L., and Pastan, I. (1971). *J. Biol. Chem.* **246**, 2419.
73. DeCrombrugge, B., Chen., B., Anderson, W. B., Nissley, P., Gottesman, M., Perlman, R. L., and Pastan, I., (1971). *Nature (London) New Biol.* **231** 139.
74. Nissley, S. P., Anderson, W. B., Gottesman, M., Perlman, R. L., and Pastan, I. (1971). *J. Biol. Chem.* **246**, 4671.
75. Eron, L., Arditti, R., Zubay, G., Connaway, S., and Beckwith, J. (1970). *Proc. Nat. Acad. Sci. U.S.* **68**, 215.
76. Eron, L., and Block, R. (1971). *Proc. Nat. Acad. Sci. U.S.* **68**, 1828.
77. Beckwith, J., Grodzicker, T., and Arditti, R. (1972). *J. Mol. Biol.* **69**, 155.
78. Allfrey, V. G., Teng, C. S., and Teng, C. T. (1971). *In* "Nucleic Acid-Protein Interaction—Nucleic Acid Synthesis in Viral Infection" (D. W. Ribbons, J. F. Woessner, and J. Schultz, eds.), p. 144. North-Holland Publ., Amsterdam.
79. Allfrey, V. G., Johnson, E. M., Karn, J., and Vidali, G. (1973). *In* "Protein Phosphorylation in Control Mechanisms" (F. Huijing and E. Y. C. Lee, eds.), p. 219. Academic Press, New York.
80. Teng, C. S., Teng, C. T., and Allfrey, V. G. (1971). *J. Biol. Chem.* **246**, 3597.
81. Shea, M., and Kleinsmith, L. J. (1973). *Biochem. Biophys. Res. Commun.* **50**, 473.
82. Rickwood, D., Threlfall, G., MacGillivray, A. J., Paul, J., and Riches, P. (1972). *Biochem. J.* **129**, 50P.
83. Johnson, E. M., and Allfrey, V. G. (1972). *Arch. Biochem. Biophys.* **152**, 786.
84. Cukier-Kahn, R., Jacquet, M., and Gros, F. (1972). *Proc. Nat. Acad. Sci. U.S.* **69**, 3643.
85. Alberts, B. M., and Frey, L. (1970). *Nature (London)* **277**, 1313.
86. Alberts, B. M., Amodio, F., Jenkins, M., Gutman, E. D., and Ferris, F. L. (1970). *Cold Spring Harbor Symp. Quant. Biol.* 33, 289.
87. Sigal, N., Delius, H., Kornberg, T., Gefter, M. L., and Alberts, B. M. (1972). *Proc. Nat. Acad. Sci. U.S.* **69**, 3537.
88. Alberts, B. M., and Herrick, G. (1971). *In* "Methods in Enzymology" (L. Grossman and K. Moldave, eds.), Vol. 21, Part D, p. 198. Academic Press, New York.
89. Siegel, R. B., and Summers, W. C. (1970). *J. Mol. Biol.* **49**, 115.
90. Chamberlin, M., McGrath, J., and Waskell, L. (1970). *Nature (London)* **228**, 227.
91. Hyman, R. W., (1971). *J. Mol. Biol.* **61**, 369.
92. Simon, M. N., and Studier, F. W. (1973). *J. Mol. Biol.* **79**, 249.
93. Studier, F. W. (1973). *J. Mol. Biol.* **79**, 237.
94. Goff, C., and Weber, K. (1970). *Cold Spring Harbor Symp. Quant. Biol.* **35**, 101.
95. Travers, A. A. (1970). *Cold Spring Harbor Symp. Quant. Biol.* **35**, 241.
96. Schachner, M., and Zillig, W. (1971). *Eur. J. Biochem.* **22**, 513.
97. Snyder, L. (1973). *Nature (London), New Biol.* **243**, 131.
98. Losick, R., Shorenstein, R. G., and Sonenshein, A. L. (1970). *Nature (London)* **227**, 910.

99. Losick, R. (1972). *Annu. Rev. Biochem.* **41**, 409.
100. Greenleaf, A. L., Linn, T. G., and Losick, R. (1973). *Proc. Nat. Acad. Sci. U.S.* **70**, 490.
101. Widnell, C. C., and Tata, J. R. (1964). *Biochim. Biophys. Acta* **87**, 531.
102. Pogo, A. O., Littau, V. C., Allfrey, V. G., and Mirsky, A. E. (1967). *Proc. Nat. Acad. Sci. U.S.* **57**, 743.
103. Roeder, R. G., and Rutter, W. J. (1969). *Nature (London)* **224**, 234.
104. Kedinger, C., Nuret, P., and Chambon, P. (1971). *FEBS Lett.* **15**, 169.
105. Chesterton, C. J., and Butterworth, P. H. W. (1971). *FEBS Lett.* **12**, 301.
106. Chambon, P., Geisinger, F., Mandel, J. L., Kedinger, C., Gniazdowski, M., and Meihlac, M. (1970). *Cold Spring Harbor Symp. Quant. Biol.* **35**, 693.
107. DiMauro, E., Hollenberg, C. P., and Hall, B. D. (1972). *Proc. Nat. Acad. Sci. U.S.* **69**, 2818.
108. Lee, S. C., and Dahmus, M. E. (1973). *Proc. Nat. Acad. Sci. U.S.* **70**, 1383.
109. Stein, H., and Hausen, F. (1970). *Eur. J. Biochem.* **14**, 270.
110. Seifart, K. H. (1970). *Cold Spring Harbor Symp. Quant. Biol.* **35**, 719.
111. Higashinakagawa, T., Onishi, T., and Muramatsu, M. (1972). *Biochem. Biophys. Res. Commun.* **48**, 937.
112. Bergstrand, A., Eliasson, N. A., Hammarsten, E., Norberg, B., Reichard, P., and von Ubisch, H. (1948). *Cold Spring Harbor Symp. Quant. Biol.* **13**, 22.
113. Daly, M. M., Allfrey, V. G., and Mirsky, A. E. (1952). *J. Gen. Physiol.* **36**, 173.
114. Allfrey, V. G., Daly, M. M., and Mirsky, A. E. (1955). *J. Gen. Physiol.* **38**, 415.
115. Steele, W. J., and Busch, H. (1963). *Cancer Res.* **23**, 1153.
116. Mirsky, A. E., and Ris, H. (1951). *J. Gen. Physiol.* **34**, 175.
117. Dingman, C. W., and Sporn, M. B. (1964). *J. Biol. Chem.* **239**, 3483.
118. Frenster, J. H., Allfrey, V. G., and Mirsky, A. E. (1963). *Proc. Nat. Acad. Sci. U.S.* **50**, 1026.
119. Frenster, J. H. (1965). *Nature (London)* **206**, 680.
120. Dolbeare, F., and Koenig, H. (1970). *Proc. Soc. Exp. Biol. Med.* **135**, 636.
121. Marushige, K., and Bonner, J. (1971). *Proc. Nat. Acad. Sci. U.S.* **68**, 2941.
122. Reeck, G. R., Simpson, R. T., and Sober, H. (1972). *Proc. Nat. Acad. Sci. U.S.* **69**, 2317.
123. Loeb, J. E., and Cruezet, C. (1970). *Bull. Soc. Chim. Biol.* **52**, 1007.
124. Platz, R. D., Kish, V. M., and Kleinsmith, L. J. (1970). *FEBS Lett.* **12**, 38.
125. Teng, C. T., Teng, C. S., and Allfrey, V. G. (1970). *Biochem. Biophys. Res. Commun.* **41**, 690.
126. Chytil, F., and Spelsberg, T. C. (1971). *Nature (London), New Biol.* **233**, 215.
127. Spelsberg, T. C., Steggles, A. W., Chytil, F., and O'Malley, B. W. (1972). *J. Biol. Chem.* **247**, 1368.
128. Zardi, L., Lin, J. C., and Baserga, R. (1973). *Nature (London), New Biol.* **245**, 211.
129. Wang, T. Y. (1971). *Exp. Cell Res.* **69**, 217.
130. Richter, K. H., and Sekeris, C. E. (1972). *Arch. Biochem. Biophys.* **148**, 44.
131. Marushige, K., and Ozaki, H. (1967). *Develop. Biol.* **16**, 474.
132. Hill, R. J., Poccia, D. L., and Doty, P. (1971). *J. Mol. Biol.* **61**, 445.

133. Seale, R. L., and Aronson, A. I. (1973). *J. Mol. Biol.* **75**, 633.
134. Spelsberg, T. C., Mitchell, W. M., Chytil, F., Wilson, E. M., and O'Malley, B. W. (1973). *Biochim. Biophys. Acta.* **312**, 765.
135. LeStourgeon, W. M., and Rusch, H. P. (1971). *Science* **174**, 1233.
136. Vidali, G., Boffa, L. C., Littau, V. C., Allfrey, K. M., and Allfrey, V. G. (1973). *J. Biol. Chem.* **248**, 4065.
137. LeStourgeon, W. M., and Rusch, H. P. (1973). *Arch. Biochem. Biophys.* **155**, 144.
138. Gershey, E. L., and Kleinsmith, L. J. (1969). *Biochim. Biophys. Acta* **194**, 519.
139. Shelton, K. R., and Allfrey, V. G. (1970). *Nature (London)* **228**, 132.
140. Teng, C. S., and Hamilton, T. H. (1970). *Biochem. Biophys. Res. Commun.* **40**, 1231.
141. Enea, V., and Allfrey, V. G. (1973). *Nature (London)* **242**, 265.
142. Helmsing, P., and Berendes, H. (1971). *J. Cell Biol.* **50**, 893.
143. Stein, G., and Baserga, R. (1970). *J. Biol. Chem.* **245**, 6097.
144. Levy, R., Levy, S., Rosenberg, S. A., and Simpson, R. T. (1973). *Biochemistry* **12**, 224.
145. Johnson, E. M., Karn, J., Vidali, G., and Allfrey, V. G. (1974). *J. Biol. Chem.* September, 1974.
146. Rovera, G., and Baserga, R. (1971). *J. Cell. Physiol.* **77**, 201.
147. Buck, M. D., and Schauder, P. (1970). *Biochim. Biophys. Acta* **224**, 644.
148. Swaneck, G. E., Chu, L., and Edelman, I. (1970). *J. Biol. Chem.* **245**, 5382.
149. Ruddon, R. W., and Rainey, C. H. (1970). *Biochem. Biophys. Res. Commun.* **40**, 152.
150. Platz, R. D., and Hnilica, L. S. (1973). *Biochem. Biophys. Res. Commun.* **54**, 222.
151. Kleinsmith, L. J., Allfrey, V. G., and Mirsky, A. E. (1966). *Science* **154**, 780.
152. Ahmed, K. (1971). *Biochim. Biophys. Acta* **243**, 38.
153. Ahmed, K., and Ishida, H. (1971). *Mol. Pharmacol.* **7**, 323.
154. Karn, J., Johnson, E. M., Vidali, G., and Allfrey, V. G. (1974). *J. Biol. Chem.* **249**, 667.
155. Platz, R. D., Stein, G. S., and Kleinsmith, L. J. (1973). *Biochem. Biophys. Res. Commun.* **51**, 735.
156. Benjamin, W. B., and Goodman, R. M. (1969). *Science* **166**, 629.
157. Langan, T. A. (1967). *In* "Regulation of Nucleic Acid and Protein Biosynthesis" (V. V. Koningsberger and L. Bosch, eds.), p. 233. Elsevier, Amsterdam.
158. Kleinsmith, L. J., Allfrey, V. G., and Mirsky, A. E. (1966). *Proc. Nat. Acad. Sci. U.S.* **55**, 1182.
159. Kleinsmith, L. J., and Allfrey, V. G. (1969). *Biochim. Biophys. Acta* **175**, 123.
160. Kleinsmith, L. J., and Allfrey, V. G. (1969). *Biochim. Biophys. Acta* **175**, 136.
161. Riches, P. G., Harrad, K. R., Sellwood, S. M., Rickwood, D., and MacGillivray, A. J. (1973). *Biochem. Soc. Trans.* **1**, 70.
162. Rickwood, D., Riches, P. G., and MacGillivray, A. J. (1973). *Biochim. Biophys. Acta.* **299**, 162.
163. Benjamin, W. B., and Gellhorn, A. (1968). *Proc. Nat. Acad. Sci. U.S.* **59**, 262.

164. Schiltz, E., and Sekeris, C. E. (1971). *Experientia* **27**, 30.
165. Jungmann, R. A., and Schweppe, J. S. (1972). *J. Biol. Chem.* **247**, 5535.
166. Kleinsmith, L. J., Heidema, J., and Carroll, A. (1970). *Nature (London)* **226**, 1025.
167. Bekhor, I., Kung, G. M., and Bonner, J. (1969). *J. Mol. Biol.* **39**, 351.
168. van den Broek, H. W. J., Nooden, L. D., Sevall, S., and Bonner, J. (1973). *Biochemistry* **12**, 229.
169. Kleinsmith, L. J. (1973). *J. Biol. Chem.* **248**, 5648.
170. Patel, G. L., and Thomas, T. L. (1973). *Proc. Nat. Acad. Sci. U.S.* **70**, 2524.
171. Chaudhuri, S., Stein, G., and Baserga, R. (1972). *Proc. Soc. Exp. Biol. Med.* **139**, 1363.
172. Inoue, A., Littau, V. C., and Allfrey, V. G. (1974). *J. Mol. Biol.* (in press).
173. Huang, R. C. C., and Huang, P. C. (1969). *J. Mol. Biol.* **39**, 365.
174. Gilmour, R. S., and Paul, J. (1969). *J. Mol. Biol.* **40**, 137.
175. Carlsson, U., Henderson, L. E., and Lindskog, S. (1973). *Biochim. Biophys. Acta* **310**, 376.
176. Pusztai, A. (1966). *Biochem. J.* **101**, 265.
177. Levy, S., Simpson, R. T., and Sober, H. A. (1972). *Biochemistry* **11**, 1547.
178. Barrett, T., and Gould, H. J. (1973). *Biochim. Biophys. Acta* **294**, 165.
179. Britten, R. J., and Kohne, D. E. (1968). *Science* **161**, 529.
180. Britten, R. J., and Smith, J. (1969). *Carnegie Inst. Wash., Yearb.* **68**, 378.
181. Kohne, D. E., and Britten, R. J. (1971). *Procedures Nucl. Acid Res.* **2**, 500.
182. McCarthy, B. J., and Church, R. B. (1970). *Annu. Rev. Biochem.* **39**, 131.
183. Walker, P. M. B. (1971). *Progr. Biophys. Mol. Biol.* **23**, 143.
184. Paul, J., and Gilmour, R. S. (1968). *J. Mol. Biol.* **34**, 305.
185. Kamiyama, M., and Wang, T. Y. (1970). *Biochim. Biophys. Acta* **228**, 563.
186. Kamiyama, M., Dastugue, B., Defer, N., and Kruh, J. (1972). *Biochim. Biophys. Acta* **277**, 576.
187. Kostraba, N. C., and Wang, T. Y. (1972). *Biochim. Biophys. Acta* **262**, 162.
188. Kostraba, N. C., and Wang, T. Y. (1972). *Cancer Res.* **32**, 2348.
189. Tan, C. H., and Miyagi, M. (1970). *J. Mol. Biol.* **50**, 641.
190. Reeder, R. H. (1973). *J. Mol. Biol.* **80**, 229.
191. Church, R. B., and McCarthy, B. J. (1968). *Biochem. Genet.* **2**, 55.
192. Melli, M., and Bishop, J. O. (1969). *J. Mol. Biol.* **40**, 117.
193. Jelinek, W., and Darnell, J. E. (1972). *Proc. Nat. Acad. Sci. U.S.* **69**, 2537.
194. Ryskov, A. P., Church, R. B., Bajszar, G., and Georgiev, G. P. (1973). *Mol. Biol. Rep.* **1**, 119.
195. Georgiev, G. P. (1972). *Curr. Top. Develop. Biol.* **7**, 1.
196. Dina, D., Crippa, M., and Beccari, E. (1973). *Nature (London) New Biol.* **242**, 101.

2

Isolation of the Nuclear Acidic Proteins, Their Fractionation, and Some General Characteristics

GORDHAN L. PATEL

I. Introduction	30
II. Isolation of Cell Nuclei and Subnuclear Components	32
A. Nuclei	32
B. Chromatin	33
C. Nucleoli	34
III. Acidic Proteins of the Chromatin	35
A. General Considerations	35
B. Methods for Isolation of Acidic Proteins	36
C. Fractionation and Some Characteristics of the Chromatin Acidic Proteins	43
IV. Acidic Proteins of the Nucleolus	47
A. Isolation of Proteins	47
B. Fractionation and Some Characteristics of the Nucleolar Acidic Proteins	48
V. Acidic Proteins of the Nucleoplasm	50
A. Isolation of Acidic Proteins of the Nucleoplasm	50
B. Fractionation and Some Characteristics of the Nucleoplasmic Acidic Proteins	52
VI. Concluding Remarks	52
References	53

I. INTRODUCTION*

The ultimate goal of studies on proteins of the nucleus is to elucidate their role in the structure of the nucleus and subnuclear components and in the nuclear metabolism, specifically, in the expression and regulation of the genetic information. Their isolation, fractionation, and characterization, therefore, is an important prerequisite for these objectives. These accomplishments would enable studies on biochemical and biophysical interactions among all nuclear macromolecules, from which total elucidation of the biology of the nucleus would emerge.

The proteins of the cell nucleus have been categorized as histone and nonhistone proteins. Histones, which comprise the majority of basic proteins of the nucleus and are integral components of the genetic apparatus, have been investigated intensely during the past two decades. Numerous monographs and reviews have been written on this subject (e.g., Phillips, 1971; DeLange and Smith, 1971; Elgin et al., 1971; Wilhelm et al., 1971; Stellwagen and Cole, 1969; Butler et al., 1968; Busch, 1965). In most of these only passing reference is made to nonhistone proteins which constitute a large majority of total proteins of the nucleus. But for a few exceptions, the majority of nonhistone proteins are considered acidic on the basis of their amino acid composition and electrophoretic mobility. Even though the importance of at least some of the nuclear acidic proteins in chromosomal structure and function and in nuclear metabolism was indicated by the works of Stedman and Stedman (1943, 1944) and Mirsky and co-workers (Mirsky and Pollister, 1946; Mirsky and Ris, 1947, 1950; Daly et al., 1952) several years ago, investigations on this class of proteins have been lacking for the following reasons. (1) Unlike histones, which can be solubilized in aqueous solvents suitable for biochemical studies, many of the acidic proteins are insoluble in aqueous buffered solvents approximating physiological conditions of ionic strength, pH, etc. Even those acidic proteins that are directly extractable or can be rendered soluble in such solvents, following their dissociation from other nuclear macromolecules, behave as notorious self-aggregating systems. These properties have made their fractionation and characterization by conventional methods difficult. (2) While histones are DNA-associated proteins of chromosomes, the nonhistone pro-

---

* Abbreviations used in this chapter: AP-NH, acidic proteins with nucleohistone affinity; CAP-I, irreversibily dissociated chromatin acidic protein; CM-, carboxymethyl-; DEAE, diethylaminoethyl-; DNA, deoxyribonucleic acid; DNH, deoxynucleohistone; DNP, deoxynucleoprotein; DOC, deoxycholate; HAP, hydroxyapatite; RNA, ribonucleic acid; SDS, sodium dodecyl sulfate.

tein class is comprised of acidic proteins associated with all subnuclear components: chromosomes, nucleoli, nucleoplasm, and nuclear membrane. They may have structural, enzymatic, gene regulatory, and many other, as yet unknown, nuclear functions. At least in part, therefore, meaningful studies of nuclear acidic proteins have had to await the development of methods for isolation of subnuclear components in high purity and yield and for dissociation and separation of their macromolecules.

In early studies, proteins of the different fractions, remaining insoluble after isolated nuclei were extracted with (1) dilute HCl or H_2SO_4, (2) 0.14 M NaCl and dilute HCl or H_2SO_4, (3) 0.14 M and 1–2 M NaCl, or (4) 0.14 M and 1–2 M NaCl and 0.05 M NaOH; and the acid-insoluble proteins of the deoxynucleoprotein extracted with 1–2 M NaCl, were solubilized with dilute alkali, and referred to as nuclear acidic or residual proteins. In all cases proteins with different properties and intranuclear locations were obtained. They were considered acidic and distinguished from histones by the predominance of acidic over basic amino acid residues in their composition and by the presence of tryptophan which histones characteristically lack. However, since the content of glutamine and asparagine residues has not been carefully determined for these proteins, some of the acidic proteins may actually carry a net positive charge in their native state. The exposure of these proteins to acid and base in their isolation and solubilization clearly resulted in their denaturation, which made such preparations unsuitable for many biochemical experiments in the past. Hnilica (1967) and Fambrough (1969) have reviewed the early work on nuclear acidic proteins in detail.

Recently reported approaches, which avoid harsh conditions, for the isolation of acidic proteins can be classified into two groups. In some laboratories they have been isolated from purified subnuclear components, while in others purified nuclei have been extracted, successively, with different solvents to obtain protein fractions distinguished on the basis of their differential solubilities. In the former, one has the advantage of assigning proteins to specific, morphologically identifiable, intranuclear components. But, one also runs the risk of losing some acidic proteins of the nuclear component in question by manipulations involved in its isolation and/or of adsorbing nonspecific proteins from the nucleoplasm and the cytoplasm (Johns and Forrester, 1969). Since many of the interesting nuclear acidic proteins may not be static structural and functional components and may in fact exist in a dynamic equilibrium among chromatin, nucleoli, nucleoplasm, and even cytoplasm, the consideration of their precise intranuclear localization is academic. This is substantiated by the gross similarities among the various nuclear acidic protein fractions (Hnilica, 1967). Successive extraction of nuclei with varying solvents to obtain protein fractions on the basis of their solubility allows complete accounting of total nuclear proteins. Cytological correlation of such fractions to nuclear components has been established (Georgiev, 1958; Zbarskii and Georgiev, 1959; Georgiev and Chentsov, 1962; Zbarskii et al., 1962). This approach is also of obvious advantage. Combinations of these two approaches will certainly prove valuable in furthering our understanding of all nuclear acidic proteins.

The purpose of this chapter is to bring together various methods that have been reported in recent years for isolating nuclear acidic protein fractions in aqueous solvents and to describe some of the characteristics of the proteins. Methods for the isolation of acidic proteins soluble in aqueous phenol and of acidic phosphoproteins are described elsewhere in this volume (Chapters 3, 4, and 5). The reader is referred to some excellent recent reviews describing other aspects of nuclear acidic proteins (Paul, 1970; MacGillivray et al., 1972a; Stein and Baserga, 1972; Spelsberg et al., 1972).

II. ISOLATION OF CELL NUCLEI AND SUBNUCLEAR COMPONENTS

A. Nuclei

Ideally, studies of isolated nuclei require that they be identical to their state *in vivo* and free of any cytoplasmic contamination and that they have not lost any natural components. None of the methods that have been described for the isolation and purification of cell nuclei meet all of these criteria. In actual practice the choice of the method in a particular study is a compromise dictated by the nature of the nuclear function under investigation. The most commonly used current methods combine the procedures of Hogeboom et al. (1952, 1953) and Chauveau et al. (1956). Cells are first disrupted by homogenization in dilute sucrose solution (usually 0.25 M) supplemented with Ca^{2+} or Mg^{2+} (1–50 mM), and the released nuclei are sedimented by low speed centrifugation (700–1000 g, 10 minutes) of the filtered homogenate. The crude nuclear preparation is then further purified by sedimentation of nuclei from dense sucrose solutions (2.2–2.4 M). Minor modifications of these basic methods have been reported by many for isolation of nuclei of satisfactory purity from various cell types. Alternately, nuclei also have been isolated by homogenization of cells in dilute citric acid (Dounce, 1963). This procedure has been particularly useful in isolating nuclei from tumor cells and for isolation of nuclei with undegraded nucleic acids. Because of its ability to solubilize cytoplasmic material the citric acid procedure yields morphologically pure nuclei. Detergents have also been employed in aqueous methods to obtain pure nuclei (Penman, 1966).

Although these aqueous methods yield nuclei in sufficient quantity and of morphological purity and integrity, translocation of nuclear and cytoplasmic components soluble in aqueous media is unavoidable. In this respect, the method for isolating nuclei in nonaqueous solvents first described by Behrens (1932) has clear advantages. In modifications of this method frozen and lyophilized cells are disrupted in organic solvents and the released nuclei are purified by partition in mixtures of organic solvents. Among disadvantages of this approach are loss of nuclear lipids and denaturation of nuclear proteins.

For the details and discussions of the various methods for isolating cell nuclei the reader is referred to Busch (1967), Busch et al. (1972), Siebert (1967), Wang (1967), and Dounce (1963).

B. Chromatin

Chromatin is defined as chromosomes in a dispersed state in the interphase nucleus. It is a macromolecular complex of DNA, RNA, and proteins. The method of Zubay and Doty (1959) with some modifications has been used for isolation of the chromatin from a variety of sources. In general, the tissue or isolated nuclei are disrupted by homogenization in $0.075\ M$ NaCl–$0.024\ M$ EDTA (pH 8), the homogenate filtered through cheesecloth, and a particulate fraction pelleted by centrifugation (1000–4000 g) of the filtrate. The pellet is washed several times with dilute Tris buffer (0.01–0.002 M) and then sedimented across a 1.7 M sucrose barrier. The final gelatinous pellet is the purified chromatin. For details of methods and criteria of purity, see Bonner et al. (1968). Alternatively, isolated nuclei are extensively washed with 0.14 M NaCl and dilute buffers and solubilized in 1–2 M NaCl (Wang, 1967) containing 5 M urea (Gilmour and Paul, 1969). A small amount of insoluble material is pelleted by centrifugation (40,000 g; 30–60 min) and the viscous supernatant is referred to as the chromatin solution.

Insofar as the exact composition of the chromatin as it exists *in vivo* is not known, evaluation of its nativeness in isolated preparations is difficult. In consideration of its acidic proteins one is never certain whether some native molecules are not lost and/or nonspecific ones adsorbed during the isolation process (Johns and Forrester, 1969; Wilhelm et al., 1972). In general, chromatin preparations exhibiting reproducible chemical composition and transcriptional template restriction similar to that observed *in vivo* have been considered satisfactory.

C. Nucleoli

The nucleolus played a central role in cytological investigations for many years because of its prominence in the nucleus and because of the dynamic morphological changes it undergoes during the life of a cell. The correlation between high protein synthetic activity of a cell and high "activity" of the nucleolus was recognized sometime ago. As a result of new methodological developments in the isolation and characterization of nucleic acids and in their analysis by molecular hybridization it has now become clear that the nucleolus is a specialized compartment of the genome for manufacturing ribosomal ribonucleic acids.

Biochemical characterization of the nucleolus became possible with its bulk isolation first reported by Vincent (1952) and Monty et al. (1956). Further improvements in the isolation procedures have been made since, and it is now possible to obtain large quantities of nucleoli from a variety of tissues. Nucleoli have been isolated from purified nuclei in aqueous and nonaqueous media. In the most commonly used aqueous procedures purified nuclei are disrupted in 0.34 M sucrose solutions containing Ca^{2+} by vigorous homogenization or sonication, and nucleoli are purified by differential centrifugation through a 0.88 M sucrose barrier. The details of the isolation procedures have been reviewed elsewhere (Maramatsu and Busch, 1967; Busch, 1967; Busch et al., 1972). Zalta et al. (1972) have substituted sucrose with 2.1% Ficoll in the disruption of nuclei. This concentration of Ficoll is equivalent in viscosity to 0.3 M sucrose and it prevents the swelling of nuclei which is common with sucrose media. Ca^{2+} has also been substituted with Mg^{2+}; acetic acid-extractable proteins of nucleoli isolated in the presence of these divalent cations are electrophoretically similar (Higashinakagawa et al., 1972). Since some enzymes are inhibited by Ca^{2+} while others are activated by Mg^{2+} the choice of the divalent ion in the media will depend on the nature of the intended biochemical investigations. Nucleoli isolated by these procedures retain their *in vivo* morphology and contain undegraded RNA. However, some soluble proteins and other molecular components, which may be natural constituents of nuclei, are undoubtedly lost in the aqueous media. Isolation of nucleoli in nonaqueous media was recently described by Shakoori et al. (1972). In their method rat liver nuclei isolated in nonaqueous organic solvent (Siebert, 1967) were disrupted by vigorous homogenization in organic solvents and nucleoli were purified by zonal centrifugation. The maximum purity of nucleoli thus prepared was 70%, indicating considerable contamination by extranucleolar material. The loss of natural nucleolar components in nonaqueous procedure, however, is minimal.

In addition to the morphologically identifiable nucleoli isolated by the above procedures the nuclear "residual" fraction, which remains insoluble after purified nuclei are successively extracted with 0.14 M NaCl and 1–2 M NaCl, has also been equated to nucleoli. Zbarskii and Georgiev (1959), Georgiev and Chentsov (1962), and Zbarskii et al. (1962) showed by direct microscopic observations that these solvents remove nuclear sap and deoxynucleoprotein, respectively, leaving insoluble a residue of nucleoli and nuclear membranes. Solubilization of the majority of this residue has been accomplished by use of sodium deoxycholate (Wang, 1966a, 1967; Patel et al., 1968).

Proteins represent the largest molecular component of isolated nucleoli. The small amount of DNA contains the ribosomal RNA cistrons, and the RNA must be the transcription product. The relative amount of these components will undoubtedly vary depending on the physiological fluctuations of the nucleolus.

III. ACIDIC PROTEINS OF THE CHROMATIN

A. General Considerations

Among all the acidic proteins of the nucleus those of the chromatin have received by far the greatest attention during the past six years. Special interest in this class of proteins is related to their possible role in structure, function, and regulation of the genome. Mirsky and Pollister (1946) first reported the presence of acidic proteins in the chromatin fraction solubilized with 1 M NaCl from isolated nuclei. Most of the acidic proteins in this extract can be separated from DNA and histones by precipitation of deoxynucleohistone at 0.14 M NaCl. Partial fractionation and characterization of the chromatin acidic proteins thus isolated were reported from the laboratory of Wang (1967; Patel and Wang, 1964). Steele and Busch (1963) also reported the separation of basic and acidic proteins in the chromatin extract on the basis of their differential solubility in dilute H_2SO_4 and in 6% p-aminosalicylate–phenol, respectively.

Recent studies on the chromatin acidic proteins were spurred by the initial reports of Paul and Gilmour (1968), Gilmour and Paul (1970), and Spelsberg et al. (1971) and later of others (Stein et al., 1972; Kostraba and Wang, 1973) who showed that, although histones indeed restrict the template activity of DNA for transcription in the chromatin, the tissue and species specificity of the restriction was conferred by the acidic proteins. This conclusion was based on qualitative comparisons of RNA synthesized on native and reconstituted chromatin and on hybrid reconstituted chromatin which contained histones and acidic proteins from different tissues. The results showed that RNA transcribed from hybrid chromatin were similar to those synthesized on native or reconstituted chromatin which served as the source of acidic proteins in the hybrid chromatin. Furthermore, Wang (1970) and Kamayama and Wang (1971) showed that a fraction of acidic proteins purified by ion-exchange chromatography activated *in vitro* transcription of new RNA species from chromatin. Later, this activation was shown to be specific to the tissue of origin of the acidic protein fraction (Wang, 1971; Kostraba and Wang, 1972a,b), which indicated selective derepression of the eukaryotic chromatin. Acidic phosphoproteins of chromatin also were shown to enhance the template activity of pure DNA in a species-specific manner (Teng *et al.*, 1971; Rickwood *et al.*, 1972; Shea and Kleinsmith, 1973). The chromosomal regulatory function of the acidic proteins suggested by these observations is further supported by the relative species specificity of their binding to DNA *in vitro* (Kleinsmith *et al.*, 1970; C. T. Teng *et al.*, 1970; C. S. Teng *et al.*, 1971; Patel and Thomas, 1973). However, only a very small fraction of the chromatin acidic proteins may be involved in a gene regulatory function. A majority of them may be enzymes (polymerases, nucleases, proteases, etc.), in a transient or stable association with the chromatin, and structural components. The isolation and separation of all the chromatin acidic proteins, therefore, are of utmost importance.

B. Methods for Isolation of Acidic Proteins

The acidic proteins have been isolated from chromatin preparations by (1) selective extraction of histones followed by dissociation and separation of acidic proteins from the nucleic acids and by (2) dissociation of the total chromatin macromolecules followed by (a) separation of histones, acidic proteins, and nucleic acids by HAP column chromatography, (b) removal of nucleic acids by ultracentrifugation, exclusion chromatography, or selective precipitation and subsequent separation of the acidic proteins from histones by ion-exchange chromatography, (c) separation of the acidic proteins from histones and DNA by reconstitution and precipitation of nucleohistone at 0.14 M NaCl, or (d) a combination of the above.

1. Isolation of Acidic Proteins following "Dehistonation"

The most commonly used method for removal of histones from nuclei or subnuclear fractions is their extraction with dilute HCl or H_2SO_4 (0.1–0.5 N). Readily extracted histones are separated from the insoluble material by low speed centrifugation. The pellet contains nucleic acids and different acidic proteins depending on the starting material. Sonnenbichler and Nobis (1970) have cautioned, however, that exposure of nuclear material to HCl or H_2SO_4 renders a significant amount (30%) of histones insoluble in acid, and these will contaminate the acidic proteins derived from the acid-insoluble fraction. Seale and Aronson (1973) reported that about 10% of purified histones added to DNA, indeed, were not recovered during subsequent acid extraction of the mixture. On the other hand the presence of nonhistone proteins in the acid-extracted histones of various tissues has also been reported (Sadgopal and Bonner, 1970a,b; Smith and Stocken, 1973). These observations indicate incomplete separation of histones and acidic proteins by this method. Exposure of acidic proteins to acids may also have deleterious denaturation effects limiting their usefulness for subsequent biochemical characterization. Spelsberg *et al.* (1971) have shown that histones can be selectively released from the chromatin with 2 M NaCl–0.05 M sodium acetate (pH 6). The DNA–acidic protein complex can then be pelleted by ultracentrifugation. Selectivity of histone extraction depends critically upon the pH, since at higher pH in 2 M NaCl all macromolecular components of chromatin begin to dissociate.

The isolation of acidic proteins from nucleic acids after the removal of histones has been achieved by the following methods:

(a) For studies where the objective is limited to determining either total quantity or some general parameters, such as chemical composition, isotope content, etc., the nucleic acids are hydrolyzed by digestion with 5% trichloroacetic acid or 0.5 N perchloric acid at 90°C and the insoluble acidic proteins solubilized in alkali or various solubilizers used in liquid scintillation counting. The proteins thus obtained are completely denatured and cannot be used for most biochemical investigations. Using this method the acidic proteins have been reported to vary quantitatively and metabolically during development (Dingman and Sporn, 1964) and in chromatin subfractions (Frenster, 1965; Frenster *et al.*, 1963). Recently, Chanda and Cherian (1973) have shown mercury-binding proteins in HCl- and trichloroacetic acid-insoluble residue of the chromatin.

(b) Separation of the acidic proteins from nucleic acids is also achieved by equilibrium centrifugation in CsCl. "Dehistoned" chromatin solubilized in 4 M CsCl containing 5 mM Na-EDTA–0.01 M mercaptoethanol–0.02 M lysine (pH 11.6) and centrifuged for 60–70 hours at 45,000 rpm (Spinco SW 50 rotor) yielded acidic proteins in the upper region of the centrifuge tube (Benjamin and Gellhorn, 1968). Although such proteins were amenable to acrylamide gel electrophoretic analysis, which showed them to be very heterogeneous, quantitative data for the efficiency of protein dissociation and recovery were not given.

(c) Alternatively, the acidic proteins have been dissociated by solubilization of the "dehistoned" chromatin in 0.05 M Tris (pH 8)–0.1% SDS at 37°C (Marushige et al., 1968; Elgin and Bonner, 1970) and in 8 M urea–1% SDS–5 mM mercaptoethanol (pH 11.5) (Seale and Aronson, 1973) and separated from DNA by centrifugation for 16–18 hours at 36,000 rpm (Spinco SW 39 or SW 50 rotor). Over 90% of the acidic proteins in the "dehistoned" chromatin were recovered in the supernatant; the upper two-thirds and lower one-third of the tubes contained 41 and 49% of protein, respectively. Electrophoresis in SDS–acrylamide gels was reported to show that all major components of the chromatin acidic proteins were recovered and that their separation from histones was complete (Elgin and Bonner, 1970). The disadvantage of using SDS is the protein denaturation caused by this detergent. It is known to bind tightly to proteins and to markedly influence their physical state. Although SDS can be removed partially by precipitation with KCl or adsorption to ion-exchange resin, concomitant precipitation of at least some of the acidic proteins results in reduced and selective yields. Furthermore, the completeness of SDS removal and reversibility of protein denaturation effect still remains to be established.

(d) Recently Wilson and Spelsberg (1973) have described the use of DNase digestion for quantitative recovery of acidic proteins. "Dehistoned" chromatin was suspended in 0.1 M Tris-HCl (pH 7.5)–2 mM MgCl$_2$–2 mM CaCl$_2$ and incubated with DNase I (25 μg DNase/1 mg chromatin in 2 ml). Upon completion of DNA digestion the acidic proteins were precipitated by the addition of excess 0.4 N HC1O$_4$ and collected by centrifugation at 2000 g for 10 minutes. This procedure is rapid and yields >95% of the acidic proteins. MacKay et al. (1968) also used DNase digestion to prepare insoluble "residual" proteins.

2. Isolation of the Acidic Proteins by Dissociation and Separation of the Chromatin Components

Methods for the isolation of acidic proteins following dissociation of the chromatin have their origin in the studies of Mirsky and Pollister (1942, 1946) who used solvents of high ionic strength to solubilize deoxynucleoproteins. Presumably, the disruption of interactions among various macromolecular components under these conditions allows their separation from one another. In current methods urea, guanidine hydrochloride, formic acid, and SDS are also used to maximize the dissociation of proteins and nucleic acids. Under these conditions covalently linked nucleoproteins (Huang and Bonner, 1965) would remain intact.

The dissociation of chromatin macromolecules has been achieved in 1–3 M NaCl (Patel and Wang, 1964; Wang, 1967; Shaw and Huang, 1970; Graziano and Huang, 1971; van den Broek et al., 1973); in buffered 2–3 M NaCl containing 5–7 M urea (Gilmour and Paul, 1969, 1970; Bekhor et al., 1969; Huang and Huang, 1969; MacGillivray et al., 1971, 1972b; Spelsberg et al., 1971; Shaw and Huang, 1970; Patel, 1972; Umanskii et al., 1971; Yoshida and Shimura, 1972; Richter and Sekeris, 1972; Monahan and Hall, 1973); in 2–2.5 M guanidine hydrochloride–0.1 M Tris, pH 8 (Arnold and Young, 1972; Hill et al., 1971); in 0.4 M guanidine hydrochloride–6 M urea–0.1% β-mercaptoethanol–0.1 M sodium phosphate, pH 7 (Levy et al., 1972); 1% SDS (Shirey and Huang, 1969); and in 60% HCOOH made 0.2 M NaCl and 8 M urea (Elgin and Bonner, 1972). In all of these cases the chromatin fraction was dispersed in the solvent and stirred until solubilized. Any undissolved matter was removed by centrifugation and the supernatant represented the dissociated chromatin macromolecules.

a. One-Step Separation of Chromatin Macromolecules. MacGillivray et al. (1971, 1972b) have reported that histones, acidic proteins, and nucleic acids of the chromatin dissociated in 2 M NaCl–5 M urea–1 mM sodium phosphate (pH 6.8) can be separated by column chromatography on hydroxylapatite (HAP). HAP allows chromatographic separation in the presence of NaCl–urea (Bernardi and Kawasaki, 1968). Differential elution of various macromolecules was effected by stepwise increase of phosphate buffer concentration in the elution solvent. At 1 mM phosphate histones did not adsorb to HAP and appeared in the breakthrough. Bulk of the acidic proteins were eluted by 0.05 M phosphate; additional, smaller amounts eluted with the RNA at higher phosphate concentrations (0.1 M and 0.2 M). Virtually all of the DNA applied to the column was eluted at 0.5 M phosphate. Depending on the method of chromatin preparation 41–100% of chromatin proteins were recovered. Thus, this method has the advantage of total recovery of proteins with certain chromatin preparations and of the ease of separating all chromatin macromolecules in a single operation.

b. Sequential Separation of Chromatin Macromolecules. The nucleic acids have been separated from the proteins by three basic methods. Ultracentrifugation of the dissociated chromatin extract results in the recovery of proteins in the supernatant (Gilmour and Paul, 1969;

Elgin and Bonner, 1972; Shirey and Huang, 1969; Shaw and Huang, 1970; Arnold and Young, 1972; Umanskii et al., 1971; Levy et al., 1972; Richter and Sekeris, 1972; Monahan and Hall, 1973). The yield of total proteins, about 90%, is in general similar for the various dissociating solvents. The absence of urea in the NaCl solvent or the use of guanidine hydrochloride reduces the time and speed of ultracentrifugation for pelleting nucleic acids. Shirey and Huang (1969) have reported that the small amount (about 10%) of protein that cosediments with nucleic acids is tightly bound and cannot be separated by exclusion chromatography of the ultracentrifuge pellet dissolved in SDS. Levy et al. (1972), on the other hand, reported that pellet-associated proteins can be released by SDS and separated from DNA by centrifugation. The different chromatin dissociation conditions employed in these studies may explain this disparity. Since the chromatin proteins that cosediment with nucleic acids have not been characterized fully, it remains to be determined whether they are qualitatively similar to or different from the bulk of chromatin proteins recovered. Holoubeck et al. (1966) have reported that acidic proteins tightly bound to DNA exhibit different metabolic activity than the dissociable proteins.

In other studies the high molecular weight nucleic acids have been separated from the dissociated total proteins by gel filtration on Bio-Gel A-50m (Shaw and Huang, 1970; Graziano and Huang, 1971; van den Broek et al., 1973) or on Sepharose 4B (Hill et al., 1971). Fractionation of the chromatin dissociated in 3 M NaCl resulted in the recovery of 66.6 and 13.2% of applied DNA and protein, respectively, in the high moleclar weight fraction and 72 and 5.8% of proteins and DNA, respectively, in the low molecular weight fraction (Graziano and Huang, 1971). The cross-contamination indicated by these values may be due to incomplete dissociation of the chromatin in the absence of urea or due to limitations of the chromatography. Chromatin dissociated in 27% guanidine hydrochloride and chromatographed on Sepharose 4B yielded 2–15% of total proteins in the excluded nucleic acid fraction. However, at higher temperature (37°C) and guanidine hydrochloride concentration (37%), dissociation of the chromatin was nearly complete and only a negligible amount of protein eluted with nucleic acids (Hill et al., 1971). The low ratio of sample/gel-bed required for satisfactory resolution of macromolecules by gel filtration and the slow flow rates limit the use of these methods for large-scale processing of the chromatin.

Alternately, Yoshida and Shimura (1972) have precipitated the nucleic acids of the calf thymus chromatin dissociated in 2 M NaCl–5 M urea–0.01 M Tris (pH 7.9) by addition of $LaCl_3$. They have shown that the selectivity of nucleic acid precipitation depends on the presence of both

NaCl and urea in high concentration, and upon pH and $LaCl_3$ concentration. Under optimal conditions (500 μg DNA/ml; pH 7.9; 8 mM $LaCl_3$) no protein is precipitated and the supernatant gives a typical protein absorption spectrum (A275 nm/260 nm = 1.66). This procedure has the advantage of speed in separating nucleic acids from proteins.

Following the separation of total proteins from nucleic acids, the acidic proteins have been separated from the histones by chromatography on a variety of cation- and anion-exchange resins. Since histones and acidic proteins form insoluble complexes at low ionic strength (Wang, 1967; Wang and Johns, 1968), their separation requires continued presence of urea to maintain the proteins in solution. The order of acidic and basic protein elution depends on the choice of resin and starting conditions of sample application. With cation-exchange resins Bio-Rex 70 (Levy et al., 1972; Richter and Sekeris, 1972; van den Broek et al., 1973), CM-Sephadex (Hill et al., 1971), CM-cellulose (Yoshida and Shimura, 1972) or SP-Sephadex (Graziano and Huang, 1971; Arnold and Young, 1972; Elgin and Bonner, 1972), the acidic proteins elute first in the unadsorbed fraction or at low ionic conditions. They are also recovered in the unadsorbed fraction when total chromatin proteins in buffered 0.5 M NaCl are chromatographed on Amberlite CG-50 (Umanskii et al., 1971). Conversely, the acidic proteins are adsorbed to QAE-Sephadex (Gilmour and Paul, 1969; Arnold and Young, 1972) or DEAE-cellulose (Monahan and Hall, 1973) at low ionic conditions and can be eluted at increased ionic strength, after washing the column with the starting solvent to remove unadsorbed basic proteins. Trace amounts of nucleic acids that may contaminate the proteins can also be separated by manipulating the ionic strength of elution solvents.

The separation of acidic proteins from histones by ion-exchange chromatography has been found to be not absolute. Arnold and Young (1972) reported that the acidic proteins recovered from QAE-Sephadex were contaminated with F_3 histones and that histones also contained some acidic proteins. With SP-Sephadex (Graziano and Huang, 1971) the acidic proteins eluting in 0.23 M NaCl–7 M urea were free of histones, but about 10% that eluted with the histones at 0.3 and 0.4 M NaCl could not be separated by rechromatography. Yoshida and Shimura (1972) have reported that even in 5 M urea some of the acidic proteins aggregated to histones and were retained by CM-cellulose. No cross-contamination has been reported for histones and acidic proteins separated on Bio-Rex 70.

    c. Separation of the Acidic Proteins by Precipitation of the Nucleohistone. It has been known for long that the solubility of deoxynucleohistone (DNH) is minimal at 0.14 M NaCl. Above and below this value DNH solubility increases. Patel and Wang (1964) and Wang (1967) have exploited this property to separate acidic proteins from histones and DNA in the chromatin extract in 1–2 M NaCl. Reduction of NaCl concentration to 0.14 M results in precipitation of reconstituted DNH, leaving a majority of acidic proteins soluble. Since the chromatin extract is prepared from nuclei washed with 0.14 M NaCl, it is clear that acidic proteins thus isolated are rendered soluble in this solvent only after dissociation of the chromatin at high ionic strength. Partial fractionation and characterization of these proteins (Patel and Wang, 1964; Wang, 1967), their interactions with histones (Wang, 1967; Wang and Johns, 1968), their effect on *in vitro* transcription (Wang, 1966c), and isolation from them of a protein fraction that stimulates transcription from chromatin in a tissue-specific manner (Wang, 1970; Kamayama and Wang, 1971; Kostraba and Wang, 1972a,b) have been reported. Separation of acidic proteins by precipitation of the nucleohistone was employed also for isolation of phosphoproteins from the chromatin (Langan, 1967; Kleinsmith and Allfrey, 1969).

d. Separation of the Acidic Proteins on the Basis of Their Differential Reconstitution with Deoxynucleohistone. In my laboratory the acidic proteins of the chromatin are isolated into two classes on the basis of their differential properties of reconstitution with deoxynucleohistone. The procedure involves a sequential combination of the methods of Wang (1967) and MacGillivray *et al.* (1971, 1972b) described above in Sections III,B,2,a and c. Purified nuclei are exhaustively washed with 0.14 M NaCl–0.05 M Tris (pH 7.5) to remove the nucleoplasm and extracted with 2 M NaCl–5 M urea–0.01 M Tris (pH 8). The insoluble residue is removed by centrifugation at 45,000 g for 1 hour and the viscous clear supernatant representing chromatin solution is dialyzed overnight against 13 volumes of distilled water to reduce NaCl concentration to 0.14 M. The reconstituted deoxynucleoprotein that precipitates is separated from the soluble phase, which contains the bulk of the chromatin acidic proteins, by centrifugation at 45,000 g for 1 hour. The supernatant is referred to as irreversibly dissociable chromatin acidic proteins (CAP-I). The DNP pellet is redissociated and reconstituted once again and the CAP-I mixed with the first. The twice reconstituted DNP contains all of the DNA and histones in the original chromatin extract and a small amount of acidic proteins that reconstituted with the DNA and histones upon reversal of dissociating conditions. These latter acidic proteins are then separated from histones and nucleic acids by dissociation of the DNP in 2 M NaCl–5 M urea–1 mM phosphate (pH 8) and chromatography on hydroxylapatite as described by MacGillivray *et al.* (1971, 1972b). They are referred to as acidic proteins with high affinity for nucleohistone (AP-NH). The small amount of nucleic acids in CAP-I and AP-NH fractions is removed by passing the proteins, in 0.3 M NaCl–8 M urea–0.005 M phosphate (pH 8), through a DEAE-Sephadex A-25 column also equilibrated in the same solvent. Under these conditions nucleic acids are retained by the resin while the proteins are recovered in the breakthrough.

The nonhistone character of the AP-NH fraction has been established by amino acid composition and solubility properties (Patel, 1972). Amino acid composition of a CAP-I subfraction that does not adsorb to DEAE-Sephadex in the absence of NaCl indicates it to be slightly acidic and not contaminated with histones. These methods for isolating acidic proteins are simple and mild, yield quantitative recovery of proteins, and provide a basis for some functional distinctions among acidic proteins of the chromatin. Differential binding of CAP-I and AP-NH to DNA *in vitro* and partial species specificity of AP-NH binding to DNA have been shown (Patel, 1972; Patel and Thomas, 1973).

C. Fractionation and Some Characteristics of the Chromatin Acidic Proteins

The chromatin acidic proteins tend to aggregate and, in many cases, form insoluble precipitates in solvents appropriate for most conventional chromatographic and electrophoretic methods for fractionation of proteins. For example, their electrophoretic separation on starch gels (Patel and Wang, 1964; Wang, 1967) and on polyacrylamide disc gels (Loeb and Creuzet, 1969) did not give satisfactory resolution of protein bands. A large portion of the sample remained at the origin and considerable smearing and streaking was evident in the gels. Likewise, in their fractionation by ion-exchange chromatography (Patel and Wang, 1964; Wang, 1968; Howk and Wang, 1969), proteins eluted in a polydisperse peak rather than in discrete fractions. The chromatographic profiles observed perhaps reflected a gradual nonspecific dissociation of protein aggregates as well as ion-exchange process by increasing ionic strength of the eluant rather than elution by a true ion-exchange process. Wang (1967) fractionated chromatin acidic proteins into four groups: a high molecular weight nucleoprotein fraction which was pelleted by ultracentrifugation and three fractions from the ultracentrifuge supernatant by sequential precipitation at pH 6 and pH 5, and with ammonium sulfate. Electrophoretic separation of these subfractions on starch gels was also unsatisfactory. Greater complexity of the chromatin acidic proteins as compared to histones was apparent, nevertheless, in all of these early attempts to fractionate them.

The use of urea and reducing agents in sample preparation and acrylamide gel systems greatly improved the resolution of acidic proteins (e.g., Benjamin and Gellhorn, 1968). The best resolution, however, was seen when sodium dodecyl sulfate (SDS) (e.g., Elgin and Bonner, 1970) and SDS–urea (e.g., MacGillivray et al., 1971) were used in the electrophoresis. Extensive molecular heterogeneity of the acidic proteins from many different cell types was demonstrated by these electrophoretic systems. (For references see those for protein isolation in the previous Section B.) Both the lack and presence of tissue, species, and developmental specificity of these proteins have been claimed by different investigators (e.g., Kostraba and Wang, 1970; Loeb and Creuzet, 1969; MacGillivray et al., 1971; Hill et al., 1971). This discrepancy may be due to differences in the protein isolation methods as well as due to limitations of the electrophoresis systems. For example, since separation in SDS–acrylamide gel systems is primarily on the basis of the molecular size rather than charge of the native proteins, two bands of similar electrophoretic mobility may contain any number of distinct polypeptide chains which will not be distinguished. Furthermore, since SDS is a strong dissociating agent, these peptides *in vivo* may be subunits of different proteins. In view of these considerations direct comparison of total acidic proteins by SDS–acrylamide gel electrophoresis cannot elucidate their specificities.

Recently Barret and Gould (1973) have compared chromatin acidic proteins from three different tissues by their separation in a two-dimensional electrophoresis system. The method combined isoelectric focusing in the first dimension and SDS electrophoresis in the second dimension. Several minor spots and 11 to 18 prominent spots were revealed in the proteins examined. Only one protein was shown to be common to the three tissues. The patterns of liver proteins from rat and chicken were closer than those of liver and reticulocytes of the chicken. Application of this separation method may prove of great value in analysis of all nuclear acidic proteins.

Isoelectric focusing of total chromatin acidic proteins has shown that the majority of them band in pH 5–7 range; a few also band as low as pH 3.7 and as high as pH 9 (Elgin and Bonner, 1970; Arnold and Young, 1972; Barret and Gould, 1973).

Fractionation of the complex acidic protein mixture based on different parameters, prior to their electrophoretic analysis, may also aid in their characterization. Levy et al. (1972) have reported that rabbit liver chromatin acidic proteins yielded two major peaks of proteins when chromatographed on DEAE-cellulose, in the presence of urea, with a linear gradient of NaCl. With a more complex NaCl gradient, however, they were able to separate the proteins into many distinct peaks. Richter and Sekeris (1972) have compared the acidic proteins from rat liver, thymus, and kidney chromatin by ion-exchange chromatography on QAE-Sephadex in 5 M urea. The elution patterns of proteins from the three tissues were complex but generally similar. Comparison of the fractions by SDS–acrylamide gel electrophoresis showed many similarities and some tissue specific bands. Tissue specific differences were more apparent by electrophoresis in the absence of SDS. Patel and Kellar (1974) also have shown qualitative and quantitative variables in acidic proteins of rat liver and brain fractionated on DEAE-Sephadex and compared by SDS–acrylamide gel electrophoresis. In general, the brain proteins were more heterogeneous than those from liver.

One of the problems in evaluating the fractionation of chromatin acidic proteins has been the lack of knowledge of their functions. With the exception of some enzymes, specific identification tags are not available to monitor the extent of protein purification. The property that can prove useful in meaningful fractionation of these proteins is their ability to interact with DNA. It is reasonable to expect that at least those that have DNA-related functions (enzymatic, regulatory, and structural) will exhibit affinity for DNA. Binding of some acidic proteins to DNA *in vitro* has been reported (Wang, 1967; Marushige *et al.*, 1968; C. T. Teng *et al.*, 1970; C. S. Teng *et al.*, 1971; Chaudhuri *et al.*, 1972; Patel, 1972; Patel and Thomas, 1973). This binding property has been exploited for fractionation of chromatin acidic proteins by affinity chromatography on heterologous and homologous DNA–cellulose/polyacrylamide columns. By successive chromatography on salmon sperm and rat liver DNA columns Kleinsmith *et al.* (1970) showed that a very small fraction of chromatin phosphoproteins bind specifically to homologous DNA at physiological ionic strength. Using this general approach van den Broek *et al.* (1973) fractionated total chromatin acidic proteins of rat liver on *Escherichia coli* and rat liver DNA–cellulose columns. Proteins in 0.05 M NaCl–0.01 M Tris (pH 7) were first applied to *E. coli* and rat liver DNA–cellulose columns in tandem. About 31% of the starting protein did not adsorb to either DNA. Separation and elution of the loaded DNA columns with buffered 0.6 M NaCl showed that 31.8 and 3.9% of protein had bound to heterologous and homologous DNA, respectively. SDS–acrylamide gel electrophoresis of protein fractions showed that *E. coli* DNA-binding proteins were similar to the proteins not binding to either DNA. They were heterogenous with a wide range of molecular weights. Those binding to rat liver DNA were less heterogeneous and comprised of components with lower than 66,000

MW; a majority of these proteins banded in 25,000 MW range. In rechromatography of this latter fraction 13% of the proteins bound to *E. coli* DNA, 40% bound to rat liver DNA, and 7% was recovered in the unbound runoff peak. Additional studies by these investigators indicated that the rat liver DNA-binding proteins did not have absolute specificity for homologous DNA.

In my laboratory the chromatin acidic proteins from rat liver have been separated into two fractions (Patel and Thomas, 1973). While the CAP-I fraction is extremely heterogeneous with a wide range of molecular sizes as indicated by SDS–acrylamide gel electrophoresis, the AP-NH fraction is less heterogenous and is comprised, predominantly, of low molecular weight proteins (Patel, 1972). The latter fraction is similar to the DNA-affinity proteins from rat liver reported recently by Wakabayashi *et al.* (1973). Affinity chromatography of the AP-NH fraction (Thomas and Patel, 1974) on highly purified salmon sperm and rat liver DNA–polyacrylamide columns in tandem showed that, with 0.14 M NaCl–5 M urea–0.01 M Tris (pH 8) as the starting solvent, 18% of the applied protein did not bind to either DNA and was recovered in the runoff peak from the rat liver DNA column. Upon separation and individual elution of the loaded columns with the starting solvent containing 2 M NaCl, 19 and 2.5% of protein was recovered from the sperm and liver DNA columns, respectively. In subsequent elution at 50°C with 2.5 M NaCl–1 M guanidine-HCl–7 M urea–1% β-mercaptoethanol–0.01 M Tris, 4 and 20% of the original sample was eluted from the sperm and liver DNA columns, respectively. The results indicated that 84% of the protein that eluted from the heterologous DNA had bound electrostatically. Conversely, 89% of the protein that eluted from homologous DNA was bound by interactions other than electrostatic. Affinity chromatography of the CAP-I fraction showed that 87% of the sample protein did not adsorb to either DNA column. Upon elution of the loaded columns all the recoverable protein, 3.6 and 2.7% from sperm and liver DNA–polyacrylamide, respectively, was eluted with 2 M NaCl. These results confirm the high DNA affinity of the AP-NH fraction and suggest differences in modes of interaction with homologous and heterologous DNA.

The average molecular weight of the chromatin acidic proteins extracted in 0.1% SDS was determined by their sedimentation analysis to be 14,300 (Marushige *et al.*, 1968). Electrophoresis in SDS–acrylamide gels which is also used for molecular weight determinations, however, has shown them to have molecular weights in the range of 5000–200,000. Their size heterogeneity has also been shown by gel filtration studies (Umanskii *et al.*, 1971; Patel and Allan, 1974).

IV. ACIDIC PROTEINS OF THE NUCLEOLUS

A. Isolation of Proteins

1. From Purified Nucleoli

Nucleoli isolated from many cell types are composed of protein (70–85%), RNA (5–15%), and DNA (3–10%) (data summarized by Maramatsu and Busch, 1967). Vincent (1952), who first isolated nucleoli from starfish oocytes, reported that they did not contain histones and that the bulk of their proteins were phosphorylated. Subsequent studies with nucleoli isolated from various sources have shown, however, that they contain histones as well as nonhistone basic and acidic proteins. Grogan et al. (1966) extracted rat liver nucleoli successively with 0.14 M NaCl–0.01 M sodium citrate and with 0.25 M HCl to obtain saline and acid-soluble fractions, respectively, and an insoluble residue. The HCl-soluble proteins, which represented about 30% of total nucleolar proteins (also Liau et al., 1965), were considered to be histones on the basis of their amino acid composition. The saline-soluble fraction contained about 31% of total nucleolar proteins. Of these, 87% were soluble also in HCl. Their amino acid composition, however, clearly showed them to be acidic proteins. Similarly, the saline- and HCl-insoluble residue proteins were also shown to be acidic.

Birnstiel et al. (1964) solubilized protein fractions of pea nucleoli by successive extractions with deoxycholate (DOC), dilute sodium citrate buffer, and 2 M NaCl. The DOC-soluble fraction, upon ultracentrifugation, yielded particulate ribonucleoproteins and a soluble supernatant fraction. A ribonucleoprotein fraction was also derived from the sodium citrate extract by precipitation with magnesium. And the 2 M NaCl extraction of the DOC and sodium citrate-insoluble residue gave a soluble deoxynucleoprotein fraction and an insoluble residue of a ribonucleoprotein complex. They reported that 0.2 N HCl extraction of nucleoli did not give a typical histone fraction. Rather, the acid extractable fraction was similar to the residue in its amino acid composition, which indicated only a small amount of histones in pea nucleoli. Histones, nonhistone basic proteins, and acidic proteins were reported to comprise 6.4, 21.6, and 55.2% of the total nucleolus, respectively. Recently, Wilhelm et al. (1972) separated the basic and acidic proteins of rat liver nucleoli by successive extraction with 2 M NaCl–5 M urea–0.01 M sodium acetate (pH 6) and 0.1% SDS–0.1% mercaptoethanol–0.05 M Tris–HCl (pH 8), respectively. The NaCl–urea solvent was shown previously to extract histones from extranucleolar chromatin (Spelsberg et al., 1971). Electrophoretic analysis of the basic proteins showed the presence of many proteins in addition to the typical histones. Since many nucleoprotein particles, which contain proteins similar to ribosomal basic proteins, have been isolated from nucleoli (Liau and Perry, 1969; Tsurugi et al., 1973), the detection of these nonhistone basic proteins is not surprising. The additional bands may also be acidic protein contaminants. By analogy and comparison to extranucleolar chromatin the SDS extract was presumed to contain acidic proteins. The quantitative distribution of proteins in these two fractions and their amino acid composition were not reported by these investigators.

Purified nucleoli have also been extracted with 0.4 M H_2SO_4 (Orrick et al., 1973) and 67% acetic acid (Higashinakagawa et al., 1972) to isolate proteins. However, the distinction of histones, nonhistone basic proteins, and acid-soluble acidic proteins in such extracts cannot be made easily.

2. From Nuclear "Residue"

The fraction of isolated nuclei that remains insoluble after their successive extraction with dilute saline and 1–2 M NaCl has been equated to nucleolar and nuclear membrane material (Georgiev and Chentsov, 1962; Zbarskii et al., 1962) and has been referred to as the residual fraction. Wang (1966a) reported that it was composed of 89.6% protein, 10.3% RNA, and 0.83% DNA. The low value of DNA was not surprising since the 1–2 M NaCl extraction would remove most of the deoxynucleoprotein. He solubilized 95% of the residual proteins with 1% sodium deoxycholate and separated the extract into four fractions. A particulate ribonucleoprotein fraction (R-RNP) was pelleted by ultracentrifugation, and the supernatant was separated into three fractions by sequential precipitation at pH 6 and pH 5 and with ammonium sulfate. All four fractions were shown to be acidic by their amino acid composition.

B. Fractionation and Some Characteristics of the Nucleolar Acidic Proteins

Meaningful fractionation of nucleolar acidic proteins suffers from the same drawbacks as pointed out for the chromatin acidic proteins; namely, the lack of identification tags. Except for a few enzymes, the functions of these proteins remain unknown. Rigorous fractionation of nucleolar acidic proteins by chromatographic and other procedures has not yet been attempted. Grogan and Busch (1967) reported preliminary fractionation of saline-soluble nucleolar acidic proteins by chromatography of DEAE-cellulose. Protein subfractions were eluted with stepwise increases in NaCl concentration (0–0.6 M) in the elution solvent (0.05 M Tris, pH 8) and with 0.2 N HCl and 0.2 N NaOH. About 10 and 20%, respectively, of total proteins eluted from DEAE-cellulose were obtained in the last two solvents. The majority of NaCl-eluted proteins appeared in 0.1, 0.2, and 0.4 M NaCl eluates. Amino acid analysis of all subfractions showed them to be acidic proteins. The attempts to characterize these protein subfractions by electrophoresis were reported to be unsatisfactory.

Patel et al. (1968) have also reported on chromatographic fractionation of residual nucleolar proteins solubilized with deoxycholate and subfractionated by isoelectric precipitation at pH 6 and pH 5 and with ammonium sulfate. They were subjected to ion-exchange chromatography on DEAE-cellulose. Proteins were eluted with a 0–1 M NaCl gradient, followed by 0.1 N NaOH elution. Of the total proteins applied to the resin, depending on the subfraction, 32–85% were eluted with NaCl, and 15–51% were eluted with 0.1 N NaOH. Starch-gel electrophoresis of the chromatographic fractions gave only a limited resolution of the proteins. Nevertheless, considerable molecular heterogeneity and the absence of large amounts of basic proteins were apparent.

The four subfractions of the DOC-solubilized nuclear "residue" have also been shown to contain 0.1 to 0.27% phosphorus and to be soluble in 0.25 N HCl to a variable extent (60–91%). The proteins that precipitate at pH 6 form spherical particles, with 100–400 Å diameter, upon removal of the detergent. It is not known if this property is related to the nucleoprotein particles of the nucleolus.

The electrophoretic resolution of nucleolar acidic proteins is greatly improved in SDS–acrylamide gels. Wilhelm et al. (1972) have shown that acidic proteins extracted with 1% SDS are heterogeneous with respect to their molecular size. Knecht and Busch (1971) have purified an acidic protein of 65,000 MW by separation of the 0.4 M H_2SO_4 extract of Novikoff hepatoma nucleoli on SDS–acrylamide gels and elution of gel portion containing the protein. It has a single N-terminal residue, which is serine, and a ratio of acidic/basic amino acids equal to 1.75. Further resolution of proteins is indicated by two-dimensional electrophoresis. Orrick et al. (1973) have resolved the 0.4 M H_2SO_4-soluble proteins of nucleoli into more than 100 spots. The presence of histones and ribosomal precursor basic proteins in the acid extract notwithstanding, these observations indicate extensive heterogeneity of nucleolar acidic proteins.

V. ACIDIC PROTEINS OF THE NUCLEOPLASM

The nucleoplasm contains nuclear ribosomes, nonribosomal ribonucleoprotein particles, and a heterogeneous class of soluble proteins. The nuclear ribosomes were first isolated and characterized by Wang (1961b, 1962, 1963a,b) and Mirsky and co-workers (Frenster et al., 1960; Pogo et al., 1962; Allfrey, 1963). The proteins of nuclear ribosomes of calf thymus are basic and similar to those of cytoplasmic ribosomes from various sources (Wang, 1962). Although nuclei and nuclear ribosomes have been shown to incorporate amino acids into protein, the question of protein synthesis in the cell nucleus is still debated (Goldstein, 1970). The identity of nuclear ribosomes is further complicated by the discovery in the nucleus of ribonucleoprotein (RNP) particles of size similar to ribosomes and ribosomal subunits but of discrete chemical characteristics (Samarina et al., 1967, 1968; Moulé and Chauveau, 1968; Parsons and McCarty, 1968; Faiferman et al., 1970, 1971). Some of these RNP particles are considered to function in the transport of messenger RNA to the cytoplasm and have been referred to as "informofers" (see Georgiev, 1972, for a recent review). The soluble proteins of the nucleoplasm are the least characterized of all proteins of the nucleus. They must contain many enzymes, proteins active in nucleocytoplasmic interactions, specific receptors of hormones and other effectors, and many other proteins with unknown nuclear functions. It is curious that although the nucleoplasmic fraction is relatively easy to isolate, much of the current work on nuclear acidic proteins has concentrated on those of the chromatin complex, which have been more difficult to isolate and investigate.

A. Isolation of Acidic Proteins of the Nucleoplasm

When isolated nuclei are disrupted in 0.14 M NaCl and dilute buffers (Tris, phosphate, citrate, etc., pH 6–7.5), the nucleoplasmic fraction is readily solubilized. The cytological confirmation of this is provided by the studies of Georgiev (1958), Zbarskii and Georgiev (1959), and Georgiev and Chentsov (1962). The saline-extractable fraction is not solubilized by disruption of nuclei in sucrose which indicates that the nuclear membrane is not responsible for their retention within the nucleus (Barton, 1960; Poort, 1961). Kirkham and Thomas (1953) described the nucleoplasm extracted by 0.14 M NaCl from calf thymus and liver nuclei as predominantly composed of "globulins." These proteins were shown to migrate as a single component in moving boundary electrophoresis. It has since become evident, however, that the saline extract of nuclei contains many different proteins, some of which are ribonucleoproteins. Sedimentation analysis of the total saline extract has shown the presence of multiple components of 2, 4, 6, 8, 14, 18, and 41 S and two diffuse groups of 18–41 S and greater than 41 S (Bakay and Sorof, 1964).

Many of the nonribosomal RNP particles in the nucleoplasm contain rapidly labeled messenger-like RNA and 70 to 80% protein by weight. They are extracted from nuclei with dilute saline buffered at pH 8. Many reports of their isolation have dealt with physicochemical characterization of the particles and their RNA moiety, but detailed information on their proteins is still lacking. The protein of the 30 S particles (informofers; Samarina et al., 1968) from rat liver nuclei has been isolated and characterized (Krichevskaya and Georgiev, 1969). The particles were dissociated in 6 M urea–0.01 M phosphate (pH 7.1), and the protein was separated from RNA by DEAE-cellulose chromatography. More than 95% of the protein did not adsorb to the resin. Upon chromatography of this fraction on CM-cellulose, the protein eluted in a single polydisperse peak. Polyacrylamide gel electrophoresis of the protein showed 3 main components and 3–5 minor bands. However, pretreatment of the protein with β-mercaptoethanol converted the three main components into a single electrophoretic band. The molecular weight of the protein subunit has been estimated to be about 40,000. Although the chromatographic behavior of this protein indicated a basic nature, amino acid analysis showed the ratio of acidic to basic residues to be 1.25. The protein of other RNP particles has not been fully characterized, but it has been shown by polyacrylamide gel electrophoresis to be distinct from histones and ribosomal proteins (Faiferman et al., 1971; Parsons and McCarty, 1968; McParland et al., 1972).

The soluble proteins of the nucleoplasm are obtained in the supernatant by ultracentrifugation of the saline and buffer extracts of nuclei at 105,000 g. Wang (1961a) has distinguished proteins in this fraction on the basis of their solubility in distilled water and their precipitation with ammonium sulfate. This method of fractionation has been applied to saline-soluble proteins of rat brain nuclei (Jokela and Piha, 1972). With the exception of a very small fraction, which contains metabolically active basic proteins (Wang, 1966b), the majority of the nucleoplasmic proteins and their subfractions are indicated to be acidic by their amino acid composition (Busch, 1965; Jokela and Piha, 1972; Patel, 1972; Steele and Busch, 1963).

The phenomenon of proteins shuttling between the nucleus and the cytoplasm has been investigated by Goldstein and co-workers (see Goldstein and Prescott, 1967, for a review). Recently Jolinek and Goldstein

(1973) have isolated some of these proteins and shown them to be heterogeneous by gel filtration chromatography. A fraction representing 17% of these proteins was shown to be a single acidic protein with a molecular weight of 2300 and with proline at its amino terminus. When injected into whole cells it was shown to concentrate rapidly in the nucleus.

B. Fractionation and Some Characteristics of the Nucleoplasmic Acidic Proteins

There are only a few reports on the fractionation and characterization of the nucleoplasmic acidic proteins. In early attempts the total nucleoplasmic fraction was characterized by paper and agar-gel (Poort, 1961) and moving boundary electrophoresis (Bakay and Sorof, 1964). It is evident now that these methods were not adequate to show the full extent of molecular heterogeneity of the nucleoplasmic fraction. SDS–polyacrylamide gel electrophoresis has shown this fraction to be extremely heterogeneous (Patel, 1972; LeStourgeon and Rusch, 1973).

About half of the nucleoplasm consists of soluble proteins (Wang, 1961a). Patel and Wang (1964) fractionated the soluble proteins of thymus nuclei by ion-exchange chromatography into ten subfractions. Only the breakthrough subfraction contained a small amount of proteins migrating toward the cathode in starch-gel electrophoresis at pH 8.2. Proteins of the other subfractions migrated toward the anode and showed considerable heterogeneity. Their comparison with analogous subfractions of the soluble cytoplasmic proteins showed qualitative and quantitative differences in many of the components.

VI. CONCLUDING REMARKS

The research on the nuclear acidic proteins is at the stage of answering the basic questions of what they are, where they are, how complex they are, etc. The methods for their isolation that have evolved during the past few years and have contributed to these investigations have been described in the foregoing presentation. None of them, however, can be recommended as the standard method applicable to all systems. Since even a simple parameter, such as the method for isolation of nuclei, affects the proteins of the various nuclear fractions (Dounce and Ickowicz, 1969), the different extraction procedures are certain to influence the characteristics of the isolated proteins. Nevertheless, some basic information on this class of proteins has begun to accumulate.

The quantitative, qualitative, and metabolic modulations in the acidic proteins, coincident with developmental changes, hormonal stimulation, etc., that have been noted in many studies have suggested a dynamic role for them. The biological significance of these observations, however, remains to be uncovered.

The difficulty in elucidation of the biology of the acidic proteins is inherent in the methodology of their investigations. The isolation of many of these proteins requires the use of conditions which may render them unsuitable for functional characterizations. Furthermore, the molecular heterogeneity of the complex mixtures of proteins used in most investigations and the limitations of the characterization methods add to these problems. These points are illustrated by the observations of Wilhelm et al. (1972) who failed to find any differences in the acidic proteins of the nucleoli and extranucleolar chromatin. The proteins were compared by SDS–polyacrylamide gel electrophoresis, a method that has been used frequently. Given the genetic differences in these components of the chromosomes, these observations indicate either a real lack of difference in acidic proteins associated with different segments of the genome or insensitivity of the characterization method to detect important differences in the unique proteins. The studies of the acidic proteins are also complicated by the possibilities of contaminations (Sonnenbichler and Nobis, 1970; Harrow et al., 1972). All of these considerations point to the need for further improvements in the methodology.

That some of the acidic proteins are important in the expression and regulation of the genome is now generally accepted. In most studies supporting this view, however, the effect of complex mixtures of proteins on general transcription have been investigated. In this connection the isolation of a nonhistone protein that specifically interacts with the ribosomal cistrons and represses their expression (Crippa, 1970) is an exciting development. Clearly the answers to the exact regulatory role of the acidic proteins and the molecular mechanism of the regulation lie in exploitation of such well-defined systems.

ACKNOWLEDGMENT

The work in the author's laboratory described in this article was supported by the U.S. Atomic Energy Commission [contract #AT (38-1)-644].

REFERENCES

Allfrey, V. G. (1963). *Exp. Cell Res.*, Suppl. 9, 183.
Arnold, E. A., and Young, K. E. (1972). *Biochim. Biophys. Acta* 257, 482.
Bakay, B., and Sorof, S. (1964). *Cancer Res.* 24, 1814.

Barrett, T., and Gould, H. J. (1973). *Biochim. Biophys. Acta* **294**, 165.
Barton, A. D. (1960). In "The Cell Nucleus" (L. S. Mitchell, ed.), p. 142. Butterworth, London.
Behrens, M. (1932). *Hoppe-Seyler's Z. Physiol. Chem.* **209**, 59.
Bekhor, I., Kung, G. M., and Bonner, J. (1969). *J. Mol. Biol.* **39**, 351.
Benjamin, W., and Gellhorn, A. (1968). *Proc. Nat. Acad. Sci. U.S.* **59**, 262.
Bernardi, G., and Kawasaki, T. (1968). *Biochim. Biophys. Acta* **160**, 301.
Birnstiel, M. L., Chipchase, M. I. H., and Flamm, W. G. (1964). *Biochim. Biophys. Acta* **87**, 111.
Bonner, J., Dahmus, M. E., Fambrough, D., Huang, R. C., Marushige, K., and Tuan, Y. H. (1968). *Science* **159**, 47.
Busch, H. (1965). "Histones and Other Nuclear Proteins." Academic Press, New York.
Busch, H. (1967). In "Methods in Enzymology" (L. Grossman and K. Moldave, eds.), Vol. 12, Part A, p. 448. Academic Press, New York.
Busch, H., Choi, Y. C., Daskal, I., Inagaki, A., Olson, M. O. J., Reddy, R., Ro-choi, T. S., Shibata, H., and Yeoman, L. C. (1972). *Acta Endocrinol. (Copenhagen)*, Suppl. **168**, 35.
Butler, J. A. V., Johns, E. W., and Phillips, D. M. P. (1968). *Progr. Biophys. Mol. Biol.* **18**, 209.
Chanda, S. K., and Cherian, M. G. (1973). *Biochem. Biophys. Res. Commun.* **50**, 1013.
Chaudhuri, S., Stein, G., and Baserga, R. (1972). *Proc. Soc. Exp. Biol. Med.* **139**, 1363.
Chauveau, J., Moulé, Y., and Rouiller, C. H. (1956). *Exp. Cell Res.* **11**, 317.
Crippa, M. (1970). *Nature (London)* **227**, 1138.
Daly, M. M., Allfrey, V. G., and Mirsky, A. E. (1952). *J. Gen. Physiol.* **36**, 173.
DeLange, R. J., and Smith, E. L. (1971). *Annu. Rev. Biochem.* **40**, 279.
Dingman, W. C., and Sporn, M. B. (1964). *J. Biol. Chem.* **239**, 3483.
Dounce, A. L. (1963). *Exp. Cell Res.*, Suppl. **9**, 126.
Dounce, A. L., and Ickowicz, R. (1969). *Arch. Biochem. Biophys.* **131**, 359.
Elgin, S. C. R., and Bonner, J. (1970). *Biochemistry* **9**, 4440.
Elgin, S. C. R., and Bonner, J. (1972). *Biochemistry* **11**, 772.
Elgin, S. C. R., Froehner, S. C., Smart, J. E., and Bonner, J. (1971). *Advan. Cell Mol. Biol.* **1**, 2.
Faiferman, I., Hamilton, M. G., and Pogo, A. O. (1970). *Biochim. Biophys. Acta* **204**, 550.
Faiferman, I., Hamilton, M. G., and Pogo, A. O. (1971). *Biochim. Biophys. Acta* **232**, 685.
Fambrough, D. M. (1969). In "Handbook of Molecular Cytology" (Lima-de-Faria, ed.), p. 437. North-Holland Publ., Amsterdam.
Frenster, J. H. (1965). *Nature (London)* **206**, 680.
Frenster, J. H., Allfrey, V. G., and Mirsky, A. E. (1960). *Proc. Nat. Acad. Sci. U.S.* **46**, 432.
Frenster, J. H., Allfrey, V. G., and Mirsky, A. E. (1963). *Proc. Nat. Acad. Sci. U.S.* **50**, 1026.
Georgiev, G. P. (1958). *Biokhimiya* **23**, 700.
Georgiev, G. P. (1972). *Curr. Top. Develop. Biol.* **7**, 1.
Georgiev, G. P., and Chentsov, J. S. (1962). *Exp. Cell Res.* **27**, 570.
Gilmour, R. S., and Paul, J. (1969). *J. Mol. Biol.* **40**, 137.

2. Isolation of Nuclear Acidic Proteins

Gilmour, R. S., and Paul, J. (1970). *FEBS Lett.* **9**, 242.
Goldstein, L. (1970). *Advan. Cell Biol.* **1**, 211.
Goldstein, L., and Prescott, D. M. (1967). *In* "The Control of Nuclear Activity" (L. Goldstein, ed.), p. 273. Prentice-Hall, Englewood Cliffs, New Jersey.
Graziano, S. L., and Huang, R. C. C. (1971). *Biochemistry* **10**, 4770.
Grogan, D. E., and Busch, H. (1967). *Biochemistry* **6**, 573.
Grogan, D. E., Desjardins, R. P., and Busch, H. (1966). *Cancer Res.* **26**, 775.
Harrow, R., Tolstoshev, P., and Wells, J. R. (1972). *Cell Differentiation* **1**, 341.
Higashinakagawa, T., Maramatsu, M., and Sugano, H. (1972). *Exp. Cell Res.* **71**, 65.
Hill, R. J., Poccia, D. L., and Doty, P. (1971). *J. Mol. Biol.* **61**, 445.
Hnilica, L. (1967). *Progr. Nucl. Acid Res. Mol. Biol.* **7**, 25.
Hogeboom, G. H., Schneider, W. C., and Striebich, M. J. (1952). *J. Biol. Chem.* **196**, 111.
Hogeboom, G. H., Schneider, W. C., and Striebich, M. J. (1953). *Cancer Res.* **13**, 617.
Hol bek, V., Fanshier, L., Crocker, T. T., and Hnilica, L. S. (1966). *Life Sci.* **5**, 1691.
Howk, R., and Wang, T. Y. (1969). *Arch. Biochem. Biophys.* **133**, 238.
Huang, R. C., and Bonner, J. (1965). *Proc. Nat. Acad. Sci. U.S.* **54**, 960.
Huang, R. C. C., and Huang, P. C. (1969). *J. Mol. Biol.* **39**, 365.
Johns, E. W., and Forrester, S. (1969). *Eur. J. Biochem.* **8**, 547.
Jokela, H. A., and Piha, R. S. (1972). *Int. J. Neurosci.* **4**, 187.
Jolinek, W., and Goldstein, L. (1973). *J. Cell. Physiol.* **81**, 181.
Kamayama, M., and Wang, T. Y. (1971). *Biochim. Biophys. Acta* **228**, 563.
Kirkham, W. R., and Thomas, L. E. (1953). *J. Biol. Chem.* **200**, 53.
Kleinsmith, L. J., and Allfrey, V. G. (1969). *Biochim. Biophys. Acta* **175**, 123.
Kleinsmith, L. J., Heidema, J., and Carroll, A. (1970). *Nature (London)* **226**, 1025.
Knecht, M., and Busch, H. (1971). *Life Sci.* **10**, 1297.
Kostraba, N. C., and Wang, T. Y. (1970). *Int. J. Biochem.* **1**, 327.
Kostraba, N. C., and Wang, T. Y. (1972a). *Biochim. Biophys. Acta* **262**, 169.
Kostraba, N. C., and Wang, T. Y. (1972b). *Cancer Res.* **32**, 2348.
Kostraba, N. C., and Wang, T. Y. (1973). *Exp. Cell Res.* **80**, 291.
Krichevskaya, A. A., and Georgiev, G. P. (1969). *Biochim. Biophys. Acta* **194**, 619.
Langan, T. (1967). *In* "Regulation of Nucleic Acid and Protein Biosynthesis" (V. V. Koningsberger and L. Bosch, eds.), p. 233. Elsevier, Amsterdam.
LeStourgeon, W. M., and Rusch, H. P. (1973). *Arch. Biochem. Biophys.* **155**, 144.
Levy, S., Simpson, R. T., and Sober, H. A. (1972). *Biochemistry* **11**, 1547.
Liau, M. C., and Perry, R. P. (1969). *J. Cell Biol.* **42**, 272.
Liau, M. C., Hnilica, L. S., and Hurlbert, R. B. (1965). *Proc. Nat. Acad. Sci. U.S.* **53**, 626.
Loeb, J. E., and Creuzet, C. (1969). *FEBS Lett.* **5**, 37.
MacGillivray, A. J., Carroll, D., and Paul, J. (1971). *FEBS Lett.* **13**, 204.
MacGillivray, A. J., Paul, J., and Threlfall, G. (1972a). *Advan. Cancer Res.* **15**, 93.
MacGillivray, A. J., Cameron, A., Krauze, R. J., Rickwood, D., and Paul, J. (1972b). *Biochim. Biophys. Acta* **277**, 384.

MacKay, M., Hilgartner, C. A., and Dounce, A. L. (1968). *Exp. Cell Res.* **49**, 533.
McParland, R., Crooke, S. T., and Busch, H. (1972). *Biochim. Biophys. Acta* **269**, 78.
Maramatsu, M., and Busch, H. (1967). *Methods Cancer Res.* **2**, 303.
Marushige, K., Brutlag, D., and Conner, J. (1968). *Biochemistry* **7**, 3149.
Mirsky, A. E., and Pollister, A. W. (1942). *Proc. Nat. Acad. Sci. U.S.* **28**, 344.
Mirsky, A. E., and Pollister, A. W. (1946). *J. Gen. Physiol.* **30**, 117.
Mirsky, A. E., and Ris, H. (1947). *J. Gen. Physiol.* **31**, 7.
Mirsky, A. E., and Ris, H. (1950). *J. Gen. Physiol.* **34**, 475.
Monahan, J. J., and Hall, R. H. (1973). *Can. J. Biochem.* **51**, 709.
Monty, K. J., Litt, M., Kay, E. R. M., and Dounce, A. L. (1956). *J. Biophys. Biochem. Cytol.* **2**, 127.
Moulé, Y., and Chauveau, J. (1968). *J. Mol. Biol.* **33**, 465.
Orrick, L. R., Olson, M. O. J., and Busch, H. (1973). *Proc. Nat. Acad. Sci. U.S.* **70**, 1316.
Parsons, J. T., and McCarty, K. S. (1968). *J. Biol. Chem.* **243**, 5377.
Patel, G., and Wang, T. Y. (1964). *Exp. Cell Res.* **34**, 120.
Patel, G., Patel, V., Wang, T. Y., and Zobel, C. R. (1968). *Arch. Biochem. Biophys.* **128**, 654.
Patel, G. L. (1972). *Life Sci.* **11**, 1135.
Patel, G. L., and Allan, M. A. (1974). In preparation.
Patel, G. L., and Kellar, K. L. (1974). In preparation.
Patel, G. L., and Thomas, T. L. (1973). *Proc. Nat. Acad. Sci. U.S.* **70**, 2524.
Paul, J. (1970). *Curr. Top. Develop. Biol.* **5**, 317.
Paul, J., and Gilmour, R. S. (1968). *J. Mol. Biol.* **34**, 305.
Penman, S. (1966). *J. Mol. Biol.* **17**, 117.
Phillips, D. M. P., ed. (1971). "Histones and Nucleohistones." Plenum, New York.
Pogo, A. O., Pogo, B. G. T., Littau, U. C., Allfrey, V. G., and Mirsky, A. E. (1962). *Biochim. Biophys. Acta* **55**, 849.
Poort, C. (1961). *Biochim. Biophys. Acta* **46**, 373.
Richter, K. H., and Sekeris, C. E. (1972). *Arch. Biochem. Biophys.* **148**, 44.
Rickwood, D., Threlfall, G., MacGillivray, A. J., and Paul, J. (1972). *Biochem. J.* **129**, 50 p.
Sadgopal, A., and Bonner, J. (1970a). *Biochim. Biophys. Acta* **207**, 206.
Sadgopal, A., and Bonner, J. (1970b). *Biochim. Biophys. Acta* **207**, 227.
Samarina, O. P., Krichevskaya, A. A., Molnar, Y., Bruskov, V. I., and Georgiev, G. P. (1967). *Mol. Biol. (USSR)* **1**, 110.
Samarina, O. P., Lukanidin, E. M., Molnar, J., and Georgiev, G. P. (1968). *J. Mol. Biol.* **33**, 251.
Seale, R. L., and Aronson, A. I. (1973). *Mol. Biol. (USSR)* **75**, 633.
Shakoori, A. R., Romen, W., Oelschlager, W., Schlatteres, B., and Siebert, G. (1972). *Hoppe-Seyler's Z. Physiol. Chem.* **353**, 1735.
Shaw, L. M. J., and Huang, R. C. C. (1970). *Biochemistry* **9**, 4530.
Shea, M., and Kleinsmith, L. J. (1973). *Biochem. Biophys. Res. Commun.* **50**, 473.
Shirey, T., and Huang, R. C. C. (1969). *Biochemistry* **8**, 4138.
Siebert, G. (1967). *Methods Cancer Res.* **2**, 287.
Smith, J. A., and Stocken, L. A. (1973). *Biochem. J.* **131**, 859.
Sonnenbichler, J., and Nobis, P. (1970). *Eur. J. Biochem.* **16**, 60.

Spelsberg, T. C., Hnilica, L. S., and Ansevin, T. (1971). *Biochim. Biophys. Acta* **228**, 550.
Spelsberg, T. C., Wilhelm, J. A., and Hnilica, L. S. (1972). *Sub-Cell. Biochem.* **1**, 107.
Stedman, E., and Stedman, E. (1943). *Nature (London)* **152**, 267.
Stedman, E., and Stedman, E. (1944). *Nature (London)* **153**, 500.
Steele, W. J., and Busch, H. (1963). *Cancer Res.* **23**, 1153.
Stein, G., and Baserga, R. (1972). *Advan. Cancer Res.* **15**, 287.
Stein, G., Chaudhuri, S., and Baserga, R. (1972). *J. Biol. Chem.* **247**, 3918.
Stellwagen, R. H., and Cole, R. D. (1969). *Annu. Rev. Biochem.* **38**, 951.
Teng, C. S., Teng, C. T., and Allfrey, V. G. (1971). *J. Biol. Chem.* **246**, 3597.
Teng, C. T., Teng, C. S., and Allfrey, V. G. (1970). *Biochem. Biophys. Res. Commun.* **41**, 690.
Thomas, T. L., and Patel, G. L., (1974). In preparation.
Tsurugi, K., Morita, T., and Ogata, K. (1973). *Eur. J. Biochem.* **32**, 555.
Umanskii, S. R., Tokarskaya, V. I., Zotova, R. N., and Migushina, V. L. (1971). *Mol. Biol. (USSR)* **5**, 270.
van den Broek, H. W. J., Nooden, L. D., Sevall, J. S., and Bonner, J. (1973). *Biochemistry* **12**, 229.
Vincent, W. S. (1952). *Proc. Nat. Acad. Sci. U.S.* **38**, 139.
Wakabayashi, K., Wang, S., Hord, G., and Hnilica, L. S. (1973). *FEBS Lett.* **32**, 46.
Wang, T. Y. (1961a). *Biochim. Biophys. Acta* **49**, 239.
Wang, T. Y. (1961b). *Biochim. Biophys. Acta* **51**, 180.
Wang, T. Y. (1962). *Arch. Biochem. Biophys.* **97**, 387.
Wang, T. Y. (1963a). *Biochim. Biophys. Acta.* **68**, 633.
Wang, T. Y. (1963b). *Exp. Cell Res., Suppl.* **9**, 213.
Wang, T. Y. (1966a). *J. Biol. Chem.* **241**, 2913.
Wang, T. Y. (1966b). *Biochim. Biophys. Acta* **114**, 620.
Wang, T. Y. (1966c). In "The Cell Nucleus: Metabolism and Radiosensitivity," p. 243. Taylor & Francis, London.
Wang, T. Y. (1967). *J. Biol. Chem.* **242**, 1220.
Wang, T. Y. (1968). *Exp. Cell Res.* **53**, 288.
Wang, T. Y. (1970). *Exp. Cell Res.* **61**, 455.
Wang, T. Y. (1971). *Exp. Cell Res.* **69**, 217.
Wang, T. Y., and Johns, E. W. (1968). *Arch. Biochem. Biophys.* **124**, 176.
Wilhelm, J. A., Spelsberg, T. C., and Hnilica, L. S. (1971). *Sub-Cell. Biochem.* **1**, 39.
Wilhelm, J. A., Ansevin, A. T., Johnson, A. W., and Hnilica, L. S. (1972). *Biochim. Biophys. Acta* **272**, 220.
Wilson, E. M., and Spelsberg, T. C. (1973). *Biochim. Biophys. Acta* **322**, 145.
Yoshida, M., and Shimura, K. (1972). *Biochim. Biophys. Acta* **263**, 690.
Zalta, J., Zalta, J. P., and Simard, R. (1972). *J. Cell Biol.* **51**, 563.
Zbarskii, I. B., and Georgiev, G. P. (1959). *Biochim. Biophys. Acta* **32**, 301.
Zbarskii, I. B., Dmitrieva, N. P., and Yermolayeva, L. P. (1962). *Exp. Cell Res.* **27**, 573.
Zubay, G., and Doty, P. (1959). *J. Mol. Biol.* **1**, 1.

3

Extraction and Characterization of the Phenol-Soluble Acidic Nuclear Proteins

WALLACE M. LESTOURGEON AND WAYNE WRAY

I.	Introduction	60
II.	Development of Procedures Using Aqueous Phenol to Dissociate and Solubilize the Nuclear Acidic Proteins	61
	A. Phenol Solutions as Protein Solvents	61
	B. Procedures Using Phenol for Solubilizing the Nuclear Acidic Proteins	62
III.	Chemical Characteristics and Biochemical Activity of the Phenol-Soluble Residual Acidic Proteins	66
	A. Composition	67
	B. Chemical Characteristics	75
	C. Phosphorylation	78
	D. Synthesis	81
	E. Affinity for Homologous DNA	86
	F. Intranuclear Protein Concentration (Molecules per Nucleus)	88
IV.	Fractionation of the Nuclear Proteins	91
	A. Saline Extraction	92
	B. Acid Extraction	93
	C. Lipid Extraction	93
	D. Solubilizing the Residual Acidic Proteins	94
	E. Notes of Considerable Interest	95
V.	Preparation of Protein for Electrophoretic Separation	96
VI.	Electrophoretic Separation of the Phenol-Soluble Proteins	98
VII.	Concluding Remarks	99
	References	100

I. INTRODUCTION

In many studies which have been designed to obtain information regarding the metabolic activity of the chromatin-associated and nucleolar proteins a "cause-and-effect" phenomenon has been the focal point of the experimental protocol. Examples can be found in (1) the experiments where cells have been stimulated to undergo various metabolic and morphological changes through the use of drugs, trauma, or the manipulation of culture techniques; (2) in the experiments where cells are observed before, during, and after a normal series of differentiative events; and (3) in studies designed to gather information concerning the metabolism of the nuclear proteins during growth-related phenomena (i.e., the interphase periods G_1, S, G_2, and mitosis). The first objective in these experiments is frequently to determine whether or not recognizable and reproducible changes such as the appearance or disappearance of specific proteins, increased phosphorylation, changes in synthesis and intranuclear turnover, etc., occur in response to a given stimulus or during a particular metabolic event. If a detectable and reproducible change is observed the second priority is logically directed at correlating the observed phenomena to a specific biochemical or cellular event. Since, however, in the absence of a cause-and-effect relationship or other observable phenomena these studies cannot be initiated, considerable emphasis must be placed on the use of rapid yet highly sensitive and dependable "screening" procedures.

This chapter considers the evolution and practical use of procedures utilizing aqueous phenol solutions to dissociate selectively nucleic acid–protein complexes. Current procedures incorporating the use of buffered phenol solutions to dissociate nucleic acid–protein complexes selectively have been shown to be highly reproducible, straightforward, rapid, thorough, and inexpensive.

A considerable portion of the available information concerning the metabolic activities and characteristics of the residual acidic proteins of chromatin has been obtained through procedures utilizing phenol solutions. These procedures have been demonstrated to be useful in studies concerned with the residual acidic proteins of chromatin and also in screening studies where the total complement of nuclear proteins must be examined. As will be pointed out below there is no conclusive evidence that extraction procedures incorporating the use of high salt concentration (i.e., 0.5 M KCl, 2.0 M NaCl, 4 M urea–2 M NaCl, 4 M guanidine-HCl) or detergents to solubilize the residual proteins of chromatin or nucleoli are less drastic than procedures incorporating phenol solutions as solvents. This statement is not however intended to attenuate continuing efforts to develop procedures for solubilizing, fractionating, and purifying DNA-bound proteins under nondenaturing conditions.

Emphasis in this chapter will be placed on the use of phenol to selectively dissociate and solubilize the nuclear proteins found to be associated primarily with DNA which are not solubilized in physiological saline or in acid. While these proteins have been termed the residual acidic or residual nonhistone proteins it should be pointed out that in many cases the "residual" nature of a particular protein reflects simply the experimental procedure used in fractionating the nuclear proteins. For example, when acidic conditions which favor the solubilization of the basic histone proteins are employed, the solubility of proteins with acidic isoelectric points is depressed. As will be discussed later certain alkaline extraction procedures favor the solubility of many proteins with acidic isoelectric points and leave the histones as "residual" material. Certain proteins of chromatin and nucleoli are however refractory to solubilization in buffered saline solutions and can be solubilized directly only with more powerful protein denaturants (i.e., 2.0 M NaCl, 4 M urea–2 M NaCl, 4 M guanidine-HCl, phenol, detergents, etc.). Since numerous studies have established that the proteins which are not solubilized in saline or acid and which remain complexed to DNA possess acidic isoelectric points and high concentrations of aspartic and glutamic acid, the proteins will be referred to here as the residual acidic proteins of nuclei, chromatin, nucleoli, or chromosomes.

II. DEVELOPMENT OF PROCEDURE USING AQUEOUS PHENOL TO DISSOCIATE AND SOLUBILIZE THE NUCLEAR ACIDIC PROTEINS

A. Phenol Solutions as Protein Solvents

While phenol was found long ago to be a good solvent for extracting protein from biological sources (Tsvett, 1899), the most extensive work on the selective partitioning of nucleic acids and protein with phenol was performed by Kirby (1957, 1959). While the studies of Kirby were primarily concerned with the purity of DNA, his observation that salts with chelating properties potentiate the ability of phenol to dissociate DNA-bound protein has been helpful to persons interested primarily in phenol as a solvent for nuclear protein.

The ability of phenol to minimize the intermolecular interactions between protein and other biological macromolecules (Craig, 1962; Pusztai,

1965, 1966a) contributes to its effectiveness as a protein solvent. The limited solubility of polysaccharide, RNA, and DNA (Pusztai, 1965, 1966a) in phenol solutions has greatly contributed to its usefulness as a selective solvent for the nuclear acidic proteins.

The use of phenol preparations for solubilizing protein has been limited since it was generally assumed to denature irreversibly proteins and abolish enzymatic activity. However, the inhibition of ribonuclease (Huppert and Pelmont, 1962) and chymotrypsinogen A (Pusztai, 1966b) by phenol is reversible. While it was observed by Herzog and Krahn (1924) that phenol did not chemically alter proteins, the most detailed physical-chemical and immunological study concerning the effects of phenol on proteins was conducted by Pusztai (1966b). In these studies the changes brought about in the properties of bovine serum albumin and chymotrypsinogen A, by varying the composition of the phenol solvent or by heat treatment in the solvents, were established to be reversible. Proteins recovered from phenol solutions were found to display essentially identical characteristics (optical rotation, viscosity, sedimentation, ultraviolet, immunochemical, and proteolytic activity) when compared to the native proteins. With bovine serum albumin, even in unbuffered phenol preparations, heating (70°–90°C) did not induce an irreversible aggregation through a thiol-disulfide exchange in contrast to that which occurs in aqueous solutions. Evidence which will be discussed later suggesting that at least portions of the nuclear acidic proteins can be renatured after solubilizing in phenol can be found in their ability to react specifically with homologous DNA and to stimulate the *in vitro* synthesis of RNA.

B. Procedures Using Phenol for Solubilizing the Nuclear Acidic Proteins

Steele and Busch (1963) were the first investigators to employ the use of phenol solutions to dissociate and solubilize a class of DNA-complexed nuclear proteins. These investigators had observed previously (Busch *et al.*, 1959, 1961) that difunctional alkylating agents and other tumor-inhibitory compounds markedly inhibit the synthesis of nonhistone proteins and that difunctional alkylating agents can cross-link acidic nuclear proteins to DNA (Steele, 1962). To further investigate these phenomena a rather complex procedure was developed for fractionating the total nuclear proteins of rat liver and Walker 256 carcinoma.

In the procedure of Steele and Busch, isolated nuclei were first extracted twice with physiological saline (0.14 M NaCl) and then with 0.1 M Tris, pH 7.6, to obtain two fractions of "nuclear sap proteins."

3. Extraction of the Nuclear Acidic Proteins

It was observed that more than 50% of the total nuclear RNA, yet only traces of DNA, were solubilized with the sap proteins. The residue remaining after the saline and Tris extractions was extensively extracted in 2.0 M NaCl and particulate material was removed by centrifugation. While the material remaining after extraction with 2.0 M NaCl was further fractionated and examined for DNA, RNA, and protein content, it was observed that most of the DNA, histone, and sufficient additional protein had been dissociated in the 2.0 M NaCl extract to account for more than half the dry mass of liver and tumor nuclei. After Mirsky and Pollister (1946) the material in the NaCl extracts was called deoxyribonucleoprotein (DNP). The DNP complex is analogous in many respects to what other investigators now describe as "purified chromatin."

The DNP complex was then treated with dilute acid to dissociate and solubilize the basic histone proteins and precipitate the remaining DNA–protein complex. It was this dehistonized DNA–protein complex which was extracted with phenol to dissociate a specific fraction of residual (salt- and acid-insoluble) DNA-bound protein. The procedure used by Steele and Busch was adopted from the p-aminosalicylate–phenol method of Kirby (1959) for obtaining essentially protein-free DNA preparations. The DNA–protein complex was neutralized and homogenized in 6% sodium p-aminosalicylate. An equal volume of 90% phenol was added and an emulsion was made by homogenization. After one minute the phases were separated by centrifugation and the extraction repeated twice. The proteins solubilized in the phenol phase were precipitated with two volumes ethanol. Through this procedure about 80% of the protein in the dehistonized deoxyribonucleoprotein complex was solubilized in the phenol phase and most of the DNA was present in the aqueous phase. While these investigators did not characterize the phenol-soluble proteins electrophoretically, it was demonstrated that these proteins were not dissociated from the deoxyribonucleoprotein complex with rather high salt concentration (2.0 M NaCl), that the proteins were acid insoluble, that greater than 20% of the amino acid residues were glutamic and aspartic acid, and that in normal rat liver the highest specific activity (after administration of [U-^{14}C]lysine for one hour) were found in the phenol-soluble proteins and in the residue remaining after phenol extraction. An additional calculation performed on figures presented by these investigators reveals that the ratio of phenol-soluble protein to DNA was about 0.74:1 (protein to DNA) for rat liver and 0.9:1 in Walker's tumor.

The first procedure in which buffered phenol solutions were used for the expressed purpose of screening the residual acidic proteins was described by Shelton and Allfrey (1970). These investigators were especially interested in the stimulatory effect of cortisol on RNA synthesis in adrenalectomized rats. In order to determine if specific nuclear acidic proteins were involved in the cortisol-induced RNA synthesis, it was necessary to examine the complement of residual acidic proteins before and at various times after the administration of the hormone. The method of Shelton and Allfrey (1970) was not a modification of the Steele and Busch (1963) procedure but rather was an adaptation of the method of Viñuela et al. (1967) for solubilizing and characterizing electrophoretically the proteins associated with bacterial virus nucleic acids.

In the procedure first described by Shelton and Allfrey (1970) the isolated rat liver nuclei were extracted in 0.15 M NaCl, 0.01 M Tris-HCl, pH 7.5, to remove "soluble proteins." This extraction is analogous to the first two extractions of Steele and Busch (1963) and is frequently used to remove considerable amounts of nuclear RNA and proteins readily soluble in saline. The nuclei were then homogenized in distilled water and salt was added to a final concentration of 0.15 M to precipitate the "chromatin." The chromatin pellet was extracted with 0.2 M HCl for one hour to remove the histones. The remaining material [analogous to the dehistonized deoxyribonucleoprotein complex of Steele and Busch (1963)] was resuspended in "TEM" buffer (0.1 M Tris-HCl, pH 8.4, 0.01 M EDTA, 0.14 M 2-mercaptoethanol). This preparation was extracted three times with phenol previously saturated with the TEM buffer.

To prepare the phenol-soluble proteins for electrophoretic characterization the phenol phases were combined and extensively dialyzed against a series of five solutions. In addition to 0.14 M 2-mercaptoethanol the solutions contained, in the order of use, 0.10 M acetic acid, 0.05 M acetic acid, 9.0 M urea, 0.1 M Tris-HCl, pH 8.4, 8.6 M urea, 0.01 M EDTA, 0.01 M phosphate buffer pH 7.2 0.1% SDS, and 0.01 M phosphate buffer, pH 7.2, 1.0% SDS. Through this rather lengthy procedure the phenol was removed and the proteins were complexed and solubilized in SDS. The SDS-complexed proteins were separated electrophoretically through 10% acrylamide gels, pH 7.2, into 16 principal bands. The characteristics of these proteins together with the observation of Shelton and Allfrey (1970) that cortisol enhanced the synthesis of a specific residual acidic protein (41,000 MW) will be discussed later.

The rather tedious procedure first described by Shelton and Allfrey (1970) was later modified (Shelton and Neelin, 1971; Shelton et al., 1972; Shelton 1973) to circumvent the lengthy dialyses described above. A second consideration in the modified procedure was the development of a method for solubilizing the remaining proteins not extracted from nuclear material suspended in TEM buffer at pH 8.4. In the modified

3. Extraction of the Nuclear Acidic Proteins

procedure the protein was recovered from the phenol phase by dissolving the phenol in a solution containing 0.10 M acetic acid and 1% 2-mercaptoethanol and sedimenting the protein by centrifugation. The precipitates were then solubilized directly in a solution containing 0.01 M sodium phosphate (pH 7.2), 1% 2-mercaptoethanol, and 3% SDS. Before electrophoresis the SDS concentration was reduced to 1% by dialysis. To solubilize the protein not dissociated and solubilized in phenol at pH 8.4, the aqueous and interphase layers were titrated to pH 9.5 with NaOH and reextracted with phenol. This procedure yields two distinct fractions of residual acidic proteins and was found to solubilize effectively all the residual acidic DNA-bound protein from goose liver, erythrocytes, and regenerating erythrocytic cells. The characteristics of the protein solubilized in phenol at pH 9.5 will be discussed in Section III of this chapter.

In studies designed to determine whether or not the residual acidic proteins of chromatin interact specifically with homologous DNA, Teng et al. (1970), like Shelton and Allfrey (1970), adapted the phenol procedure of Viñuela et al. (1967) for solubilizing the proteins of bacteriophage MS2. The procedure as first used by Teng et al. (1970) was described in detail later (Teng et al., 1971). The distinguishing feature of the procedure was the addition of a three-step extraction to remove lipids and phospholipids prior to extracting the residual acidic protein–DNA complex with phenol. The dehistonized nuclear or chromatin residues were extracted first with 10 volumes 1:1 (v/v) chloroform–methanol containing 0.2 N HCl, then with 2:1 chloroform–methanol–0.2 N HCl, and finally with ether. The lipid extraction procedure was incorporated to insure phospholipid removal prior to quantitating protein-bound phosphate. Since in the studies of Teng et al. (1970) and Teng et al. (1971) the ability of these proteins to bind DNA and to influence in vitro transcription reactions was under investigation, the lengthy dialysis procedure for reconstituting the proteins in aqueous buffers (less SDS) was retained.

Helmsing and Berendes (1971) and Helmsing (1972) used a combined phenol–SDS procedure for their studies on the nonhistone proteins associated with polytene chromosome puffs in Drosophila. In this procedure nuclei were suspended directly in Ringer's solution containing 0.16 M EDTA and 1% SDS. This preparation was extracted three times with phenol saturated with 0.25 M Tris-HCl (pH 7.8) and the protein in the phenol phase precipitated with cold acetone containing 0.1 M acetic acid. While this procedure is the most direct and rapid method for separating DNA–protein complexes and prevents DNA contamination of the protein sample, resolution problems can be encountered when essentially all the nuclear proteins are separated electrophoretically at one time.

LeStourgeon and Rusch (1971) noting that little was known concerning the natural metabolic activity of the residual acidic proteins during periods of synchronous growth and differentiation used the phenol procedure of Teng et al. (1970) to demonstrate for the first time that major quantitative and qualitative changes occur in the residual nuclear acidic proteins during differentiative periods. Taking advantage of the natural and absolute mitotic synchrony of the lower eukaryote *Physarum polycephalum* these investigators demonstrated for the first time that neither quantitative nor qualitative changes occur in the phenol-soluble proteins during interphase but that maximum protein phosphorylation occurs during the G_2 period of active growth.

In later reports (LeStourgeon and Rusch, 1973; LeStourgeon et al., 1973a,b,c), the phenol procedure was simplified and an alakaline discontinuous SDS–acrylamide gel system was used to greatly enhance protein separation and resolution. In the simplified procedure the TEM-buffered (pH 8.4) phenol extracts were prepared for electrophoresis by dialyzing directly in a "sample buffer" containing 0.1% SDS, 0.01 M sodium phosphate (pH 7.2), 0.14 M 2-mercaptoethanol, and 0.25 M sucrose. Through this procedure aliquots of the SDS–complexed protein may be separated electrophoretically on the same day as nuclear isolation.

The residual material remaining after phenol extraction and which still contained considerable amounts of protein and most of the DNA was washed briefly with 10% methanol to remove contaminating phenol then suspended in a solution containing 0.01 M sodium phosphate (pH 7.2), 5% SDS, and 0.14 M 2-mercaptoethanol. This preparation was heated 3 minutes in boiling water and dialyzed as above to prepare the protein for electrophoresis. This protein fraction, which was not solubilized in phenol at pH 8.4, was termed the residual hot SDS-soluble protein fraction and was found (LeStourgeon and Rusch, 1973) to consist mostly of nucleolar proteins. The residual hot SDS-soluble fraction is analogous to the pH 9.5 phenol-soluble fraction of Shelton and Neelin (1971) but can be expected to contain still additional protein.

III. CHEMICAL CHARACTERISTICS AND BIOCHEMICAL ACTIVITY OF THE PHENOL-SOLUBLE RESIDUAL ACIDIC PROTEINS

Since the phenol-soluble residual acidic proteins of interphase nuclei, chromosomes, or nucleoli are in no way different from similar fractions

3. Extraction of the Nuclear Acidic Proteins   67

solubilized through other procedures, the chemical characteristics (i.e., amino acid composition, isoelectric points, molecular weight, phosphorylation, etc.) and biochemical activities (i.e., effect on *in vitro* transcription, affinity for homologous DNA, etc.) described below can be expected to reflect the characteristics of the residual acidic proteins in general.

A. Composition

The number and types of residual acidic proteins solubilized in buffered phenol preparations (i.e., pH 8.4 or 9.5) can be expected to vary with the composition of the starting material. For example, if isolated nuclei or chromatin preparations are not extracted to remove the saline-soluble components (ribonucleoprotein, nuclear ribosomes, neutral proteins, etc.) or histone protein, it can be expected that proteins in addition to the acidic proteins of chromatin will be present in the phenol phase. Histones for example are soluble in phenol as is demonstrated in Fig. 1.

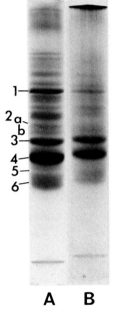

Fig. 1. Electrophoretic banding patterns of the histone proteins of *Physarum* extracted and separated by the procedure of Mohberg and Rusch (1969) (gel A). Gel B shows the histone proteins solubilized in phenol from 0.02 N HCl and reconstituted in 0.02 N HCl by dialysis. Note that the slower moving histones in gel B have aggregated and remained at the top of the gel. This aggregation does not occur if the phenol extracts are dialyzed directly against 8.0 M urea containing 0.02 N HCl. When extracting histones or other proteins from aqueous solutions with phenol essentially all the protein is recovered in the phenol phase.

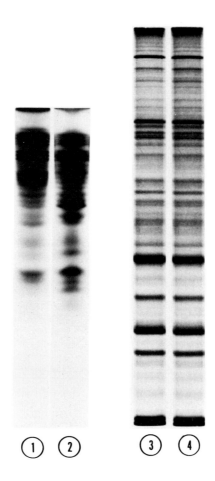

Fig. 2. Electrophoretic banding patterns of the phenol-soluble acidic proteins from intact rat liver nuclei (gel 1) and from isolated rat liver chromatin (gel 2). The similarities of the banding patterns indicate a chromosomal origin for most of the protein in the whole nuclear extracts. Gels 1 and 2 courtesy of C. S. Teng and the American Society of Biological Chemists, Inc. [C. S. Teng et al., J. Biol. Chem. 246, 3597 (1971)]. Gel 3 shows the phenol-soluble nuclear acidic proteins of Physarum and gel 4 shows the same fraction extracted from nuclei in which the nuclear envelope was previously removed (as judged by electron microscopy) by sheering in Triton. While gels 3 and 4 appear identical there are slight quantitative differences not evident in contrast photographs. The quantitative differences are judged to result from a slight solubilizing effect of Triton on certain of the nuclear acidic proteins. These gels demonstrate that the nuclear envelope proteins must constitute a very small fraction of the phenol-soluble acidic nuclear proteins of Physarum. Gels 3 and 4 courtesy of J. R. Jeter, Jr.

3. Extraction of the Nuclear Acidic Proteins

Perhaps then the best evidence for the existence of a unique class of residual acidic DNA-bound protein can be found in the observation that among highly different eukaryotes the residual acidic nuclear or chromatin proteins which can be solubilized in phenol show many gross similarities, even when varied procedures are followed for preparing the residual chromatin or nuclear material. While other examples in support of this statement will be presented below it can be mentioned first that the binding ratio (phenol-soluble protein, pH 8.4, to total DNA) is very similar among diverse eukaryotes [0.64:1 in *Physarum* (Le Stourgeon and Rusch 1973); 0.45:1 in rat liver (Teng *et al.* 1971)] and second that the electrophoretic profile of the phenol-soluble proteins isolated from intact nuclei or purified "chromatin" are essentially identical (Teng *et al.*, 1971) (Fig. 2). In *Physarum* the complement of phenol-soluble residual acidic protein is similar whether prepared from intact nuclei or from nuclei in which the nuclear envelope has been previously removed with Triton (Fig. 2).

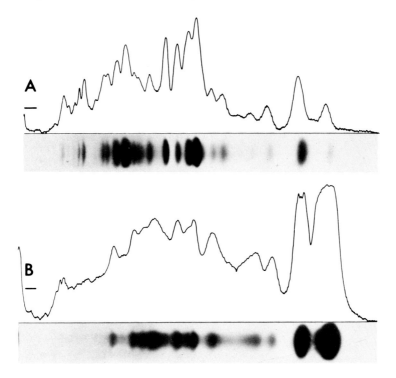

Fig. 3. Electrophoretic profiles of the phenol-soluble acidic nuclear proteins from rat liver (A) and from rat kidney (B). Courtesy of C. T. Teng and Academic Press [C. T. Teng *et al.* (1970). *Biochem. Biophys. Res. Commun.* **41**, 690].

Since molecular weight sieving through SDS–gel electrophoresis has been used almost universally for separating the phenol-soluble proteins, several observations can be made even though gel formulas have varied among laboratories. The electrophoretic profiles of the phenol-soluble residual acidic proteins (extracted from nuclei suspended in TEM buffer at pH 8.4) from rat liver and kidney (Teng et al., 1970; Teng et al., 1971) (Fig. 3), from goose liver (Shelton and Neelin, 1971) (Fig. 4), and from Physarum and HeLa cells (LeStourgeon and Rusch, 1973; LeStourgeon et al., 1973a,b,c) (Fig. 4) reflect "polarized" electro-

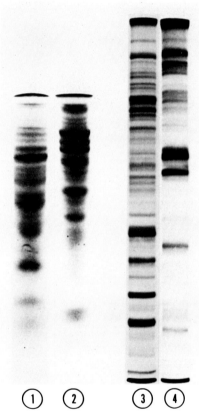

Fig. 4. Electrophoretic banding patterns of the proteins extracted with phenol from goose liver nuclei suspended first in "TEM" buffer at pH 8.4 (gel 1) and after reextracting with phenol from TEM buffer titrated to pH 9.4 with base (gel 2). Gels 1 and 2 reprinted from Shelton and Neelin, Biochemistry 10, 2342 (1971). Copyright by the American Chemical Society. Gel 3 shows the electrophoretic profile of the phenol-soluble proteins of Physarum extracted from TEM buffer at pH 8.4 and gel 4 shows the banding pattern of the residual acidic proteins remaining after phenol extraction and which have been solubilized in hot SDS as described in Section V of this chapter.

3. Extraction of the Nuclear Acidic Proteins

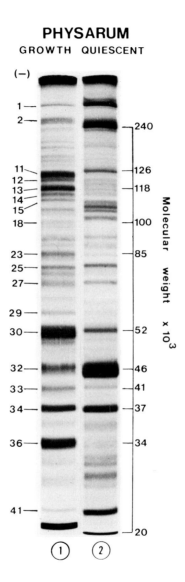

Fig. 5. Electrophoretic profiles of the phenol-soluble acidic nuclear proteins of *Physarum* extracted from actively growing plasmodia (gel 1) and after starvation induction of the nonproliferative cell state (gel 2). Significant quantitative or qualitative changes do not occur in the pattern pictured in gel 1 during the synchronous cell cycle of *Physarum* (for example, see Fig. 11). The starvation-induced changes pictured in gel 2 are reversed by refeeding and this reversal is the result of specific resynthesis and intranuclear protein accumulation. From LeStourgeon and Rusch, *Arch. Biochem. Biophys.* **159**, 861 (1973).

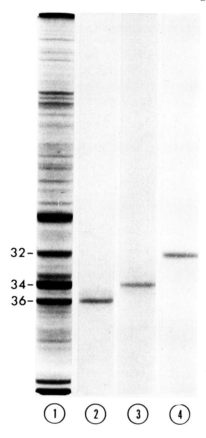

Fig. 6. Electrophoretic banding pattern of the phenol-soluble acidic nuclear proteins of *Physarum* (gel 1) and purified bands 32, 34, and 36 (gels 2–4). For amino acid analysis of the purified proteins see Table I and for the approximate molecular weights of the purified proteins see Fig. 5.

pherograms. In other words, there is usually in the low molecular weight regions (ca. 30,000 to 60,000 daltons) several large and well-separated proteins which in some cells comprise almost half of the total protein present in the phenol extract. Conversely in the high molecular weight region from 80,000 to 250,000 daltons there are numerous very fine protein bands and in high resolution gels (W. M. LeStourgeon, unpublished findings) as many as 50 discrete bands can be visualized. Four of the major low molecular weight proteins (bands 36, 34, 33, 32, Fig. 5) from the lower eukaryote *Physarum* have been isolated and purified (Fig. 6) and have been shown through amino acid analyses and isoelectric point determinations to be unique and acidic proteins (Table I and Fig. 7).

3. Extraction of the Nuclear Acidic Proteins

TABLE I

Amino Acid Composition of Several Purified Nuclear Acidic Proteins of *Physarum polycephalum* (Moles per 100 Moles)

Amino acid	Band 32[a]	Band 33[b]	Band 34[a,b]	Band 36[a]
Lysine	5.7	7.7	6.8	6.4
Histidine	2.4	3.0	4.4	2.4
Arginine	5.2	4.8	4.6	5.2
Aspartic acid	8.9	11.7	12.9	8.4
Threonine	6.3	4.3	4.5	1.6
Serine	6.3	8.2	6.2	4.8
Glutamic acid	11.4	13.4	10.0	9.6
Proline	5.3	3.8	3.7	6.9
Glycine	9.5	7.7	8.8	17.4
Alanine	7.9	9.1	5.6	8.9
Cysteine/2	—	—	—	—
Valine	5.9	6.6	7.4	8.1
Methionine	3.0	—	0.6	0.6
Isoleucine	6.1	3.9	3.9	4.9
Leucine	7.9	8.4	8.0	6.7
Tyrosine	3.7	3.0	2.9	3.0
Phenylalanine	4.6	4.5	3.5	4.4
N^τ-Methylhistidine	[c]	None	None	None
Unknown amino acids	None	None	None	[d]

[a] Average of two separate analyses.
[b] Nucleolar proteins.
[c] At least 1 mole/mole protein.
[d] A basic residue comprising about 3% of the total amino acids and which elutes in 0.35 N citrate buffer pH 5.28 from Beckman PA-35 columns equidistant between ammonia and arginine.

The residual acidic proteins which are not soluble at pH 8.4 in phenol but which can be solubilized in phenol at pH 9.5 (Shelton and Neelin, 1971) or in hot SDS (LeStourgeon and Rusch, 1973) show electrophoretic profiles different from the pH 8.4 fraction (Fig. 4). Evidence that the proteins not extracted at pH 8.4 in phenol are not artifact-induced by phenol extraction can be seen in experiments where the total residual proteins are solubilized with hot SDS or dilute alkali and compared electrophoretically to both the phenol and hot SDS-soluble fractions (Fig. 8). Further evidence that the phenol-insoluble (pH 8.4) proteins are not artifact can be seen in the demonstration (LeStourgeon and Rusch, 1973) that at least several of these proteins are in fact nucleolar in origin and that they can be solubilized in alkaline salt solutions. The significant difference between the phenol pH 8.4 fraction and the phenol pH 9.5 or hot SDS fraction is that the second fraction is mostly

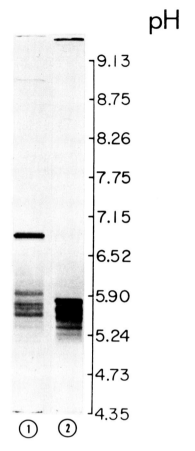

Fig. 7. The phenol-soluble acidic nuclear proteins of *Tetrahymena* (gel 1) and *Physarum* (gel 2) migrated to their isoelectric points in ampholite gels. Courtesy of J. R. Jeter, Jr.

very high molecular weight (i.e., 100,000 to 250,000 daltons) (Figs. 4 and 8).

That the phenol-soluble acidic nuclear proteins are highly heterogeneous can be seen by inspecting the various figures in this chapter. The heterogeneity and tissue specificity of the phenol-soluble proteins does not preclude the existence of functionally homologous proteins among the eukaryotes. In fact evidence for functionally homologous proteins can be found in the observation that in both HeLa cells and the lower eukaryote *Physarum polycephalum* several proteins (bands 11, 13, 36: Fig. 9) with similar molecular weights disappear from the residual chromatin while other proteins (band 32 especially; Fig. 9)

3. Extraction of the Nuclear Acidic Proteins

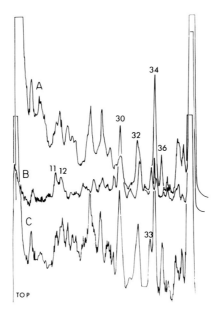

Fig. 8. Densitometer tracings of gels in which the total complement of residual acidic proteins of *Physarum* were solubilized in hot SDS (C); the phenol-soluble fraction (B); and the phenol-insoluble proteins which have been solubilized in hot SDS (A). Note that phenol selectively solubilizes only certain of the high molecular weight proteins but solubilizes essentially all of the major lower molecular weight proteins (i.e., bands 30, 32, 33, 34, and 36). Numbered bands can be compared to correspondingly numbered bands in Fig. 5. Since the proteins not solubilized in phenol (A) are present in total protein extracts (C) the phenol-insoluble proteins cannot be artifact-induced by phenol treatment.

accumulate in the nucleus in response to the stimulus of step-down culture (LeStourgeon et al., 1973b).

B. Chemical Characteristics

The amino acid composition of the residual phenol-soluble acidic nuclear proteins has been determined for several animal and plant tissues (Table II), for the phenol-soluble nucleolar proteins of *Physarum* (Table II, column 9) and for four isolated and purified proteins (two nucleolar and two chromatin) also from *Physarum* (Table I). In mass (total phenol fraction pH 8.4 or 9.5) glutamic and aspartic acid comprise 22% of the residues from all tissues and species. Glutamic acid is always present in excess of aspartic acid except in the nucleolar proteins of *Physarum* (see Table I, columns 2 and 3 and Table II, column 9). The ratio of acidic to basic residues is usually about 1.3–1.6:1 (acidic/basic) for

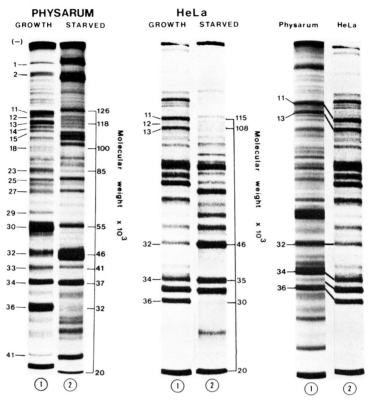

Fig. 9. Simultaneous electrophoretic separation of the phenol-soluble acidic protein fraction from actively growing plasmodia of *Physarum* (gel 1) and 48 hours after initiation of starvation (gel 2). The middle set of gels show the same protein fraction isolated from exponentially growing HeLa cells (gel 1) and 48 hours after peak exponential growth (gel 2). The gels on the right show the molecular weight relationships between the proteins present in exponentially growing *Physarum* (gel 1) and HeLa cells (gel 2). The numbered proteins from HeLa cells (11, 12, 13, 32, 34, and 36) are thought to be functionally homologous to the proteins of *Physarum* with corresponding numbers. From LeStourgeon and Wray, *Exp. Cell Res.* 79, 487 (1973).

all tissues (Table II). The phenol-soluble proteins are uniformly low in the sulfur-containing residues (1.5% methionine, trace amounts of cystine). The amino acid compositions of the total phenol-soluble proteins are thus quite similar among the various tissues; however, isolated and purified proteins from the same source reflect considerable residue variation (Table I). It is not known to what extent glutamic and aspartic acid (as analyzed after acid hydrolysis) represent glutamine or asparagine. A point of interest, however, is that if both residues were entirely in the amide form, calculated isoelectric points are still below neutrality.

TABLE II
Amino Acid Composition of the Phenol-Soluble Acidic Nuclear Proteins from Various Tissues

Amino acid	Rat liver[a]	Rat liver[b]	Rat liver[c]	Rat kidney[c]	Goose liver pH 8.4[d]	Goose liver pH 9.4[d]	Pollen mother cell[e]	Physarum nuclei[f]	Physarum nucleoli[f]
Lysine	7.3	6.62	6.04	6.72	7.0	7.6	6.18	7.65	7.37
Histidine	2.3	2.25	2.52	2.41	2.6	2.2	2.68	2.82	3.21
Arginine	7.7	5.60	5.89	7.08	6.1	6.5	5.38	5.92	4.63
Aspartic acid	9.1	10.22	10.28	9.55	9.3	10.0	9.83	10.51	12.86
Threonine	5.0	4.79	3.81	4.91	5.0	4.7	5.31	5.59	4.63
Serine	6.9	5.77	4.38	5.33	5.8	5.5	7.11	6.95	8.23
Glutamic acid	12.4	13.92	12.16	12.29	11.8	13.7	12.21	11.53	11.09
Proline	6.4	5.59	6.49	5.14	5.3	5.0	5.12	5.19	4.91
Glycine	9.0	8.60	15.28	9.62	8.2	7.5	9.69	8.60	11.91
Alanine	6.7	8.08	6.73	7.55	8.5	8.0	8.41	8.27	9.83
Cysteine/2	0.4	—	—	—	1.3	1.2	—	—	—
Valine	5.1	6.35	6.10	6.31	6.4	6.0	8.58	6.36	3.31
Methionine	2.2	1.52	1.92	2.00	1.6	1.7	0.85	1.42	1.42
Isoleucine	4.4	3.98	4.13	5.22	4.6	4.3	5.21	4.86	2.18
Leucine	8.3	9.47	6.79	8.90	9.5	9.4	7.48	8.34	8.40
Tyrosine	3.2	3.08	3.26	3.23	3.2	3.1	2.00	2.65	2.93
Phenylalanine	3.8	4.10	3.96	3.64	3.8	3.6	3.38	3.46	3.22
Acidic/basic	1.24	1.67	1.55	1.34	1.30	1.50	1.54	1.34	1.44

[a] Steele and Busch (1963).
[b] Shelton and Allfrey (1970).
[c] Teng et al. (1971).
[d] Shelton and Neelin (1971).
[e] Pipkin and Larson (1973).
[f] LeStourgeon and Rusch (1973).

Evidence that the bulk of the phenol-soluble residual acidic proteins are in fact acidic can be seen from isoelectric point determinations (Fig. 7) and from the anode-directed migration at pH 7 in 8.0 M urea gels (LeStourgeon and Rusch, 1971; Teng et al., 1971). When the phenol-soluble proteins are reconstituted in alkaline 8.0 M urea and focused in ampholite gels the experimental isoelectric points vary from pH 5 to 7 (Fig. 7).

C. Phosphorylation

Many of the residual nuclear acidic proteins which can be dissociated from DNA and solubilized in phenol are phosphorylated (Teng et al., 1970; Teng et al., 1971; Shelton and Neelin, 1971; LeStourgeon and Rusch, 1971). Most of the phosphate in the phenol-soluble proteins is alkali labile and solvent insoluble. Isotope incorporation experiments with ^{3}H-thymidine and ^{3}H-uridine (LeStourgeon and Rusch, 1973) and ^{14}C-orotic acid (Teng et al., 1971) have so far failed to demonstrate that the phosphorylated proteins are bound to nucleic acid. Hydrolysis of ^{32}P-labeled protein from rat liver and chromotographic separation of the isotopically labeled residues has demonstrated phosphate esterified to the hydroxy residues with serine being the major phosphorylated amino acid (Teng et al., 1971). When the phenol-soluble proteins are separated electrophoretically it can be demonstrated that some protein bands are more highly phosphorylated than others (Teng et al., 1971; Shelton et al., 1972; Karn et al. 1974; Johnson et al. 1973). The differential phosphorylation of these proteins can be seen in Fig. 10. In this figure the proteins of *Physarum* were labeled with ^{32}P-orthophosphate during a one-hour period during the G_2 period of exponential growth and separated electrophoretically. Observations to be noted are (1) that only about one-fourth of the proteins show ^{32}P incorporation after one hour and (2) that the most highly phosphorylated proteins are present in relatively small amounts. Evidence that protein phosphorylation is a highly specific phenomenon can be found in the observation that the extent of phosphorylation is not related to the concentration of hydroxyamino acids in the proteins. For example, the nucleolar protein (band 33) is highly phosphorylated and present in small amounts compared to proteins 32, 34, and 36 (Fig. 10), but contains essentially the same amounts of serine and threonine as the other proteins (Table I).

In the very precise mitotic synchrony of *Physarum*, maximum protein phosphorylation occurs during the G_2 period of synchronous growth (LeStourgeon and Rusch, 1971) yet sequential temporal differences in individual protein phosphorylation does not occur during the synchro-

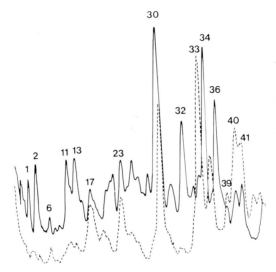

Fig. 10. Densitometer tracings of the phenol-soluble acidic nuclear proteins of *Physarum* isolated after labeling for one hour with ^{32}P-orthophosphate (upper solid tracing) during the G_2 period of active growth. The lower dashed line is a densitometer tracing of the radiogram produced from the same gel on x-ray film. Note that the proteins which incorporate the greatest amount of phosphate after a one-hour period are present in minor concentrations and that the major protein (band 30) is not phosphorylated but rather a minor component migrating in the forward shoulder of this protein.

nous cell cycle (Fig. 11). However, in HeLa S-3 cells in which synchrony was induced by the double thymidine procedure, Karn et al. (1974) observed that certain phenol-soluble proteins incorporate ^{32}P-orthophosphate at varying rates during the cell cycle. Also in difference to the observations concerning phosphorylation in *Physarum* (LeStourgeon and Rusch, 1971), maximum phosphorylation was found to occur not during the G_2 period but rather during early G_1 and S. In the studies of Karn et al. (1973) it was observed that the radioactive half-life of individual proteins from cells incubated in the presence of ^{32}P-orthophosphate during a 23-hour period varied from 5 to 12 hours, thus demonstrating differences in phosphate turnover.

The observation of LeStourgeon and Rusch (1971, 1973) and later confirmed by others (Bhorjee and Pederson, 1972; Karn et al., 1974) that significant quantitative or qualitative changes do not occur in the nuclear acidic proteins during the synchronous cell cycle that if these proteins play an active role in differential transcription then mechanisms other than quantitative changes must be operative. As pointed out by Allfrey (1970), Allfrey et al. (1971, 1973), Teng et al. (1971),

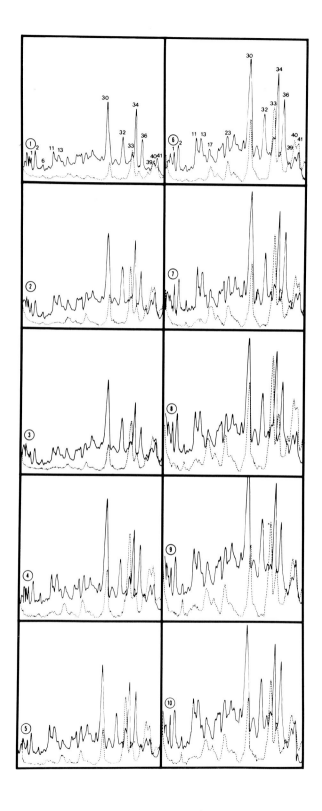

Karn et al. (1974), and Johnson et al. (1973), phosphorylation may represent an important aspect of the metabolic activity of the phenol-soluble proteins and could perhaps exist as a mechanism for specific regulatory activity. Support for this contention can be found in the experiments of Allfrey et al. (1971, 1973) where it has been observed that within 2 hours after cortisol administration to rats the pattern of protein phosphorylation in liver nuclei was completely different yet no quantitative change in the complement of phenol-soluble proteins could be detected (see Fig. 12).

Phosphorylation of the phenol-soluble proteins may be under the control of or influenced by specific histone proteins. Johnson et al. (1973) has observed that isolated and purified histones will bind in isolated nuclei and specifically influence *in vitro* protein phosphorylation. For example, histone F_1 was found to stimulate the phosphorylation of a protein fraction of about 40,000 daltons while F_{2A1} enhances phosphorylation of a protein with a molecular weight of 22,000. Both these histones inhibit the phosphorylation of certain low molecular weight proteins yet histones F_{2A2}, F_{2B}, and F_3 have little effect on phosphorylation of the phenol-soluble acidic proteins.

The information summarized here is concerned with some of the phosphorylation studies that have been performed on the phenol-soluble residual acidic proteins of chromatin. As demonstrated here phenol is not specific for extracting a unique class of highly phosphorylated residual acidic proteins; rather, the obervations reported here simply reflect characteristics unique to the residual acidic proteins of chromatin and a more comprehensive review is presented in the chapter by Kleinsmith.

D. Synthesis

In contrast to the nuclear histones the biosynthesis of the residual acidic proteins of chromatin is not necessarily coupled to periods of DNA synthesis (Zampeth-Bosseler et al., 1969; Stein and Baserga, 1970;

Fig. 11. The phenol-soluble nuclear acidic proteins of *Physarum* were allowed to incorporate ^{32}P-orthophosphate for consecutive one-hour periods during the 10-hour synchronous cell cycle; proteins were separated electrophoretically and densitometer tracings of stained gels and their corresponding autoradiograms were obtained as described in Fig. 10. While changes in phosphate turnover cannot be detected through this procedure it is clear that there is no period during interphase in which a single phosphorylated species incorporates phosphate in all-or-none fashion. It can also be observed in this figure that major changes in the complement of phenol-soluble proteins do not occur during the synchronous cell cycle of *Physarum*. Minor changes in peak height and separation probably reflect differences in gel characteristics more than protein differences. The increase in optical density of these scans reflects the increase in protein per nucleus which occurs maximally during the first 4 hours of interphase.

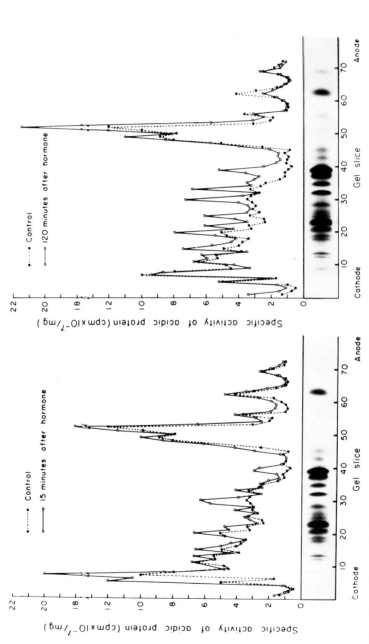

Fig. 12. The specific ^{32}P activities of the phenol-soluble proteins of control and cortisol-treated animals at 15 minutes after hormone administration and again at 120 minutes. Note that ^{32}P uptake into proteins of low mobility (high molecular weight) are evident at early times while several proteins of high mobility (low molecular weight) show a greater stimulation of phosphate uptake at later times after hormone administration. While these differences can be induced in the phosphorylation of liver protein no corresponding changes occurred in the phenol-soluble protein of kidney which is not sensitive to cortisol. From Allfrey, V. G. (1971), Changes in chromosomal proteins associated with gene activation. *In* "Nucleic Acid Protein Interactions" (D. W. Ribbons, ed.). North Holland Publishing Co., Amsterdam.

3. Extraction of the Nuclear Acidic Proteins

Stein and Borun, 1972; Seale and Aronson, 1973; LeStourgeon et al., 1973c). The residual acidic proteins of chromatin are apparently synthesized in the cytoplasm (Stein and Baserga, 1971) and increased synthesis of these proteins is frequently associated with and often precedes increased chromatin activity as demonstrated by RNA synthesis (Rovera and Baserga, 1971; Teng and Hamilton, 1969; Stellwagen and Cole, 1969; Stein and Baserga, 1970; Shelton and Allfrey, 1970; Levy et al., 1973; LeStourgeon et al., 1974).

As is to be expected the residual acidic proteins which can be dissociated and solubilized in phenol also reflect these phenomena. For example, the adrenalectomized attenuation of RNA synthesis by rat liver chromatin can be reversed by steroid administration (Feigelson et al., 1962; Yu and Feigelson, 1969); associated with this induced RNA synthesis is the synthesis of a specific phenol-soluble acidic protein (Shelton and Allfrey, 1970). In the experiments of Shelton and Allfrey (1970), cortisol was administered to adrenalectomized rats and through the use of double-label techniques with control animals it was observed that a specific protein (41,000 MW) increased in its relative specific activity by some 200% within 7 hours after cortisol administration (Fig. 13). The enhanced synthesis could actually be detected through increased isotope incorporation within 1 hour after administration of the hormone. In similar experiments (Enea and Allfrey, 1973) it has been observed that within 5 hours after administering the peptide hormone glucagon to rats two specific proteins (60,000 and 80,000 daltons), solubilized to phenol from liver nuclei, show major increases in ^{3}H-leucine incorporation.

In the lower eukaryote *Physarum polycephalum* an inactive chromatin complex (i.e., no RNA synthesis, no mitosis) can be induced through step-down culture; associated with this "inactivation" is the disappearance of numerous specific acidic proteins (LeStourgeon and Rusch, 1971, 1973; LeStourgeon et al., 1973b). If however chromatin "reactivation" is induced by transferring to fresh nutrient media, the proteins which disappeared during chromatin inactivation are specifically resynthesized and reaccumulate in the residual chromatin complex. This resynthesis and intranuclear accumulation occurs in a sequential temporal order throughout the 12-hour period of dedifferentiation leading to mitosis and DNA synthesis (LeStourgeon et al. 1973c). The quantitative and qualitative changes which result in the intranuclear concentration of the phenol-soluble proteins in response to step-down culture and chromatin inactivation can be seen in Figs. 5 and 14. The reaccumulation of the "activation" or proliferation-associated proteins can be seen in Figs. 15 and 16.

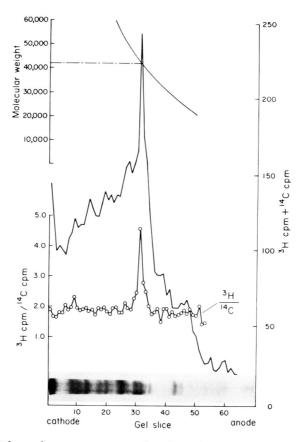

Fig. 13. Electrophoretic separation and analysis of labeling patterns of the phenol-soluble proteins from rat livers of cortisol-treated and control animals. The lower curve shows the ratio of ^{3}H-/^{14}C-leucine incorporation into the phenol-soluble proteins at different positions in the gel. The middle curve shows total radioactivity and the upper curve describes the relationship between mobility and molecular weight. Note that through the double-label procedure with control and treated animals the ratio of the distribution of radioactivity is essentially constant except at the molecular weight region corresponding to a protein of 41,000 daltons in the cortisol-treated animals. Courtesy of K. R. Shelton and *Nature*.

Through the use of the very sensitive and precise methods of acrylamide gel autoradiography (Fairbanks *et al.*, 1965) it has been possible to determine the extent of isotope (^{14}C-glutamic acid) incorporation per polypeptide per hour after initiation of refeeding and to relate these data to the actual intranuclear concentration of the various proteins. While it is beyond the scope of this chapter to discuss in detail the synthesis, intranuclear accumulation, and turnover for the various pro-

3. Extraction of the Nuclear Acidic Proteins

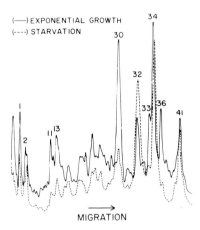

Fig. 14. Densitometer tracings of gels in which the phenol-soluble acidic nuclear proteins were isolated from actively growing plasmodia (solid tracing) and after starvation induction of the nonproliferative cell state. Note the generalized attenuation in the relative concentration of the high molecular proteins (i.e., between bands 1 and 30) in quiescent plasmodia. Note also the significant difference in bands 30, 32, 33, and 36. From LeStourgeon and Rusch, *Arch. Biochem. Biophys.* **159**, 861 (1973).

teins, several polypeptides can be taken as examples. Evidence suggesting the establishment of pooled protein and delayed intranuclear accumulation have been observed in phenomena associated with the reappearance of proteins 30 and 36 (Figs. 15 and 16). After consecutive 1-hour labeling periods during the period of chromatin reactivation, protein 30 shows its highest specific activities during the first 5 hours after initiation of refeeding (Fig. 17) yet increases in the intranuclear concentration of this protein are minimal during this period (Figs. 15 and 16). The maximum rate of intranuclear accumulation for this protein occurs between hours 7–14 of the cell state conversion period (Fig. 15) yet little isotope incorporation is associated with the increase in protein (Fig. 17). Similarly, protein 36 shows maximum specific activities midway during the refeeding period (Fig. 17) but maximum increases in the intranuclear concentration of this protein occurs just preceding mitosis and initiating of DNA synthesis (Fig. 15). The low specific activities obtained for proteins 30 and 36 once they have obtained their intranuclear concentration unique to active growth could suggest that these proteins are involved in the structural reorganization of the once-inactive chromatin. The observation that these proteins as well as proteins 11, 13, and 33 (Fig. 14) are present in maximum concentrations only during periods of active growth continued to suggest an important role for these proteins in the "reactivation" of *Physarum* chromatin.

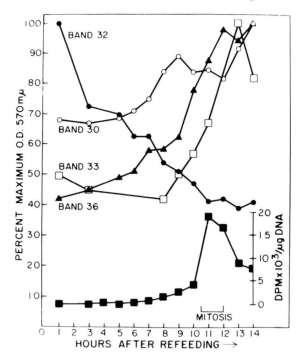

Fig. 15. Diagrammatic representation of the quantitative changes which occur in three major chromatin-associated proteins (30, 32, 36) and one nucleolar protein (33) during the period of chromatin reactivation in *Physarum*. Quantification of individual protein bands determined from integrated densitometer tracings and each point represents the percent change per hour relative to the maximum intranuclear concentration for each protein. Incorporation per hour ^{3}H-thymidine into nuclear DNA monitored as DPM/μg DNA. Nuclei in metaphase observed under phase contrast. From LeStourgeon and Rusch, *Arch. Biochem. Biophys.* **159**, 861 (1973).

The marked increase in the intranuclear concentration of the nucleolar protein 33 just preceding mitosis and DNA synthesis, together with the high specific activities of both nucleolar proteins (bands 33 and 34), points to the possibility that these proteins may be involved in ribosomal protein synthesis or that these proteins may be ribosomal components. In *Physarum* little ribosomal RNA synthesis occurs in quiescent microplasmodia; however, during periods of active growth it is the major RNA component.

E. Affinity for Homologous DNA

A unique aspect of many of the phenol-soluble residual acidic proteins is their ability to interact selectively with DNA. The specificity of the

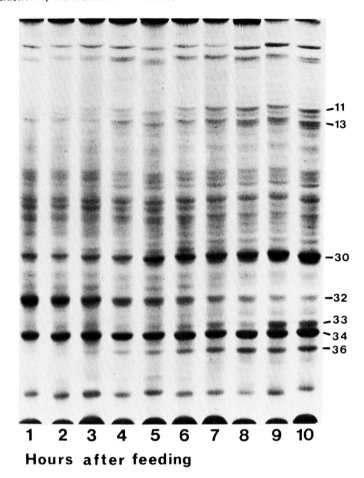

Fig. 16. Electrophoretic profiles of the phenol-soluble acidic nuclear proteins of *Physarum* extracted hourly after refeeding nonproliferative microplasmodia. Each profile can be taken to represent the complement of phenol-soluble acidic protein present per nucleus. From LeStourgeon and Rusch, *Arch. Biochem. Biophys.* **159**, 861 (1973).

DNA-binding reaction exists as further evidence for the *in vivo* association of these proteins with DNA. In the annealing experiment of Teng et al. (1970) and Teng et al. (1971), the proteins solubilized in phenol were reconstituted in 5 M urea–2.0 M NaCl–0.01 M Tris-HCl, pH 8.0, and mixed with purified DNA from various sources which had also been solubilized in the same solution. After stepwise dialysis to remove the salt and urea the DNA protein mixture (DNA-bound protein, free protein, and free DNA) was reconstituted in 0.01 M NaCl–0.01 M Tris-HCl (pH 8.0) and the DNA banded in buffered sucrose gradients

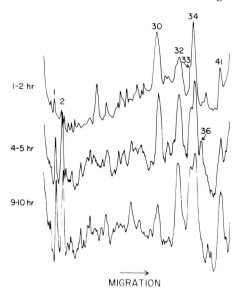

Fig. 17. Densitometer tracings of autoradiograms produced from ^{14}C-labeled phenol-soluble acidic nuclear proteins separated electrophoretically. The area under each peak represents the relative amount of isotope ^{14}C-glutamic acid incorporated per protein band per each one-hour labeling period. Autoradiograms were chosen from early, middle, and late during the period of dedifferentiation. In an order from top to bottom the tracings can be directly compared to gels 2, 5, and 10 pictured in Fig. 16. From LeStourgeon and Rusch, *Arch. Biochem. Biophys.* **159**, 861 (1973).

(5–25%). In experiments where the phenol-soluble proteins were mixed with homologous DNA (i.e., rat liver protein plus rat liver DNA), significantly more protein was found associated with the DNA than remained at the top of the gradients (Fig. 18C). However, when the proteins from rat liver were annealed with purified DNA from calf thymus or pneumococcus, essentially no protein banded with the DNA but instead remained on top of the gradients (Fig. 18A,B). Taking into consideration the limitations of the reconstitution experiments and the possibilities for protein denaturation, the observed binding ratio of 0.13:1 (protein to DNA) seems significant. In these experiments it was also demonstrated that certain proteins bind DNA to greater extents than others and that increasing the salt concentration from 0.01 M to 0.05 M would dissociate the DNA–protein complex.

F. Intranuclear Protein Concentration (Molecules per Nucleus)

Considerable evidence has accumulated in the past few years suggesting that the acidic nuclear proteins may be responsible for the regulation

3. Extraction of the Nuclear Acidic Proteins

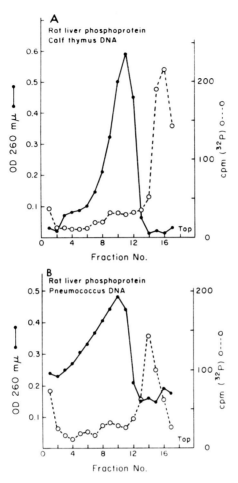

Fig. 18. Evidence for the selective binding of the phenol-soluble acidic nuclear proteins of rat liver to the DNA of the species of origin. The phenol-soluble proteins of liver were labeled *in vivo* with ^{32}P-orthophosphate, isolated, and "annealed" with various DNA's as described in the text. Each mixture was layered over 5 to 25% sucrose gradient and the DNA banded by centrifugation. The distribution of DNA is indicated by its absorption at 260 mμ. The position of the isotopically labeled protein is shown by its ^{32}P activity (- - -). Note that complexes of the type shown in (C) for rat liver protein and rat DNA are not formed with the DNA of calf thymus (A) nor with pneumococcal DNA (B). Courtesy of C. S. Teng and The American Society of Biological Chemists, Inc. [C. S. Teng *et al., J. Biol. Chem.* **246**, 3597 (1971)].

of specific gene expression in the eukaryotes (Wang, 1968; Gilmour and Paul, 1970; Spelsberg and Hnilica, 1970; Dingman and Sporn, 1964; Teng *et al.*, 1971; LeStourgeon and Rusch, 1971, 1973; LeStourgeon *et al.*, 1973b). As outlined above the residual acidic proteins which can be solubilized in phenol are highly heterogeneous (Figs. 2–5) and

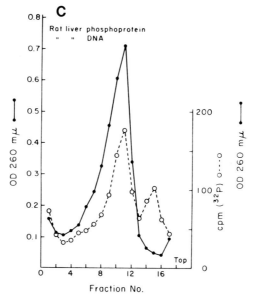

Fig. 18C.

tissue specific (Figs. 3, 4, and 9). Many of the phenol-soluble proteins are phosphorylated and increased protein phosphorylation is coincident with periods of increased gene activity (Figs. 10–12). Numerous proteins are only present in the nucleus during periods of active cell growth while others are associated only with nonproliferative cell states (Figs. 5, 9, 14). The phenol-soluble proteins also display species-specific affinity for DNA (Fig. 18), stimulate *in vitro* transcription (Teng *et al.*, 1971), and are synthesized and accumulate in the nucleus in a sequential temporal order during periods of chromatin reactivation (Figs. 15–17).

While these observations certainly indicate that the residual acidic proteins show high metabolic activity as compared to the histones, few models (Paul, 1972) have been presented attempting to outline the kinetics or mechanisms whereby these proteins may act as specific gene regulatory agents. In fact many of the observed activities (i.e., high turnover, increased synthesis, or phosphorylation) which are coincident with periods of known gene activity are also consistent with the possibility that these proteins may be intranuclear enzymes (i.e., RNA polymerase, protein kinase, etc.). The observation (LeStourgeon *et al.*, 1973b) that in both HeLa cells and *Physarum* numerous parallel proteins either disappear or are specifically synthesized in response to the stimulus of step-down culture could reflect changes in the intranuclear enzyme requirements common to both eukaryotes. More interestingly the high

TABLE III
Approximate Intranuclear Concentration of Several Proteins Present during Active Growth[a]

Protein band	Molecular weight	gm/nucleus	Moles/nucleus	Molecules/nucleus
11	126,000	2.00×10^{-14}	0.15×10^{-18}	90.3×10^{4}
13	118,000	1.62×10^{-14}	1.37×10^{-19}	8.25×10^{4}
30	52,000	9.16×10^{-14}	1.66×10^{-18}	9.99×10^{5}
32	46,000	2.93×10^{-14}	6.10×10^{-19}	3.67×10^{5}
34	37,000	6.16×10^{-14}	1.66×10^{-18}	9.99×10^{5}
36	34,000	3.08×10^{-14}	9.06×10^{-19}	5.45×10^{5}

[a] Molecular weight estimations and calculations as described elsewhere (LeStourgeon et al., 1974). The above values are based on the assumption that each protein band represents a single protein species and that all protein have the same dye-binding characteristics.

molecular weight proteins 11 and 13 (Figs. 5, 16) of *Physarum*, which are not present in a quiescent chromatin complex, have the same molecular weight and perfectly coelectrophorese with the two major components of *Physarum* RNA polymerase (W. M. LeStourgeon and A. Burgess, unpublished). That several of the major proteins such as bands 30, 32, 34, and 36 (Figs. 5, 9, 14, 16) may be proteins responsible for maintaining chromatin in a structurally "active" or "quiescent" state seems a distinct possibility. Evidence perhaps suggesting that many of the residual acidic chromatin proteins do not function singularly as specific gene regulators can be found in their rather high intranuclear concentration (Table III) (LeStourgeon et al. 1973c). For example there are about 80,000 to 90,000 copies of the polymerase-like proteins 11 and 13 (Figs. 5, 9, 16) in a single nucleus and as many as 500,000 to 1,000,000 copies of the major protein (bands 30, 32, 34, and 36) in *Physarum*.

IV. FRACTIONATION OF THE NUCLEAR PROTEINS

As pointed out in the introduction to this chapter a frequent prerequisite to studies designed to elucidate the metabolic activity and mechanism of action of a particular nuclear protein is the detection of reproducible changes (i.e., protein modification, the synthesis or catabolism of a specific protein, etc.) in response to a specific stimulus or developmental event. Initial attempts to correlate a particular cellular activity to nuclear protein metabolism can be greatly facilitated through the use of straightforward procedures for examining the total complement of nuclear proteins. The procedure described below utilizing aqueous phenol solutions to selectively dissociate, solubilize, and fractionate the residual nuclear acidic proteins has been demonstrated to be useful in studies designed to detect differences in nuclear protein metabolism as well as in studies concerned primarily with characterizing the residual acidic proteins of chromatin. While the procedure to be described is rapid, straightforward, inexpensive, and yields highly reproducible results, perhaps the most important advantage of the procedure is that potentially interesting proteins are not lost through chromatographic procedures or are never solubilized (for example, solvents such as 4.0 M guanidine-HCl, 6 M urea, or 4 M NaCl at pH less than 10 will not solubilize all the chromatin proteins from HeLa cells, *Physarum*, or rat fibroblasts). In the event that a particular protein should warrant further investigation methods for circumventing the use of phenol as a protein solvent will also be described.

Since it is beyond the scope of this chapter to discuss what is or is not "chromatin" nor the shortcomings of procedures which have been suggested to yield purified chromatin, it will be assumed that the starting material may be either purified nuclei, nucleoli, metaphase chromosomes, or chromatin together with other nuclear constituents. The procedure as outlined below has been found to yield excellent results with nuclei or nuclear components from HeLa cells (LeStourgeon *et al.*, 1973b), *Physarum* (LeStourgeon and Rusch, 1971, 1973; LeStourgeon *et al.*, 1973a,b,c), mouse embryo fibroblasts (W. M. LeStourgeon, unpublished), *Neurospora* (R. E. Totten, unpublished), *Tetrahymena* (Jeter, 1973), and chicken erythrocytes (I. L. Cameron, personal communication). In this procedure the total complement of nuclear proteins are divided into four specific solubility classes (LeStourgeon and Rusch, 1973) each of which can be characterized electrophoretically and all extractions and centrifugations are performed in a single homogenizing tube.

A. Saline Extraction

The isolated nuclear material having a wet packed volume of about 0.05 to 0.1 ml should be pelleted in a straight Potter-Elvehjem type homogenization tube (Tri R, S30; Tri-R Instruments, Inc., Rockville Center, N.Y.) by centrifugation at about 1×10^3 g for 10 minutes or longer depending on the density and viscosity of the isolation solution. After decanting add 2.0 ml cold 0.14 M NaCl and with the Teflon pestle (Tri-R, S20) disperse the nuclear material with the motor drive (about 1000 rpm) for 10 seconds (or manually for about 1 minute). Add an additional 2.0 ml cold 0.14 M NaCl and vortex briefly. Pellet by centrifugation for about 10 minutes at 1000 g, collect the supernatant, and repeat. The combined saline extracts (about 8.0 ml) should be labeled and set aside in ice. If chromatin, envelope, or nucleolar fragments are suspended in the salt extract they should be removed by high speed centrifugation.

B. Acid Extraction

The residue remaining after saline extraction should be suspended in 2.0 ml cold 0.25 N HCl as above, an additional 2.0 ml cold 0.25 N HCl should be added, and after vortexing briefly the preparation is allowed to stand in ice for one hour. The nuclear material is then pelleted by centrifugation for 15 minutes at 1000 g, the supernatant collected, and the residue reextracted with acid as above except that the preparation is not allowed to stand before pelleting. The combined acid extracts should be labeled and set aside in ice. The acid-soluble proteins have been shown to be almost entirely histone proteins and a third extraction with acid removes no additional protein (LeStourgeon and Rusch, 1973). As with the saline extraction only traces of DNA are dissociated during acid extraction (Table IV).

C. Lipid Extraction

Excess liquid above the dehistonized nuclear material should be removed by aspiration if necessary since large amounts of water can cause

TABLE IV
The Quantitative Distribution of Nuclear Proteins and DNA during Protein Fractionation

	% total nuclear protein and DNA dissociated during nuclear protein fractionation		Weight ratio of dissociated protein and DNA
	Protein	DNA	
% total nuclear weight	30	6.5	4.6:1
0.14 M NaCl extraction	23	1.2	—
0.25 N HCl extraction	32	0.5	—
Phenol extraction	14	16.6	3.9:1
Hot SDS extraction	31	81.7	1.8:1

phase separation during the first lipid extraction. The dehistonized residue should be gently suspended in about 2.0 ml chloroform–methanol 1:1 (v/v)–0.25 N HCl, pelleted by centrifugation (about 5 minutes at 1000 g), suspended in 2.0 ml chloroform–methanol 2:1 (v/v)–0.25 N HCl, again pelleted, and finally washed with about 4.0 ml diethyl ether. The nuclear material in the ether suspension can be pelleted by brief centrifugation (about 1 minute at 1000 g). If this procedure is to be performed in a closed centrifuge, the Potter-Elvehjem tube should be corked or capped with aluminum foil and over wrapped in Parafilm before spinning. Ether remaining after decanting can be evaporated rapidly with a gentle stream of air. Since lipids frequently incorporate isotope from many sources, contain appreciable amounts of phosphate, and migrate with the "front" in most SDS gel systems, it is advisable not to eliminate this procedure.

D. Solubilizing the Residual Acidic Proteins

The residual DNA–protein complex (usually containing about 95% of the total nuclear DNA and about 45% of the total nuclear protein) remaining after the above procedure should be gently suspended in about 1.5 ml TEM buffer [0.1 M Tris-HCl (pH 8.2), 0.01 M EDTA, 0.14 M 2-mercaptoethanol]. To this suspension is added 1.5 ml aqueous phenol previously saturated with the TEM buffer and, with the Teflon pestle and motor drive, an emulsion is made by homogenizing at about 1500 rpm for several passes. For convenience sake this preparation can be allowed to stand overnight under refrigeration. However, the phenol-soluble proteins are effectively solubilized within 30 minutes with occasional vortexing. Neither the electrophoretic profile nor protein yields are altered by allowing the phenol preparation to stand for periods up to 24 hours. If the phenol–TEM emulsion is allowed to stand overnight it should be allowed to come to room temperature with occasional vortexing before phase separation. The phenol and aqueous phases should be separated by centrifugation at 1000 g for 10 minutes and the lower phenol phase should then be drawn out with a small syringe and long flat tipped 22–24 gauge hypodermic needle. The nuclear material (which bands at the phenol–buffer interphase) and the aqueous TEM buffer should be reextracted with an equal volume of fresh phenol. In the second extraction the emulsion should be allowed to stand at least 10 minutes with occasional vortexing. The phenol phases should be combined and at this stage it is not necessary to keep the phenol cold.

The residual acidic proteins not solubilized in phenol (which remained complexed with DNA at the interphase) together with small traces of protein in the aqueous TEM buffer can be solubilized and prepared for electrophoresis as follows: to the TEM buffer and residual material add about 3 volumes cold methyl alcohol to precipitate the residual material. This procedure also solubilizes contaminating traces of phenol. After centrifugation at 1000 g for 10 minutes decant the supernatant and add 1.5–3 ml 3% SDS containing 0.14 M 2-mercaptoethanol. Suspend the residual material gently in the SDS solution and heat in boiling water for 3 minutes. Particulate material remaining should be removed by centrifugation.

E. Notes of Considerable Interest

1. Crystalline AR-grade phenol contains some hydrocarbon, heavy metals, and approximately 0.15% phosphine (H_3PO_2) as a preservative. Since various analytical tests may be performed on the phenol-soluble proteins it is suggested that the phenol be purified by distillation. Since phosphine is liberated during distillation and is spontaneously flammable distillation should be carried out in a closed system still using an asbestos heating mantle.

2. The purified phenol can be liquified by warming and saturating with TEM buffer in a separatory funnel of suitable size. The phenol should be shaken (taking care to release pressure) at least 3 times with equal volumes TEM buffer and stored under refrigeration in the dark. During saturation safety glasses and gloves should be used. The phenol solution will contain at room temperature 73% phenol and should have a pH very near 7.6 after saturation.

3. A pink or violet color may appear when the residual nuclear material is suspended in TEM buffer. This color is produced from an iron–EDTA–mercaptoethanol chelate and indicates the presence of $FeCl_2$ produced by the action of 0.25 N HCl on metal stemmed pestles. The presence of color does not affect the electrophoretic separation of protein nor quantitative yields.

4. While phenol is a poor solvent for nucleic acids, traces of this material can appear in the phenol phase. Nucleic acid contamination in the phenol phase is almost entirely low molecular weight material and migrates with the front in most SDS gel systems.

5. The rather high concentrations of 2-mercaptoethanol do not seem to be necessary since the electrophoretic profiles obtained for *Physarum* and HeLa cells are not affected even in the absence of 2-mercaptoethanol.

6. If considerable amounts of material remain after phenol extraction a gel-like solution can result on solubilizing the material in hot SDS. Since this indicates the presence of considerable amounts of free DNA

and since high molecular weight DNA interferes with electrophoresis, the SDS-complexed protein can be selectively solubilized in phenol by extracting as described above.

7. DNA in the aqueous TEM buffer after phenol extraction is (as a result of sheer forces) of low molecular weight and in alkaline sucrose gradients yields an S value of 13.

8. Often when intact nuclei or nucleoli are extracted, DNA-free and protein-free "ghosts" will remain after the hot SDS extraction. This material usually appears as a clear gel after high speed centrifugation and if insufficient volumes of 3% SDS are used it will be impossible to obtain a clean supernatant. As previously reported (LeStourgeon and Rusch, 1973), the nuclear ghost material is essentially pure polysaccharide and its presence would seem to suggest a polysaccharide nuclear superstructure in the eukaryotes.

V. PREPARATION OF PROTEIN FOR ELECTROPHORETIC SEPARATION

When the foregoing procedure is followed for extracting and fractionating the protein from small amounts of nuclear material four protein fractions (i.e., saline-soluble, acid-soluble, phenol-soluble, and the hot SDS-soluble) will be obtained. In preliminary studies it is important to characterize the proteins in each fraction since the possibility exists that the appearance or disappearance of a protein from one fraction may simply reflect a solubility modification leading to the appearance of the protein in a different fraction. It is also important for the investigator to get a "feel" for the nuclear proteins in general. In order that a direct comparison can be made (for monitoring reproducibility, specificity, changes, etc.), it is important to use a single high resolution electrophoretic system for separating the various protein fractions. It will be assumed here that the system of choice will be that used by LeStourgeon and Rusch (1973) and LeStourgeon et al. (1973a,b,c) for increased resolution of the nuclear acidic proteins. The development and use of the electrophoretic procedure will be described elsewhere.

Although the volume of the saline and acid extracts is relatively small (about 8.0 ml), frequently when small amounts of nuclear material are used the protein in these solutions is too dilute for direct electrophoresis or protein determination. The protein can be concentrated by various procedures (precipitated with trichloroacetic acid, alcohol, lyophilization, etc.) although perhaps the most straightforward procedure is to concentrate the protein by solvent reduction directly in small dialyzing tubes. In this procedure the saline and acid extracts, containing no particulate material, are transferred into dialysis tubing (i.e., Union Carbide, visking 1 cm flat) and covered with absorptive agents (such as polyethylene glycol or Sephadex G-100). Through this procedure the volume can be reduced by half usually within 45 minutes and the salt concentration is not appreciably elevated. Several drops of 5% SDS should be added to the solution before dialysis to denature the proteins before dialyzing at room temperature. After concentrating (if necessary), the preparation should be dialyzed in 1 liter 0.1% SDS for about 4 hours and aliquots can then be taken for protein determinations. SDS-complexed protein in 0.1% SDS yields essentially identical color values with the Lowry et al. (1951) procedure, providing exceptionally large volumes are not required. It is however good practice to prepare standards in 0.1% SDS. The protein can also be reconstituted in 0.1 N NaOH before protein determination.

If protein determinations are not necessary the saline and acid extracts can be denatured by adding several drops 5% SDS heated briefly in boiling water then dialyzed directly in "sample buffer" [0.01 M sodium phosphate buffer (pH 7.2), 0.1% SDS, 0.14 M 2-mercaptoethanol, 0.25 M sucrose]. Dialysis should be carried out overnight in at least 50 volumes of sample buffer. Through this procedure the proteins are effectively complexed with SDS and aliquots of these preparations can be separated electrophoretically.

The proteins solubilized in the phenol phase can also be prepared for electrophoretic separation by simple dialysis in the "sample buffer." During the first 2 hours of dialysis it is good practice to mix the solutions in the dialyzing tubes by passing the tubes through closed fingers. When high concentrations of protein and large amounts of phenol (4–10 ml) are used it may be more expedient to precipitate the protein out of the phenol phase with about 3 volumes of alcohol or ether. If this procedure is used, the precipitate should be thoroughly washed with solvent before solubilizing in the sample buffer. For protein determinations these preparations should be extensively dialyzed in either 0.1% SDS or 0.1 N NaOH since phenol gives a strong reaction with the Lowry procedure.

The phenol-insoluble proteins in the clear 3% SDS solution can be prepared for electrophoresis also by dialyzing in the sample buffer. If the solution is overly viscous the protein can be selectively solubilized in phenol as previously described leaving most of the DNA in the aqueous phase. Protein determinations can be performed after reconstituting in 0.1% SDS as previously described.

VI. ELECTROPHORETIC SEPARATION OF THE PHENOL-SOLUBLE PROTEINS

If the procedures outlined above are followed, all four protein fractions will be complexed with SDS and reconstituted in the same "sample buffer." In order to make comparisons based on electrophoretic mobility it is important that all protein samples be in the same sample buffer since differences in electrolyte concentration can affect the speed of migration and distribution of polypeptide bands. It is also important that the protein concentration (1.0 mg/ml as ideal) be adjusted such that all samples contain similar amounts of protein per unit volume. Adding an increased volume of dilute sample to a gel will result in a slower migration although the protein distribution will not be appreciably affected. The same volume should be added to all tubes in a single electrophoresis run. If Coomassie brilliant blue is to be used as the protein stain, protein per gel should be kept between 50 and 100 μg. If aniline blue black or similar stain is to be used, ideal protein per gel will be in the range between 80 and 150 μg. These amounts are based on the assumption that either 5 or 6 mm gels will be used and that the protein will be migrated over a distance of 10–15 cm. Bromphenol blue (0.005 ml of a 0.1% solution) should be added per ml of sample and stacking should be carried out at a current of 1 mA/gel until the marker dye just enters the lower gel; the current should then be adjusted to 2 mA/gel. Under these conditions the marker dye should reach the bottom of the gels in about 5 hours.

While electrophoretic procedures are continually being refined, the discontinuous alkaline-SDS system used by Laemmli (1970) for demonstrating the presence of many previously unreported proteins in bacteriophage T4 seems to be superior to date for separating the residual nuclear acidic proteins. The procedure as used for many of the gels pictured in this chapter can easily resolve as many as 100 polypeptides from the residual acidic protein fractions and by manipulating the concentration of acrylamide and N,N^1-methylenebisacrylamide attention can be focused on proteins in specific molecular weight regions between 10,000 and 250,000 or perhaps even broader. A problem frequently encountered however is the need for more sophisticated densitometric and narrow-angle photographic equipment.

A procedure for preparing 12 gels 10.5 cm long with a diameter of 6 mm follows. To 18.0 ml distilled water add 10.0 ml of the following solution: Tris base 18.17 gm, 4 ml 10% SDS, distilled water, and enough 6.0 N HCl (about 5 ml) to make a final volume of 100 ml at pH 8.8.

Next add 11.7 ml of the following solution: acrylamide 30.0 gm, N,N^1-methylenebisacrylamide 0.8 gm, and enough distilled water to make a final volume of 100 ml. The combined solutions should be deaerated by heating in a 1000-ml round bottom flask under vacuum until bubbles first appear, cooled to room temperature, and to this preparation 0.12 ml 10% (w/v) ammonium persulfate prepared just before use and 0.01 ml N,N,N^1,N^1-tetramethylethylenediamine (TEMED) should be added. The solution should be mixed thoroughly yet gently and filled into pre-marked glass tubes. A flat interface should be made by overlaying with 0.1% SDS or by adding a few drops of isobutyl alcohol. The stacking gel is prepared by adding to 6.5 ml distilled water, 2.5 ml of the following solution: Tris base 6.0 gm, 4.0 ml 10% SDS, and enough 6 N HCl (about 7 ml) and distilled water to make a final volume of 100 ml at pH 6.8. Then add 1.0 ml of the above acrylamide solution and 0.03 ml of the above ammonium persulfate solution. Since it is not necessary to deaerate the stacking gel, add 0.01 ml TEMED and polymerize 0.25 ml of the solution over the lower gel. The electrode buffer contains 0.025 M Tris, 0.192 M glycine, and 0.1% SDS and should have a pH of 8.4.

VII. CONCLUDING REMARKS

As described in this chapter, procedures incorporating the use of phenol for solubilizing certain fractions of the residual nuclear acidic proteins have been designed primarily to be used as screening techniques. Considering the complexity of the eukaryotic nucleus and the apparent fact that hundreds of polypeptides are associated with the genetic material the development of rapid and thorough methods for fractionating the total complement of nuclear protein has been especially valuable.

In the hypothetical case where as a result of one's initial "screening studies," a specific protein in the phenol fraction is observed to undergo a change (i.e., phosphorylation, synthesis, catabolism, etc.) in response to a given stimulus or developmental event considerable information can be obtained even though the protein may have been denatured by its exposure to phenol. This statement is supported by the information presented in Section III of this chapter. Often an understanding of the physical-chemical characteristics of a given protein can point to methods for purifying the protein under nondenaturing conditions. Since proteins in phenol can be reconstituted in aqueous solutions, specific polypeptides can be isolated and purified by contemporary procedures and characterized as to their molecular weight, isoelectric point, amino acid composition, degree of phosphorylation, sequence of amino acids, rate of synthesis, intranuclear turnover, etc.

Should attempts be initiated to determine the metabolic activity of a specific phenol-soluble protein, the phenol fraction can serve as a control for monitoring the ability of other procedures to solubilize the protein of interest under nondenaturing conditions. For example, it has been observed that extracting isolated nuclei in 0.5 M sodium citrate buffered to pH 7.9 will solubilize the phosphorylated nucleolar protein band 33 (see Fig. 5) yet leave bands 34 and 36 still bound in the nuclear material. Extracting isolated nuclei in 1.0 M KCl buffered to pH 7.9 will solubilize many of the high molecular weight proteins as well as bands 32 and 36 (see Fig. 5) yet leaves the nucleolar protein bands 33 and 34 essentially unsolubilized. Chromatographic procedures which would not normally resolve proteins with such similar molecular weights and isoelectric points become useful when gross fractionation based on the solubility characteristics of certain proteins are first employed. Additional methods of isolating, purifying, and identifying proteins in the phenol fraction found to undergo changes during periods of physiological differentiation will be described in Chapter 6.

In summary it should be pointed out that the primary rationale for using phenol as a nuclear protein solvent lies in its ability to exclude nucleic acids and polysaccharide both of which can severely reduce the ability of electrophoretic procedures to separate proteins. It should also be evident through the information presented in this chapter that phenol is not specific for solubilizing only phosphorylated acidic proteins but that it is simply an efficient and selective solvent for protein.

REFERENCES

Allfrey, V. G. (1970). *Fed. Proc., Fed. Amer. Soc. Exp. Biol.* **29**, 1447–1460.
Allfrey, V. G., Teng, C. S., and Teng, C. T. (1971). *In* "Nucleic Acid-Protein Interactions—Nucleic Acid Synthesis in Viral Infection" (D. W. Ribbons, J. F. Woessner, and J. Schultz, eds.), pp. 144–167. North-Holland Publ., Amsterdam.
Allfrey, V. G., Johnson, E. M., Karn, J., and Vidali, G. (1973). *In* "Protein Phosphorylation and Control Mechanisms" (F. Huijing and E. Y. C. Lee, eds.), pp. 34–36. Academic Press, New York.
Bhorjee, J. S., and Pederson, T. (1972). *Proc. Nat. Acad. Sci. U.S.* **69**, 3345–3349.
Busch, H., Amer, S. M., and Nyhan, W. L. (1959). *J. Pharmacol. Exp. Ther.* **127**, 195–199.
Busch, H., First, D. C., Lipsey, A., Kohen, E., and Amer, S. (1961). *Biochem. Pharmacol.* **7**, 123–134.
Craig, L. C. (1962). *Arch. Biochem. Biophys., Suppl.* **1**, 112–118.
Dingman, C. W., and Sporn, M. B. (1964). *J. Biol. Chem.* **239**, 3483–3492.

Enea, V., and Allfrey, V. G. (1973). *Nature (London)* **242**, 265–267.
Fairbanks, G., Levinthal, C., and Reeder, R. H. (1965). *Biochem. Biophys. Res. Commun.* **20**, 393–399.
Feigelson, M., Gross, P. R., and Feigelson, P. (1962). *Biochim. Biophys. Acta* **55**, 495–504.
Gilmour, R. S., and Paul, J. (1970). *FEBS Lett.* **9**, 242–244.
Helmsing, P. J. (1972). *Cell Differentiation* **1**, 19–24.
Helmsing, P. J., and Berendes, H. D. (1971). *J. Cell Biol.* **50**, 893–896.
Herzog, R. D., and Krahn, E. (1924). *Hoppe-Seyler's Z. Physiol. Chem.* **134**, 290–295.
Huppert, J., and Pelmont, J. (1962). *Arch. Biochem. Biophys.* **98**, 214–223.
Jeter, J. R., Jr. (1973). Doctoral Dissertation, University of Texas Medical School, San Antonio.
Johnson, E. M., Vidali, G., Littau, V. C., and Allfrey, V. G. (1973). *J. Biol. Chem.* **248**, 7595–7600.
Karn, J., Johnson, E. M., Vidali, G., and Allfrey, V. G. (1974). *J. Biol. Chem.* **249**, 667–677.
Kirby, K. S. (1957). *Biochem. J.* **66**, 495–504.
Kirby, K. S. (1959). *Biochim. Biophys. Acta* **36**, 117–124.
Laemmli, U. K. (1970). *Nature (London)* **227**, 680–685.
LeStourgeon, W. M., and Rusch, H. P. (1971). *Science* **174**, 1233–1236.
LeStourgeon, W. M., and Rusch, H. P. (1973). *Arch. Biochem. Biophys.* **155**, 144–158.
LeStourgeon, W. M., Goodman, E. M., and Rusch, H. P. (1973a). *Biochim. Biophys. Acta* **317**, 524–528.
LeStourgeon, W. M., Wray, W., and Rusch, H. P. (1973b). *Exp. Cell Res.* **79**, 487–490.
LeStourgeon, W. M., Nations, C., and Rusch, H. P. (1973c). *Arch. Biochem. Biophys.* **159**, 861–872.
LeStourgeon, W. M., Nations, C., and Rusch, H. P. (1974). (In press.)
Levy, R., Levy, S., Rosenberg, S. A., and Simpson, R. T. (1973). *Biochemistry* **12**, 224–228.
Lowry, O. H., Rosebrough, N. J., Farr, A. L., and Randall, R. J. (1951). *J. Biol. Chem.* **193**, 265–275.
Mirsky, A. E., and Pollister, A. W. (1946). *J. Gen. Physiol.* **30**, 117–147.
Mohberg, J., and Rusch, H. P. (1969). *Arch. Biochem. Biophys.* **134**, 577–589.
Paul, J. (1972). *Nature (London)* **238**, 444–446.
Pipkin, J. L., and Larson, D. A. (1973). *Exp. Cell Res.* **79**, 28–42.
Pusztai, A. (1965). *Biochem. J.* **94**, 611–616.
Pusztai, A. (1966a). *Biochem. J.* **99**, 93–101.
Pusztai, A. (1966b). *Biochem. J.* **101**, 265–273.
Rovera, G., and Baserga, R. (1971). *J. Cell. Physiol.* **77**, 201–211.
Seale, R. L., and Aronson, A. I. (1973). *J. Mol. Biol.* **75**, 633–645.
Shelton, K. R. (1973). *Can. J. Biochem.* **51**, 1442–1447.
Shelton, K. R., and Allfrey, V. G. (1970). *Nature (London)* **228**, 132–134.
Shelton, K. R., and Neelin, J. M. (1971). *Biochemistry* **10**, 2342–2348.
Shelton, K. R., Seligy, V. L., and Neelin, J. M. (1972). *Arch. Biochem. Biophys.* **153**, 375–383.
Spelsberg, T. C., and Hnilica, L. S. (1970). *Biochem. J.* **120**, 435–437.
Steele, W. J. (1962). *Proc. Amer. Ass. Cancer Res.* **3**, 364.
Steele, W. J., and Busch, H. (1963). *Cancer Res.* **23**, 1153–1163.

Stein, G. S., and Baserga, R. (1970). *J. Biol. Chem.* **245**, 6097–6105.
Stein, G. S., and Baserga, R. (1971). *Biochem. Biophys. Res. Commun.* **44**, 218–223.
Stein, G. S., and Borun, T. W. (1972). *J. Cell Biol.* **52**, 292–307.
Stellwagen, R., and Cole, R. D. (1969). *J. Biol. Chem.* **244**, 4878–4887.
Teng, C. S., and Hamilton, T. (1969). *Proc. Nat. Acad. Sci. U.S.* **63**, 465–472.
Teng, C. S., Teng, C. T., and Allfrey, V. G. (1971). *J. Biol. Chem.* **246**, 3597–3609.
Teng, C. T., Teng, C. S., and Allfrey, V. G. (1970). *Biochem. Biophys. Res. Commun.* **41**, 690–696.
Tsvett, M. (1899). *C. R. Acad. Sci.* **129**, 551–552.
Viñuela, E., Algranati, I. D., and Ochoa, S. (1967). *Eur. J. Biochem.* **1**, 3–11.
Wang, T. Y. (1968). *Exp. Cell Res.* **53**, 288–291.
Yu, F. L., and Feigelson, P. (1969). *Biochem. Biophys. Res. Commun.* **35**, 499–504.
Zampeth-Bosseler, F., Malpoix, P., and Fievez, M. (1969). *Eur. J. Biochem.* **9**, 21–26.

4

Acidic Nuclear Phosphoproteins

LEWIS J. KLEINSMITH

I. Introduction	103
II. Isolation and Fractionation of Acidic Nuclear Phosphoproteins	104
A. Salt Extraction Technique	105
B. Other Extraction Techniques	108
C. Electrophoretic Fractionation	110
III. Phosphate Metabolism of Acidic Nuclear Phosphoproteins	111
A. Protein-Bound Phosphate Metabolism in Intact Nuclei	112
B. Phosphorylation Reaction: Protein Kinase Activities	113
C. Dephosphorylation Reaction: Phosphatase Activities	117
D. Summarizing Model	118
IV. Functional Properties of Acidic Nuclear Phosphoproteins	119
A. Cell, Tissue, and Species Specificity	119
B. Localization and Quantitation in Chromatin	121
C. Changes in Protein Phosphorylation and Gene Activity	121
D. Specificity of DNA Binding	123
E. Effects on RNA Synthesis *in Vitro*	125
F. Effects of Cyclic AMP on Protein Phosphorylation	126
G. Binding of Steroid Hormones	126
H. Histone–Phosphoprotein Interactions	127
V. Role of Acidic Phosphoproteins in Nuclear Function	129
VI. Concluding Remarks	132
References	133

I. INTRODUCTION[*]

For many years it has been known that tissues rapidly incorporate radioactive phosphate into proteins, and more recently it has been discovered that such incorporation reflects the fact that the structure and function of many types of proteins can be altered by phosphorylation

[*] The following abbreviations have been used in this chapter: SDS, sodium dodecyl sulfate; cyclic AMP, cyclic adenosine 3′,5′-monophosphate.

and dephosphorylation reactions. Such protein modification reactions have been shown to be important regulatory mechanisms for a variety of different types of proteins, including enzymes, membrane proteins, contractile proteins, and ribosomal proteins (Krebs, 1972; Segal, 1973; Roses and Appel, 1973; Pratje and Heilmeyer, 1972; Kabat, 1970). In the context of our present concern with acidic nuclear proteins, however, it is of special interest to note that the highest concentration of phosphorylated proteins in eukaryotic cells is found localized in the cell nucleus.

The first definitive experiments on the existence of nuclear phosphoproteins were performed by Langan (1967), who purified a protein fraction from isolated rat liver nuclei which was extensively phosphorylated, containing about 1% phosphorus by weight. About 90% of the phosphate was found to be esterified to serine residues in the protein, with the remaining 10% esterified to threonine. Amino acid analyses of this protein fraction indicated that it was acidic, and subsequent analyses of acidic chromatin protein fractions prepared by other techniques confirmed that they contained large amounts of phosphorylated amino acids (Kleinsmith et al., 1966a; Benjamin and Gellhorn, 1968). Although it has now been clearly established that some histone fractions are phosphorylated (Kleinsmith et al., 1966a; Ord and Stocken, 1966; Langan, 1968a,b; Sherod et al., 1970; Shepherd et al., 1971), it is now known that over 95% of the nuclear protein-bound phosphorus is located on acidic chromatin proteins.

In view of the general importance of protein phosphorylation as a cellular regulatory mechanism, this finding of extensive phosphorylation of acidic nuclear proteins is of considerable interest. As the study of the phosphorylation of these proteins has proceeded over the years, it has become increasingly apparent that this process exhibits many of the characteristics of a regulatory phenomenon, specifically in relationship to the control of gene expression. In this chapter, we will first review the general approaches employed for the isolation and fractionation of phosphorylated acidic proteins and will then examine the metabolic and functional properties of these proteins which have led to the conclusion that their phosphorylation is involved in the regulation of gene activity.

II. ISOLATION AND FRACTIONATION OF ACIDIC NUCLEAR PHOSPHOPROTEINS

The procedure originally developed by Langan for the purification of acidic nuclear phosphoproteins basically involved salt extraction of nuclei and selective absorption of phosphorylated proteins on calcium phosphate

4. Acidic Nuclear Phosphoproteins

gel. This procedure has some distinct advantages, in that it avoids harsh treatment with acids and/or phenol and thus yields proteins which are useful as enzymes and substrates. The final product is considerably enriched in phosphorylated proteins and has therefore been referred to as the acidic nuclear phosphoprotein fraction. It is thus a subclass of the acidic nuclear proteins, although it should be emphasized that the distinctions are not absolute. It is possible, and even likely, that the acidic nuclear phosphoprotein fraction still contains some unphosphorylated proteins, while some phosphorylated proteins are probably lost during purification. The situation is even further complicated by the fact that these acidic proteins are continually being metabolically phosphorylated as well as dephosphorylated, so that a given protein molecule may be phosphorylated at one moment and not phosphorylated at another. Thus, the distinction between phosphorylated and nonphosphorylated acidic proteins is not an absolute one, but is subject to both methodological and theoretical limitations. The best one can do in practice is to refer to protein fractions which are enriched in phosphorus as acidic nuclear phosphoproteins, realizing the limitation inherent in such a designation.

One implication of this lack of a clear distinction between phosphorylated and nonphosphorylated acidic proteins is that it is often difficult to know whether various studies on acidic nuclear proteins are in fact dealing with phosphorylated proteins or not. The only way to answer this question directly is to analyze for phosphorus, and in much of the literature on acidic nuclear proteins this type of analysis has unfortunately not been done. Thus, many of the studies reviewed in the rest of this book on acidic nuclear proteins are in fact studying the behavior of at least some phosphorylated proteins. Keeping these limitations in mind, in this chapter we will focus on the isolation and characterization of acidic protein fractions which are enriched in phosphorus and on studies where phosphorylation of acidic proteins has been examined specifically by using radioactive labeling with ^{32}P.

Since the salt extraction technique pioneered by Langren has been employed in so many of the basic studies on the physical, chemical, and enzymological characterization of nuclear phosphoproteins, it will be described first in some detail. We will then examine some of the other procedures which have been more recently employed for isolating nuclear phosphoproteins, following which we will examine the electrophoretic approach to fractionation of these various protein preparations.

A. Salt Extraction Technique

This technique for purification of acidic nuclear phosphoproteins usually employs as starting material nuclei which have been isolated by purification through dense sucrose, although other types of nuclei can often be substituted. The procedure will be described for rat liver, but it can easily be adapted to a wide variety of tissues and cell types. The method is based on the procedure of Langan, but includes several modifications and improvements which have been developed in our laboratory (Gershey and Kleinsmith, 1969a; Kleinsmith and Kish, 1974).

All operations are performed at 4°C. Thirty grams of fresh tissue are finely minced with scissors and added to 300 ml of 0.32 M sucrose–3 mM MgCl$_2$. The tissue is homogenized for 2 minutes at 6000 rpm in a Sorvall Omni-Mixer with the small-bladder chamber. The resulting homogenate is filtered through double-napped flannelette and centrifuged for 7 minutes at 1000 g. The resulting pellet is resuspended in 225 ml of 2.4 M sucrose–1 mM MgCl$_2$ by homogenizing 2 minutes at 2000 rpm in the Sorvall Omni-Mixer using the large-bladed chamber. The resulting suspension is centrifuged for 60 minutes at 70,000 g. The nuclear pellets are collected and washed twice by resuspending in 0.01 M Tris-HCl, pH 7.5–0.25 M sucrose–4 mM MgCl$_2$ and centrifuging for 7 minutes at 1000 g.

These purified nuclei serve as the starting material for subsequent purification of phosphorylated acidic chromatin proteins. The soluble proteins of the nuclear sap are removed by suspending the nuclei in 30 ml of 0.14 M NaCl per ml of packed nuclei for 20 minutes. The nuclei are collected by centrifugation at 10,000 g for 10 minutes and are then suspended in 1.0 M NaCl–0.02 M Tris-HCl, pH 7.5, to a final protein concentration of 2 mg/ml. The suspension is dispersed using a Polytron homogenizer (Brinkmann) for 20 seconds at setting 3. The resulting viscous solution is mixed with 1.5 volumes of 0.02 M Tris-HCl (pH 7.5) to lower the salt concentration to 0.4 M, and the precipitated nucleohistone is removed by centrifugation at 200,000 g for 2 hours. Bio-Rex 70 (Na$^+$), which has previously been equilibrated with 0.4 M NaCl–0.02 M Tris-HCl, pH 7.5, is then added to the supernatant at a ratio of 20 mg Bio-Rex per mg protein. After stirring slowly for 10 minutes, the suspension is centrifuged for 10 minutes at 6000 g and the supernatant withdrawn. The resin is washed by suspending it in 10–15 ml of 0.4 M NaCl–0.02 M Tris-HCl, pH 7.5, and centrifuging again for 10 minutes at 6000 g. The two supernatants are then combined and calcium phosphate gel added at a ratio of 0.46 mg of gel per mg protein. After slowly stirring for 20 minutes the suspension is centrifuged for 5 minutes at 6000 g and the supernatant discarded. The gel is washed by resuspension in 10–15 ml of 1.0 M (NH$_4$)$_2$SO$_4$–0.05 M Tris-HCl, pH 7.5, followed by centrifugation at 6000 g for 5 minutes. The supernatant is again discarded, and the gel then dissolved by gentle

4. Acidic Nuclear Phosphoproteins

homogenization with a Teflon Potter-Elvehjem tissue grinder in 0.3 M EDTA, pH 7.5–0.33 M $(NH_4)_2SO_4$ in a ratio of 0.2 ml of solution per mg gel. The suspension is allowed to stand for one hour in the cold with occasional rehomogenization, and the insoluble residue is then removed by centrifugation for 15 minutes at 33,000 g. The supernatant is then dialyzed overnight against 0.05 M Tris-HCl, pH 7.5, to yield the final purified acidic nuclear phosphoprotein fraction. The overall purification scheme is outlined in a flow diagram in Fig. 1.

A comparison of the yields of phosphoprotein obtained by this procedure from several different tissues is given in Table I. As can be seen, the purification runs from approximately 10- to 20-fold, yielding products heavily enriched in phosphorus (from 1.0 to 1.3% phosphorus by weight). It is important to note at this point that the phosphorus contents of nuclear proteins are routinely measured by determining alkali-labile phosphate (Kleinsmith et al., 1966a). This alkali treatment releases phosphate which is esterified to serine and threonine hydroxyl groups in proteins, but does not hydrolyze phosphate present in nucleic acids. Thus, this technique is not significantly affected by any nucleic acid contamination which might be present.

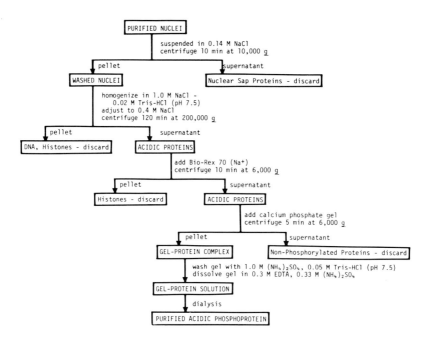

Fig. 1. Flow diagram of the salt extraction method for purification of acidic nuclear phosphoproteins.

TABLE I
Yields of Saline-Extractable Acidic Phosphoproteins from Various Tissues

	Calf thymus	Rat liver	Beef kidney
Phosphorus content			
Total nuclear protein	0.07%	0.14%	0.12%
Purified phosphoprotein	1.2%	1.3%	1.0%
Purification	17×	9×	8×
Yield	12%	25%	20%
Phosphoprotein content expressed as % of			
Dry weight nucleus	5%	9%	10%
Total acidic proteins	15%	50%	42%

Amino acid analyses of the purified phosphoprotein fraction confirm that the phosphorus occurs primarily as phosphoserine (about 90%), with the remainder of the phosphorus present as phosphothreonine. It can be calculated that there is enough phosphorus present to phosphorylate 4–5 amino acids out of every 100 residues, thus emphasizing the fact that these proteins must be extensively phosphorylated at multiple sites. Examination of Table I also points to the conclusion that these acidic nuclear phosphoproteins are major components of the cell nucleus. It is estimated that they represent between 5 and 10% of the total dry weight of the nucleus and from 15 to 50% of the total acidic proteins. Considering that some phosphorylated proteins may be lost during this purification procedure, the real values may be even higher.

B. Other Extraction Techniques

Many other techniques are currently employed for extraction and fractionation of acidic nuclear proteins, and some of these no doubt yield fractions which are enriched in phosphorylated proteins. Although in some cases analyses of phosphorus content have been made (Richter and Sekeris, 1972; Kostraba and Wang, 1972a), such measurements are not usually available as a basis for comparison. One approach which has been shown to be specifically useful in the preparation of acidic nuclear protein fractions enriched in phosphorus is phenol extraction (see Chapter 2), which has been extensively employed by Teng *et al.* (1971) in studies on phosphorylated acidic proteins. Since the phosphorus contents of these phenol-soluble phosphoproteins are similar (around 1% phosphorus by weight) to those of the salt-extracted phosphoproteins, it is tempting to conclude that these are analagous protein preparations. However, amino acid analyses show that in spite of overall

4. Acidic Nuclear Phosphoproteins

similarities, there are some basic differences in the composition of the two types of preparations. As can be seen in Table II, there are major differences between the two types of phosphoprotein fractions in at least 3 amino acids, namely arginine, serine, and glycine, as well as in the overall ratio of acidic to basic residues. Table II also gives the amino acid composition of a typical histone fraction for comparative purposes, showing that these phosphoproteins are easily distinguishable from histones.

In addition to differences in amino acid composition, electrophoretic analysis has also indicated significant differences between these two types of phosphoproteins, with major differences in banding patterns evident when the two types of preparations are compared from the same tissue

TABLE II
Amino Acid Compositions of Various Nuclear Protein Fractions

	Acidic phosphoproteins			Histone F_{2A2}
	Saline-extracted		Phenol-extracted	
	Calf thymus[a]	Rat liver[b]	Rat liver[c]	Calf thymus[d]
Lysine	9.4	8.2	6.0	12.9
Histidine	1.9	2.0	2.5	2.7
Arginine	8.5	9.1	5.9	9.4
Aspartic acid	10.5	9.8	10.3	5.6
Glutamic acid	14.9	14.1	12.2	9.0
Threonine	3.8	4.5	3.8	5.0
Serine	10.3	10.2	4.4	5.0
Proline	6.2	6.8	6.5	4.0
Glycine	6.8	8.0	15.3	9.1
Alanine	6.2	6.2	6.7	13.2
Cysteine	0.6	0.4		
Valine	4.5	5.0	6.1	6.0
Methionine	1.9	1.4	1.9	0.2
Isoleucine	3.0	3.0	4.1	4.3
Leucine	6.2	6.4	6.8	10.2
Tyrosine	2.4	1.9	3.3	2.3
Phenylalanine	2.8	2.9	4.0	1.0
Total acidics	25.4	23.9	22.5	14.6
Total basics	19.8	19.3	14.4	25.0
Acidics/basics	1.28	1.24	1.56	0.58

[a] From Kleinsmith and Allfrey (1969a).
[b] From Kleinsmith (1973).
[c] From Teng et al. (1971).
[d] From Hnilica et al. (1971).

(Kleinsmith and Kish, 1974). It is thus dangerous to assume that phosphorus-enriched acidic protein fractions prepared by different procedures are in fact the same.

In addition to these two basic procedures for preparation of acidic protein fractions highly enriched in phosphorus, other techniques for isolation of acidic proteins (see Chapter 3) may be used in conjunction with labeling with radioactive phosphorus in order to study phosphorylation in acidic protein fractions which have not been extensively enriched in terms of phosphoproteins. For example, such an approach has been extensively employed in the analysis of phosphorylation of acidic protein fractions isolated by chromatography on hydroxylapatite and QAE-Sephadex (Rickwood et al., 1973; MacGillivray and Rickwood, 1973).

C. Electrophoretic Fractionation

Acidic nuclear phosphoprotein fractions isolated by any of the above techniques can be shown to be still heterogeneous when analyzed by electrophoresis in acrylamide gels in the presence of the detergent SDS (Fig. 2). Typically one obtains between 25 and 50 bands, depending on the resolving power of the electrophoretic system employed. Electrophoretic comparison of acidic nuclear phosphoprotein fractions prepared by different procedures has shown them to be all highly heterogeneous, although major differences in banding patterns can be detected in materials prepared in different ways (Kleinsmith and Kish, 1974).

Thus far SDS–acrylamide gels have been the only electrophoretic technique which has provided reproducible and sensitive subfractionation of the acidic nuclear phosphoproteins. This is unfortunate because this technique has several inherent limitations. First of all, the presence of the detergent SDS causes protein denaturation, thus inhibiting enzyme activities and breaking up proteins into subunits. Second, due to differences in staining properties of different types and concentrations of proteins, it is difficult to apply rigorous quantitation to this technique. Third, because this technique separates only on the basis of molecular weight, it can obviously not separate different types of proteins of approximately

Fig. 2. SDS–acrylamide gels of salt-extracted acidic nuclear phosphoproteins prepared from nuclei of rat liver and HeLa cells. Gels were stained for protein with Coomassie Brilliant Blue.

the same size. This is especially critical in the case of acidic nuclear phosphoproteins, where a single protein backbone may exist in varying states of phosphorylation. Finally, because of these limitations it is not possible to easily compare materials prepared under different conditions in different laboratories. For example, if two different procedures of phosphoprotein isolation each result in an SDS–acrylamide gel pattern with a band at 40,000 molecular weight, there is no way of knowing whether that is the same protein or different proteins with the same molecular weight.

These factors thus impose serious limitations on gathering and communicating information about specific acidic nuclear phosphoproteins, and as a result other techniques for subfractionation are badly needed. Unfortunately due to the tendency of these proteins to aggregate, other types of electrophoresis, even in the presence of urea, have not been very successful. One possible new approach is hydroxylapatite chromatography in the presence of SDS, which has been reported to separate proteins differently than SDS–acrylamide gels (Moss and Rosenblum, 1972). However, since this technique also employs detergent, it would be subject to some of the same limitations discussed previously. Preliminary reports (MacGillivray and Rickwood, 1973) have suggested that two-dimensional acrylamide gel fractionation, with isoelectric focusing in one dimension, may be another useful approach to this problem, and the development of this technique will be awaited with interest.

III. PHOSPHATE METABOLISM OF ACIDIC NUCLEAR PHOSPHOPROTEINS

The unique, distinguishing feature of the acidic nuclear phosphoproteins is of course the presence of large numbers of protein-bound phosphate groups. Molecular weight estimates based on SDS–acrylamide gel electrophoresis for the family of phosphorylated proteins cover a broad range of from about 7000 to over 150,000. Using the average phosphate content of 1.0% as a basis for calculation, this means that individual protein chains may contain from 3 to 60 or more covalently bound phosphate groups. If individual protein chains assemble together as subunits as part of protein quaternary structure, the number of phosphate groups per protein molecule would be even larger. Although these estimates are based on an average calculation, and there may be wide differences in phosphate content between different proteins in this fraction, the overall conclusion that we are dealing with an extremely high concentration of phosphorylated sites is still valid.

Thus, the metabolic characteristics of this large number of protein-bound phosphate groups must have extremely important implications for the structural and functional properties of the acidic nuclear proteins. In this section we will review the metabolism of these phosphate groups, first examining studies with radioactive phosphate tracers in intact nuclei and then focusing attention on the enzymes involved in these metabolic reactions.

A. Protein-Bound Phosphate Metabolism in Intact Nuclei

The first studies which demonstrated that protein phosphorylation occurs in intact nuclei were performed by incubating calf thymus nuclei with ^{32}P-orthophosphate (Kleinsmith et al., 1966a). In such a system it was shown that the site of phosphorylation is mainly the hydroxyl group of serine residues, but small amounts of radioactive phosphothreonine could also be detected. The phosphorylation reaction was found to be energy dependent, since it was blocked by inhibitors of ATP formation such as iodoacetate and 2,4-dinitrophenol. Since puromycin, which inhibits protein synthesis in these nuclei, was found to have no effect on the phosphorylation reaction, it could be concluded that the phosphate group is put on the protein after the polypeptide chain is completed.

Not only are the phosphate groups attached to the protein independently of protein synthesis, but the phosphate groups are also being continually removed. During a two-hour cold-chase experiment, 70–80% of previously incorporated ^{32}P is lost from the protein independent of any protein breakdown (Fig. 3). In these studies it was also observed that most of ^{32}P radioactivity is found associated with acidic chromatin proteins, although a small amount of phosphorylation of histones and nuclear sap proteins was also observed.

These studies, then, focused attention on the fact that acidic proteins are continually being phosphorylated and dephosphorylated inside the nucleus in extremely active reactions. The basic findings of these original studies have since been confirmed in a number of other *in vitro* systems involving nuclei, chromosomes, or chromatin (Schiltz and Sekeris, 1969, 1971; Benjamin and Goodman, 1969; Ahmed, 1971; Kamiyama and Dastugue, 1971; Rickwood et al., 1973). However, since these other types of nuclei and subnuclear fractions do not retain the capacity to synthesize ATP, it has been necessary to incubate them with ^{32}P-labeled ATP in order to observe protein phosphorylation. It should be pointed out that in addition to radioactive phosphoserine and phosphothreonine, small

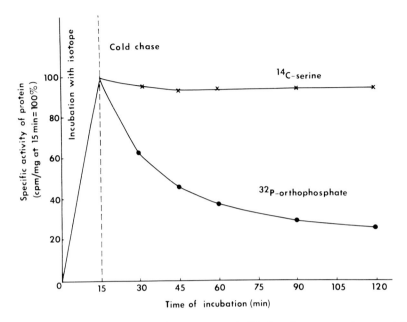

Fig. 3. Turnover of previously incorporated ^{32}P-phosphate in nuclear proteins. Calf thymus nuclei were incubated for 15 minutes with ^{32}P-orthophosphate or ^{14}C-serine, following which they were washed and reincubated in radioisotope-free media. Note that the ^{14}C-serine radioactivity is stable, indicating the absence of significant protein degradation, while the ^{32}P-phosphate is rapidly lost from the protein fraction (from Kleinsmith et al., 1966a).

amounts of labeled phosphoarginine and phospholysine can also be detected in such systems if appropriate precautions are taken (Schiltz and Sekeris, 1969).

The occurrence of both phosphorylation and dephosphorylation reactions in the metabolism of acidic nuclear proteins raises the questions of the metabolic origin and fate of the phosphate groups involved. These important questions are most easily dealt with in the context of the enzymological studies described in the following sections.

B. Phosphorylation Reaction: Protein Kinase Activities

Acidic nuclear phosphoprotein fractions purified by the salt extraction procedure retain endogenous protein kinase activity (Langan, 1967; Kleinsmith and Allfrey, 1969a). Incubation of these proteins in the presence of [γ-^{32}P]ATP and Mg$_2^+$ results in rapid protein phosphorylation, again mainly in the form of phosphoserine with small amounts of phosphothreonine. A wide variety of ^{32}P-labeled substrates have been tested in this system to determine the possible range of phosphate donors. In addition to ATP, it has been found that a wide variety of other nucleoside and deoxynucleoside triphosphates can donate their terminal (γ) phosphate groups to nuclear phosphoproteins (Table III). Thus, it appears as if the natural phosphate donors may include a wide range of high energy compounds, a conclusion consistent with the previously mentioned finding of the energy dependence of protein phosphorylation in intact nuclei.

Since the salt-extracted phosphoprotein fraction is still highly heterogeneous as evidenced by SDS–acrylamide gel electrophoresis, the question arises as to whether the protein kinase activities are separate proteins which contaminate the nuclear phosphoprotein fraction or whether the protein kinase activity is inherent in the phosphoproteins themselves. Attempts to resolve this question by fractionating the acidic phosphoproteins via ion-exchange chromatography on phosphocellulose columns have succeeded in separating the protein kinase activities from the bulk of the protein. However, these purified kinases can still be phosphorylated in the absence of any exogenous protein substrate, so a complete separation of kinase activity and the protein substrates has still not been achieved.

TABLE III

Phosphorylation of Acidic Nuclear Phosphoproteins by Various Substrates during *in Vitro* Incubation[a]

Substrate	Alkali-labile ^{32}P formed	
	(pmoles)	(% of ATP activity)
[γ-^{32}P]ATP	11.3	100.0
Zero time	0.01	0.07
Minus Mg^{2+}	0.03	0.2
[γ-^{32}P]GTP	6.5	57.4
[γ-^{32}P]ITP	4.2	37.5
[γ-^{32}P]CTP	2.4	21.5
[γ-^{32}P]UTP	0.7	6.0
[γ-^{32}P]dATP	11.3	100.0
[β-^{32}P]ADP	0.3	2.2
^{32}P$_i$	0.06	0.5
^{32}PP	0.002	0.02

[a] Purified nuclear phosphoproteins were incubated in the presence of 5 mM MgCl$_2$ and various ^{32}P-labeled substrates for 10 minutes, and the formation of protein-bound alkali-labile ^{32}P was determined (from Kleinsmith and Allfrey, 1969a).

4. Acidic Nuclear Phosphoproteins

Several of the investigations on fractionation of nuclear protein kinases have shown the presence of only a small number of enzyme activities (Takeda et al., 1971; Kamiyama et al., 1971, 1972; Kamiyama and Dastugue, 1971; Ruddon and Anderson, 1972; Desjardins et al., 1972). This limitation in resolution is probably due to the relatively low specific activity of the starting materials employed, as well as the use of casein and histones as substrate, both of which inhibit many nuclear protein kinase activities. When one starts with a highly purified acidic nuclear phosphoprotein preparation as described in Section II,A, and protein kinase activity of phosphocellulose column fractions is measured by incubation with [γ-^{32}P]ATP and magnesium in the absence of any exogenous protein substrate, a much greater degree of heterogeneity is observed (Kish and Kleinsmith, 1972, 1974a,b). When beef liver nuclear phosphoprotein is employed as starting material in such a system, at least 12 distinct enzyme activities are reproducibly observed (Fig. 4). These

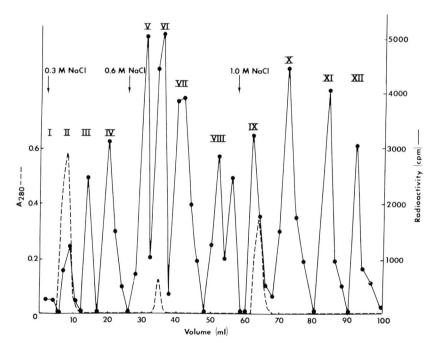

Fig. 4. Fractionation of nuclear protein kinases via column chromatography on phosphocellulose. The salt-extracted nuclear phosphoprotein fraction prepared from beef liver was employed as starting material, and protein kinase activity in the fractions was determined by incubation in the presence of Mg^{2+} and [γ-^{32}P]ATP, followed by measurement of protein-bound radioactivity (see Kish and Kleinsmith, 1974a,b).

individual enzyme fractions can be further fractionated into multiple components by acrylamide gel electrophoresis, thus demonstrating a remarkable degree of heterogeneity amongst the chromatin-associated protein kinases. The various protein kinase fractions exhibit different specificities for casein, histone, and acidic proteins as substrates, as demonstrated both by acidic precipitable radioactivity and analysis of ^{32}P-labeled substrates by SDS–acrylamide gel electrophoresis. The latter type of analysis has shown that different protein kinase fractions specifically phosphorylate different components of the total acidic nuclear phosphoprotein fraction. Further evidence which suggests that we are dealing with distinctly different protein kinases is shown by the differing sensitivity of these fractions to the presence of cyclic AMP. At a concentration of 10^{-6} M, this cyclic nucleotide stimulates protein phosphorylation in some cases and inhibits it in others, depending both on the choice of substrate and the particular protein kinase fraction employed (Fig. 5).

Data have also been obtained which suggest that the profile of these

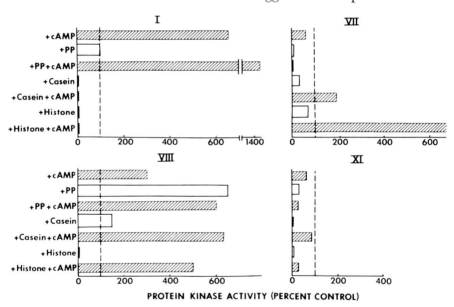

Fig. 5. Histogram showing the effectiveness of 4 different nuclear protein kinase fractions (indicated by Roman numerals referring to peaks in Fig. 4) in catalyzing the phosphorylation of various substrates in the presence or absence of cyclic AMP. Note that each enzyme fraction has a uniquely identifying profile of activity. Dotted lines represent control activity in the absence of any added substrate or cyclic AMP. cAMP = cyclic AMP, PP = total acidic nuclear phosphoprotein fraction prepared by salt extraction technique and having its own protein kinase activity destroyed by heat treatment.

protein kinase activities is different in different tissues and that the enzymes are distinct from the soluble protein kinases which phosphorylate histones (Kish and Kleinsmith, 1974a). The existence of such a complex set of enzymes which phosphorylate acidic proteins and which respond in different ways to cyclic AMP is of obvious potential importance in helping us understand how the phosphorylation of these proteins is regulated, and has implications for the possible functions of these proteins which will be discussed later.

C. Dephosphorylation Reaction: Phosphatase Activities

Less is known about the enzymology of the dephosphorylation reaction of acidic proteins than was the case above for the phosphorylation reaction. In order to determine the nature of the reaction by which phosphate is removed from the acidic nuclear proteins in intact nuclei, it is necessary to follow the fate of the ^{32}P-labeled phosphate groups. Unfortunately this approach is not feasible in experiments where phosphoproteins are simply labeled and then subjected to cold-chase conditions (Fig. 3), since in this case molecules other than acidic phosphoproteins may be labeled with ^{32}P. Thus the original approach employed to attack this problem involved extraction and purification of the acidic nuclear phosphoproteins which had been previously labeled with ^{32}P-phosphate or ^{3}H-serine and subsequent addition of these radioactive proteins back to fresh unlabeled nuclei. In such experiments, autoradiographic evidence showed that these proteins entered the nuclei, and removal of ^{32}P-phosphate groups was found to occur at a rate comparable to that seen in cold-chase experiments where nuclei were labeled and chased directly. In terms of its sensitivity to various inhibitors, the dephosphorylation reaction was found to be indistinguishable from that which occurs in phosphoproteins labeled and chased in intact nuclei. Furthermore, the ^{3}H-serine label was not lost from the protein fraction, indicating that protein hydrolysis was not responsible for the loss of phosphate groups. When analyses were performed to determine the fate of the ^{32}P released from these phosphoproteins, it was discovered that the radioactivity could all be recovered in the form of inorganic phosphate (Kleinsmith and Allfrey, 1969b).

These results suggested that the enzyme(s) responsible for the dephosphorylation reaction is a phosphatase. In order to find out whether this enzyme activity is inherent in the purified phosphoprotein fraction (as was seen to be the case for the protein kinase activities), cold-chase experiments have been performed on purified nuclear phosphoproteins labeled with ^{32}P *in vitro*. In such experiments the phosphate groups have been found to be stable to subsequent incubation (Kleinsmith and Allfrey, 1969a), in marked contrast to the rapid turnover seen in intact nuclei. These results thus show that the phosphatase activity is not carried along during the purification of the acidic nuclear phosphoproteins.

Since the nuclear phosphatase(s) responsible for the dephosphorylation of acidic proteins have not yet been purified, little is known of their substrate specificities. An acid phosphatase activity has been localized cytochemically in chromatin and nucleoli (Soriano and Love, 1971) and may be involved in such reactions. A histone phosphatase has been purified by Meisler and Langan (1969), but is found localized mainly in the cytoplasm and has very low activity with phosphorylated acidic proteins as substrates. Since the dephosphorylation reaction may be an important site of control in regulating the structure and functional properties of these acidic proteins, studies on the phosphatases involved should be an important area for future investigation.

D. Summarizing Model

The data thus far present an overall picture of phosphorylation and dephosphorylation of acidic nuclear proteins which allows for continual and dynamic changes in protein structure and function. The model presented in Fig. 6 summarizes the relationships between the various substrates and enzyme activities involved in these processes.

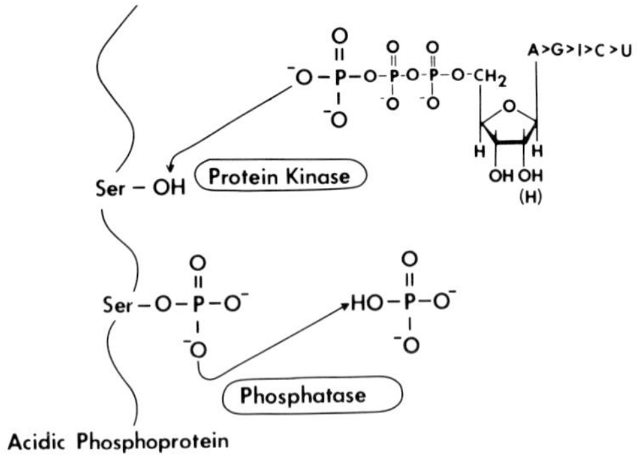

Fig. 6. Model summarizing the metabolic relationships of the acidic nuclear phosphoprotein side chains. Serine (and threonine) residues in the protein are phosphorylated in a kinase reaction utilizing the terminal phosphate group of various nucleoside and deoxynucleoside triphosphates. In a separate phosphatase reaction, phosphate groups already present in the protein are cleaved off and released as inorganic phosphate.

IV. FUNCTIONAL PROPERTIES OF ACIDIC NUCLEAR PHOSPHOPROTEINS

Since the acidic nuclear phosphoproteins are present in such high concentration in nuclei and are involved in such active metabolism, the obvious question arises as to their functional significance. When they were first discovered in the mid-1960's, preliminary data from a few systems led to the suggestion that they were somehow involved in the process of gene regulation (Kleinsmith et al., 1966a,b; Langan, 1967). If these proteins are in fact genetic regulatory molecules, a number of predictions can be made regarding their expected properties. For example, one might predict that they would exhibit tissue specificity, quantitative and metabolic differences correlating with differences in gene activity, localization in specific regions of chromosomes, specific binding to DNA, effects on RNA synthesis in vitro, and modulation by external effectors such as hormones and cyclic AMP. In this section we will examine the data which has now been accumulated in relationship to these various predictions.

A. Cell, Tissue, and Species Specificity

If the acidic nuclear phosphoproteins are specific genetic regulatory molecules, then one would expect to find differences in these proteins in different tissues and cell types where differences in gene expression occur. Analysis of nuclear phosphoproteins from different tissues by SDS–acrylamide gel electrophoresis has shown that although there are major similarities, significant and reproducible differences in specific protein bands are evident in different tissues and in the same tissue at different stages of differentiation (Platz et al., 1970; Platz, 1972; C. T. Teng et al., 1970; C. S. Teng et al., 1971; LeStourgeon and Rusch, 1971, 1973; Rickwood et al., 1973; Vidali et al., 1973). Labeling experiments with ^{32}P have demonstrated that most of the bands contain phosphorylated proteins, and the phosphorylation patterns have also been shown to be tissue specific (Platz et al., 1970; Platz, 1972; Teng et al., 1971; Rickwood et al., 1973). The observed heterogeneity of phosphorylated acidic proteins is not explained simply by the fact that individual tissues are composed of multiple cell types, since nuclear phosphoproteins isolated from pure cell lines are also highly heterogeneous (Figs. 2 and 7). Examination of phosphorylated nuclear proteins from the same tissue in different species has shown basic similarities, although major differences are evident as one compares tissues from widely diverging species (Platz, 1972).

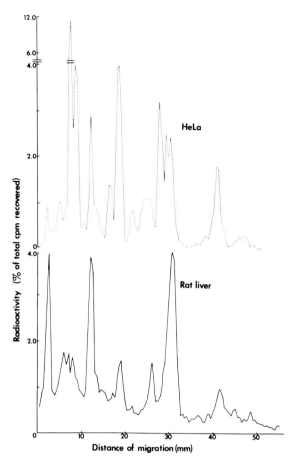

Fig. 7. Radioactivity profiles of salt-extracted acidic nuclear phosphoproteins labeled with ^{32}P and electrophoresed in SDS–acrylamide gels. Rats were injected with ^{32}P-orthophosphate, while HeLa cells were incubated with the same isotope in culture prior to isolation of nuclear phosphoproteins. The patterns obtained are reproducibly different from each other.

The observation of major similarities in nuclear phosphoproteins from different cells and tissues is not surprising, since many nuclear functions must be universal for all cell types. However, the observation of significant tissue-specific differences in these proteins is quite noteworthy and may have significant implications for the proposed role of these molecules as specific gene regulators. This observed heterogeneity and specificity is in sharp contrast to the histones, which exist in a limited number of types which do not vary substantially from tissue to tissue.

B. Localization and Quantitation in Chromatin

The design of the salt extraction technique for purification of nuclear phosphoproteins suggests that these proteins are part of the chromatin fraction. Starting with purified nuclei, this procedure involves removal of nuclear sap proteins by isotonic saline extraction and removal of membranes and nucleoli by sedimentation from the high salt extract. By the process of elimination this leaves only chromatin components, a conclusion which has been substantiated by the identification of these phosphoproteins in chromatin fractions prepared by a variety of different techniques. The phosphoproteins isolated directly from chromatin preparations have been shown to be generally indistinguishable from those isolated from intact nuclei (Teng et al., 1971).

Investigation of the distribution of the amount of protein-bound phosphate in different types of chromatin has indicated that active, euchromatin fractions contain a severalfold higher concentration of protein-bound phosphate per milligram DNA than do inactive, heterochromatin fractions. Cell types with different general levels of nuclear activity also seem to differ in their levels of chromatin phosphoprotein. For example liver nuclei, which are very active in RNA synthesis, contain about 3 times as much phosphoprotein per milligram DNA as do thymus nuclei, which are derived from an organ undergoing involution and are relatively inactive in RNA synthesis (Kleinsmith and Allfrey, 1969a). Likewise mouse brain chromatin, which has a greater amount of its genome available for transcription than other mouse tissues, has a significantly higher level of acidic protein phosphorylation (Riches et al., 1973).

The conclusion that the acidic nuclear phosphoproteins are localized in the chromatin fraction, based originally on biochemical separations, has since been independently confirmed by direct autoradiographic observation. Benjamin and Goodman (1969), working with giant insect chromosomes, have shown that incubation for short periods of time with [γ-^{32}]ATP leads to phosphorylation of acidic proteins which can be localized autoradiographically to discrete regions of the chromosomes.

C. Changes in Protein Phosphorylation and Gene Activity

If the phosphorylation of acidic proteins is related to the control of gene expression, then one would expect to observe changes in phosphorylation which correlate with changes in activity of the genome. One of the first observations along these lines was made in human lymphocytes stimulated to divide by the addition of phytohemagglutinin (PHA). Such cells are known to dramatically increase their capacity for RNA synthesis during the first 24 hours after addition of PHA, and consistent with our hypothesis, this major increase in RNA synthesis was found to be preceded by an increase in the rate of phosphorylation of nuclear proteins (Kleinsmith et al., 1966b).

Since that initial observation, similar correlations have been observed in a number of other systems. For example, during the maturation of avian erythrocytes the nuclear chromatin progresses from a diffuse configuration which is active in RNA synthesis to a dense, compact form which has a relatively inactive genome. During this maturation process phosphorylation of acidic chromatin proteins decreases progressively, with the final mature red blood cells having less than one third as much protein-bound phosphate as the immature, active cells (Gershey and Kleinsmith, 1969b).

This correlation between cell activity and nuclear protein phosphorylation has also been studied in a number of systems where different states of cell growth can be compared. During the mitotic cycle of *Physarum polycephalum,* distinct changes in the incorporation of ^{32}P-orthophosphate and content of alkali-labile phosphate occur in nuclear acidic proteins (LeStourgeon and Rusch, 1971). In Yoshida ascites sarcoma cells, the rate of phosphorylation of these proteins increases sharply as the cells move from the exponential phase to the stationary phase of growth (Riches et al., 1973). During the cell cycle of synchronously dividing HeLa cells it has been found that significant changes in the phosphorylation of specific nuclear proteins occurs (Platz et al., 1973). Similarly, the phosphorylation of a specific nonhistone protein component has been shown to change dramatically during sea urchin development (Platz and Hnilica, 1973). Finally, in rat salivary glands stimulated by isoproterenol, the prereplicative phase involves an inhibition of RNA synthesis followed by a stimulation, which correlates directly with observed changes in acidic protein phosphorylation (Ishida and Ahmed, 1973).

Another type of physiological situation where changes in acidic protein phosphorylation have been observed is during hormonal stimulation. Both steroid and polypeptide hormones are now known to have at least some effects at the level of gene activity, and it is thus of interest to note that a number of hormones have now been shown to induce alterations in the phosphorylation of acidic nuclear proteins. Insulin and prolactin have been shown to stimulate acidic protein phosphorylation in mouse mammary epithelial cells in close association with an increase in RNA synthesis (Turkington and Riddle, 1969). Chorionic gonadotropin causes an increase in protein phosphorylation and RNA synthesis in ovarian nuclei of immature rats (Jungmann ad Schweppe, 1972a,b). Testosterone has been observed to stimulate phosphorylation of acidic proteins in rat ventral prostate (Ahmed, 1971; Ahmed and Ishida, 1971), while cortisol, which is known to increase RNA synthesis in rat liver, has been shown to increase the rate and alter the pattern of phosphorylation of acidic proteins within 5 minutes after its administration (Allfrey et al., 1971).

Studies on the synthesis of acidic nuclear proteins have resulted in a large number of examples where changes in the synthesis of particular acidic proteins have been shown to be correlated with changes in gene activity. This area is thoroughly discussed in Chapters 5, 6, and 7 and so will not be dealt with here. However, it should be pointed out that many of the observed changes in synthesis of acidic proteins may in fact be changes in proteins which are, or ultimately will be, phosphorylated.

Thus, in a wide variety of different experimental situations the phosphorylation of these nuclear proteins has been found to be under dynamic control, changing in a variety of ways as gene activity is modulated in different physiological states.

D. Specificity of DNA Binding

Since specific genetic regulatory molecules which have been purified from microbial systems have turned out to be allosteric proteins which recognize and bind to specific sites on DNA, it might be expected that specific gene regulators in eukaryotic cells would also be able to bind to DNA in a specific fashion. Experiments employing DNA–cellulose chromatography have demonstrated that a small portion of the acidic nuclear proteins which are phosphorylated can bind to DNA in a specific fashion (Kleinsmith et al., 1970; Kleinsmith, 1973). When binding is carried out in 0.14 M NaCl, approximately 1% of the purified acidic nuclear phosphoprotein prepared from rat liver binds to rat DNA. Much lower levels of binding are observed when foreign DNA's are used, thus demonstrating the specificity of the reaction (Fig. 8). Under these conditions, binding sites on the DNA are saturated at a value of approximately 1 μg of phosphorylated protein per 100 μg of DNA. The material which selectively binds to the DNA is still a heterogeneous family of proteins as demonstrated via SDS–acrylamide gel electrophoresis, but the majority of them fall in the molecular weight range of 30,000 to 70,000.

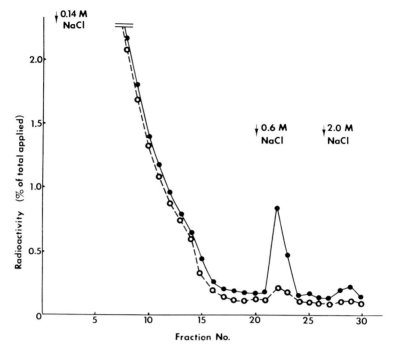

Fig. 8. Chromatography of ^{32}P-labeled rat liver nuclear phosphoproteins on columns of DNA-cellulose. Proteins which bind to DNA are eluted by raising the ionic strength to 0.6 M NaCl. Rat liver phosphoproteins bind much better to columns made from rat DNA (●——●) than to columns made from *E. coli* DNA (○---○). For details of methodology, see Kleinsmith (1973).

Specificity of binding of phosphorylated acidic proteins to DNA has also been demonstrated by sucrose gradient centrifugation (C. T. Teng *et al.*, 1970; C. S. Teng *et al.*, 1971). Under these conditions higher percentages of the phosphoprotein fraction bind to DNA, but the experiments are difficult to compare to the ones employing DNA–cellulose chromatography, because both the method of protein isolation and conditions of the binding assay are different. Since the degree of binding is dependent on a wide variety of experimental factors, one must be cautious in attempting to interpret the significance of the absolute quantity of phosphorylated protein which is bound in any given experiment.

The most significant fact to be obtained from these different experimental approaches is that they both suggest that a portion of the phosphorylated acidic proteins can specifically recognize and bind to certain types of DNA, a finding with considerable signficance for the proposed role of these proteins in specific gene regulation.

E. Effects on RNA Synthesis *in Vitro*

The first experiments showing a direct effect of nuclear phosphoproteins on RNA synthesis *in vitro* were performed by Langan (1967), who demonstrated that these proteins could partially reverse the inhibitory effects of histones on RNA synthesis when purified DNA was employed as a template with bacterial RNA polymerase. Similar results were obtained subsequently by Kamiyama *et al.* (1971, 1972), who in addition demonstrated that the ability of these phosphoproteins to increase RNA synthesis was directly related to their phosphate content.

Enhancement of template activity for RNA synthesis has also been observed in systems where phosphorylated acidic proteins are added directly to intact chromatin (Kostraba and Wang, 1972a,b). Hybridization analyses in these experiments indicated that new DNA sequences are being transcribed and that addition of phosphorylated proteins from one tissue to chromatin prepared from another can alter the pattern of transcription to one more like that of the tissue from which the acidic phosphorylated proteins were obtained.

A direct effect of phosphorylated proteins on RNA synthesis has also been observed in systems where histones and other chromosomal proteins are not present. Using isolated DNA as a template with purified bacterial or mammalian RNA polymerase, acidic phosphorylated proteins prepared

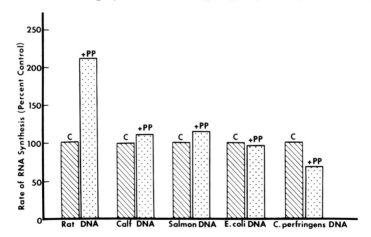

Fig. 9. Histograms showing the effects of the addition of rat liver nuclear phosphoproteins to an RNA-synthesizing system containing rat liver RNA polymerase and DNA's prepared from various species as template. Note that a stimulation of RNA synthesis by rat liver nuclear phosphoproteins only occurs when rat DNA is employed as template (see Shea and Kleinsmith, 1973). C = control, + PP = with rat liver acidic phosphoprotein added.

by a variety of techniques will directly enhance the rate of RNA synthesis (Teng et al., 1971; Rickwood et al., 1972; Shea and Kleinsmith, 1973). This ability to stimulate RNA synthesis has been shown to be dependent on the presence of protein-bound phosphate groups (Shea and Kleinsmith, 1973), and most interestingly, the effect is template specific. If foreign DNA's are employed as template, the enhancement of RNA synthesis does not occur (Fig. 9).

These data thus indicate that phosphorylated proteins can enhance RNA synthesis in a number of different systems using either intact chromatin or isolated components recombined in a variety of ways.

F. Effects of Cyclic AMP on Protein Phosphorylation

Since cyclic AMP has been shown to be generally involved in the regulation of protein phosphorylation reactions, it seems logical to ask if phosphorylation of nuclear acidic proteins is regulated by this cyclic nucleotide. Many studies on protein kinase activity in isolated nuclei or purified acidic phosphoprotein fractions have failed to show any significant effects of cyclic AMP (Kaplowitz et al., 1971; Ahmed, 1971; Kamiyama and Dastugue, 1971). However, when more highly purified kinase fractions are obtained, effects of this cyclic nucleotide become apparent (Kish and Kleinsmith, 1972, 1974a,b; Johnson and Allfrey, 1972). As was discussed in Section III,B, the nucleus contains a wide spectrum of protein kinases which phosphorylate acidic proteins, some of which are stimulated by cyclic AMP and some of which are inhibted by it. Thus, the failures to observe such effects reported earlier in the literature may have been due to the masking of these complex and divergent effects which occurs when analyses are performed on less purified preparations.

A similar problem has occurred in attempts to observe effects of cyclic AMP on phosphorylation of acidic nuclear proteins *in vivo*. Phosphorylation of total acidic protein fractions in rat liver have been reported to be unaffected by injection of dibutyryl cyclic AMP (Byvoet, 1971), but analyses of individual protein components by SDS–acrylamide gel electrophoresis indicates a small stimulation of some individual components (Johnson and Allfrey, 1972).

The present data, although relatively sketchy, thus indicate the possibility of some selective effects of cyclic AMP in modulating the phosphorylation of acidic nuclear proteins.

G. Binding of Steroid Hormones

Since steroid hormones are known to become localized in cell nuclei, where they affect gene transcription (Tomkins and Martin, 1970) and acidic protein phosphorylation (see Section IV,C), the question arises as to whether there is any direct interaction between these steroid hormones and the acidic proteins. Dastugue et al. (1971) have reported that radioactive corticosterone binds specifically to rat liver acidic chromatin proteins with a dissociation constant of 5×10^{-9}. Although the phosphorus content of their acidic protein fraction was not reported, experiments in our laboratory have suggested that the phosphorylated acidic proteins may be involved in this reaction (Fig. 10).

H. Histone–Phosphoprotein Interactions

We have already seen how a variety of extranuclear factors such as hormones and cyclic AMP can interact with and affect the nuclear phosphoproteins. Another type of molecule which has been shown to interact with these proteins and affect their properties is the histones. Langan (1967) was the first to observe that histones and nuclear phosphoproteins form insoluble complexes when mixed together, and we have already

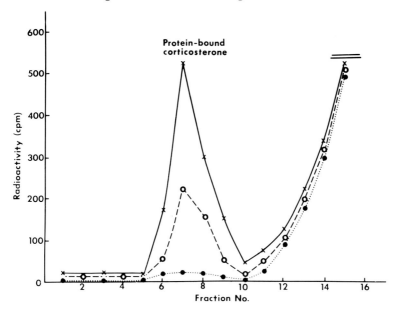

Fig. 10. Binding of ³H-corticosterone to equal amounts of salt-extracted acidic phosphoprotein (×——×), total acidic chromatin protein (○---○), and phosvitin (●·····●) as measured by gel chromatography. Incubations and gel chromatography were performed as described by Dastugue et al. (1971). The phosphorylated acidic protein fraction appears to be enriched in the steroid-binding material. The phosphorylated egg protein phosvitin is included as a control to demonstrate the specificity of the reaction (R. Andrews and L. J. Kleinsmith, unpublished observations).

seen how this interaction can lead to a partial reversal of histone-mediated inhibition of RNA synthesis (Section IV,E).

Another interesting effect of the histone–phosphoprotein interaction is that it causes an alteration in the rate of the phosphorylation of the acidic proteins. In view of the fact that histones are usually viewed as general inhibitors of enzymatic reactions, it is somewhat surprising to find that the addition of histones to the acidic phosphoproteins causes a dramatic overall increase in the rate of phosphorylation of the acidic proteins (Kaplowitz et al., 1971). In spite of the fact that addition of histones causes the phosphoproteins to precipitate, this change in solubility is not in itself responsible for the increased rate of phosphorylation. The stimulatory effect of histones is quite specific, since other small basic proteins like cytochrome c do not exhibit a comparable effect (Fig. 11).

When the effect of histones on acidic protein phosphorylation is studied by adding histone directly to isolated nuclei, specific effects

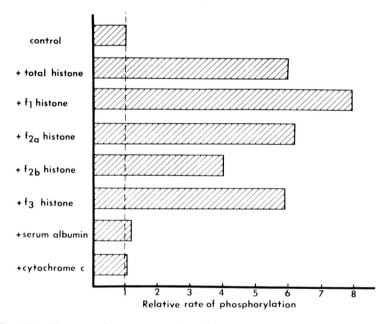

Fig. 11. Histogram demonstrating the enhancement of phosphorylation of acidic proteins induced when various histone fractions are added to the salt-extracted acidic phosphoprotein fraction and the phosphorylation rate measured *in vitro* by incorporation of radioactivity from [γ-^{32}P]ATP into acidic protein. The specificity of the observed stimulation is especially evident from the failure of cytochrome c (a small basic protein like histones) to exert any effect (from Kaplowitz et al., 1971).

of different histone fractions on the phosphorylation of different acidic proteins can be observed (Allfrey et al., 1973). Histone F_{2A1} stimulates phosphorylation of an acidic protein with molecular weight 22,000, while histone F_1 has a similar effect on an acidic protein of molecular weight 35,000. Both these histones inhibit the phosphorylation of an acidic protein of 7000 molecular weight.

V. ROLE OF ACIDIC PHOSPHOPROTEINS IN NUCLEAR FUNCTION

We have thus seen that a wide variety of evidence exists that supports the original hypothesis that the phosphorylation of acidic nuclear proteins is somehow involved in the regulation of gene activity. Although none of these lines of evidence is conclusive in itself, the total picture which emerges from these various independent approaches is strongly suggestive of some role for these phosphorylated proteins in gene control.

If the phosphorylation of acidic nuclear proteins is involved in gene regulation, then the question arises as to how the phosphorylation of these proteins in turn is regulated. Since cyclic AMP levels are known to be altered within cells in response to varying hormonal and physiological states, the observed effects of this cyclic nucleotide in stimulating and depressing the phosphorylation of specific acidic protein species are certainly of potential significance in this regard. However, the complexity of the acidic phosphoproteins and the enzymes involved in their phosphorylation and dephosphorylation makes it difficult to create a simple model of gene regulation based on these proteins.

It was originally speculated that the negatively charged phosphate groups on the acidic proteins might interact with the positively charged histones, thereby displacing the inhibitory histones from the DNA–histone complex and thus allowing the DNA to become active as a template for RNA synthesis (Kleinsmith et al., 1966a). This displacement was speculated to be correlated with a change in the structure of the chromatin from a condensed to an extended state, and such a relationship between phosphoproteins and chromatin structure has been observed in a few instances (Whitfield and Perris, 1968; Gershey and Kleinsmith, 1969b).

The more recent studies which have shown that histones stimulate the phosphorylation of acidic chromatin proteins (see Section IV,H) have led to a refinement of this hypothesis (Kaplowitz et al., 1971). According to this newer model (see Fig. 12), the increase in the phosphorylation of acidic chromatin proteins induced by the presence of

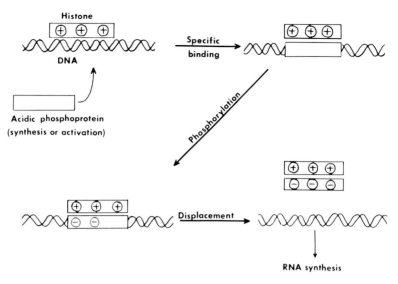

Fig. 12. One possible model of acidic phosphoprotein action capable of explaining specific gene activation. The model is based on specific binding of acidic proteins to DNA and histone displacement occurring during subsequent phosphorylation. The resulting naked DNA is capable of synthesizing RNA. See text for details.

histones would result in an increased negative charge on the acidic chromatin proteins and would therefore in turn serve to strengthen the force of ionic bonding between acidic phosphoproteins and the positively charged histones. Thus, when acidic phosphoproteins and histones come together *in vivo*, one would expect this interaction to lead to phosphorylation of the acidic chromatin proteins, resulting in a rapid increase in the strength of attraction between phosphoprotein and histone. Such an increased force of attraction might be sufficient to displace the histone from the DNA, thereby freeing the DNA template from histone inhibition and allowing gene transcription to take place. Since some of the acidic proteins have the ability to bind to specific types of DNA, the initial step in specific gene activation could thus be visualized as involving the binding of these proteins to specific gene loci in the DNA, bringing the acidic proteins in close association with histones; this would in turn lead to an increased phosphorylation of the acidic proteins and histone displacement as described above.

In spite of the attractive features of this model, it also has a number of serious limitations. Although data have already been cited which show that some of the phosphorylated acidic proteins can bind to specific types of DNA sequences in purified systems, such binding has not yet

4. Acidic Nuclear Phosphoproteins

been reported in chromatin where the DNA is covered with histones and other acidic proteins. Specific binding under such conditions, of course, is required for models such as the one described above. Another observation which is difficult to reconcile with this model is the previously mentioned stimulation of RNA synthesis by acidic phosphoproteins in the absence of any histone at all (Section IV,E). Histone displacement can obviously not explain these observations, and it is thus possible that some of these phosphorylated proteins may exert their effects on the DNA template or on RNA polymerase directly. One alternative model along these latter lines is suggested by reports that sigma factors, which regulate the specificity of bacterial RNA polymerases, can also be phosphorylated (Martelo et al., 1970). It is thus possible to speculate that phosphorylated acidic proteins may associate with RNA polymerase to activate the enzyme for the transcription of particular genes (Fig. 13). The binding of phosphorylated proteins to RNA polymerase might be regulated by their degree of phosphorylation, which in turn might be controlled by levels of cyclic AMP.

In view of the great heterogeneity present amongst the acidic nuclear phosphoproteins it is perhaps not even appropriate to think in terms of any single model which will explain all of their actions in a satisfactory

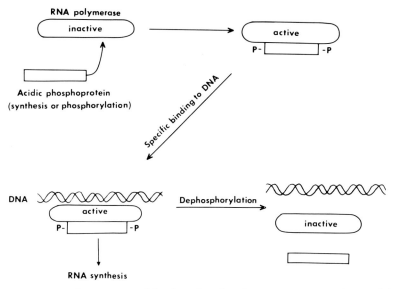

Fig. 13. An alternative model of acidic phosphoprotein action. This model is based on phosphoprotein associating with RNA polymerase, with the resulting complex being thereby activated for binding to specific families of genes and synthesizing RNA. The initial act of phosphorylation which triggers association with RNA polymerase might be regulated by cyclic AMP levels. See text for details.

way. It is more likely that this heterogeneous family of proteins is involved in nuclear metabolism in a variety of different ways. Consistent with this general view is the recent finding that a nuclear enzyme involved in histone deacetylation is in fact a phosphorylated acidic protein, containing 1.3% phosphorus by weight (Vidali *et al.*, 1972). Thus in the final analysis, phosphorylated acidic proteins may be found to play a variety of enzymatic and structural roles in the cell nucleus, the nature of which can only be determined as individual proteins are purified and their functional properties analyzed.

VI. CONCLUDING REMARKS

It is clear that a wide range of properties of the acidic nuclear phosphoproteins have now been described and that in general they support a definite correlation between the behavior of these phosphorylated proteins and the process of gene activity. Although experiments of the sort described in this chapter have been very useful in pointing out the existence of this relationship, these approaches have some serious limitations. The most critical one is that although correlation experiments can show that the various properties of the phosphorylated acidic proteins are consistent with their proposed role as gene regulators, correlations of this type cannot be employed to provide direct evidence for a cause-and effect relationship between nuclear phosphoproteins and the activity of specific genes.

It is clear that what is now required is the direct demonstration that the presence or absence of one specific phosphorylated protein (or its change from one state of phosphorylation to another) causes the activity of one identifiable gene to be altered. With the recent improvements in techniques for chromatin reconstitution, and the recent successes in the translation of messenger RNA's into specific, identifiable proteins *in vitro*, it should soon be possible to devise general methods for assaying whether particular genes are active in chromatin made up of components recombined in various ways. Although such techniques may be useful in directly demonstrating whether the class of nuclear phosphoproteins as a whole contains the molecules responsible for regulating the activity of specific genes, we would still fall short of our goal of demonstrating a relationship between a specific phosphorylated protein and a specific gene. This latter accomplishment would of course require methods for purifying single phosphorylated protein species in a native state. Due to the inherent difficulties in handling these proteins, including their tendencies to aggregate and precipitate and the requirements for detergent treatment during their subfractionation, it appears as if purification and assay of individual phosphoproteins will present the major obstacle to progress in this field in the near future.

ACKNOWLEDGMENT

Studies on this subject in our laboratory have been supported by grants from the National Science Foundation (GB-8123 and GB-23921). The figures were prepared by Sandra Beadle.

REFERENCES

Ahmed, K. (1971). *Biochim. Biophys. Acta* **243**, 38–48.
Ahmed, K., and Ishida, H. (1971). *Mol. Pharmacol.* **7**, 323–327.
Allfrey, V. G., Teng, C. S., and Teng, C. T. (1971). In "Nucleic Acid-Protein Interactions—Nucleic Acid Synthesis in Viral Infection" (D. W. Ribbons, J. F. Woessner, and J. Schultz, eds.), pp. 144–167. North-Holland Publ., Amsterdam.
Allfrey, V. G., Johnson, E. M., Karn, J., and Vidali, G. (1973). In "Protein Phosphorylation in Control Mechanisms" (F. Huijing and E. Y. C. Lee, eds.), pp. 217–249. Academic Press, New York.
Benjamin, W., and Gellhorn, A. (1968). *Proc. Nat. Acad. Sci. U.S.* **59**, 262–268.
Benjamin, W. B., and Goodman, R. M. (1969). *Science* **166**, 629–631.
Byvoet, P. (1971). *Res. Commun. Chem. Pathol. Pharmacol.* **2**, 78–86.
Dastugue, B., Defer, N., and Kruh, J. (1971). *FEBS Lett.* **16**, 121–124.
Desjardins, P. R., Lue, P. F., Liew, C. C., and Gornall, A. G. (1972). *Can. J. Biochem.* **50**, 1249–1260.
Gershey, E. L., and Kleinsmith, L. J. (1969a). *Biochim. Biophys. Acta* **194**, 331–334.
Gershey, E. L., and Kleinsmith, L. J. (1969b). *Biochim. Biophys. Acta* **194**, 519–525.
Hnilica, L. S., McClure, M. E., and Spelsberg, T. C. (1971). In "Histones and Nucleohistones" (D. M. P. Phillips, ed.), pp. 187–240. Plenum, New York.
Ishida, H., and Ahmed, K. (1973). *Exp. Cell Res.* **78**, 31–40.
Johnson, E. M., and Allfrey, V. G. (1972). *Arch. Biochem. Biophys.* **152**, 786–794.
Jungmann, R. A., and Schweppe, J. S. (1972a). *J. Biol. Chem.* **247**, 5535–5542.
Jungmann, R. A., and Schweppe, J. S. (1972b). *J. Biol. Chem.* **247**, 5543–5548.
Kabat, D. (1970). *Biochemistry* **9**, 4160–4175.
Kamiyama, M., and Dastugue, B. (1971). *Biochem. Biophys. Res. Commun.* **44**, 29–36.
Kamiyama, M., Dastugue, B., and Kruh, J. (1971). *Biochem. Biophys. Res. Commun.* **49**, 1345–1350.
Kamiyama, M., Dastugue, B., Defer, N., and Kruh, J. (1972). *Biochim. Biophys. Acta* **277**, 576–583.
Kaplowitz, P. B., Platz, R. D., and Kleinsmith, L. J. (1971). *Biochim. Biophys. Acta* **229**, 739–748.
Kish, V. M., and Kleinsmith, L. J. (1972). *J. Cell Biol.* **55**, 138a.
Kish, V. M., and Kleinsmith, L. J. (1974a). *J. Biol. Chem.* **249**, 750–760.
Kish, V. M., and Kleinsmith, L. J. (1974b). In "Methods in Enzymology" (in press).

Kleinsmith, L. J. (1973). *J. Biol. Chem.* **248**, 5648–5653
Kleinsmith, L. J., and Allfrey, V. G. (1969a). *Biochim. Biophys. Acta* **175**, 123–135.
Kleinsmith, L. J., and Allfrey, V. G. (1969b). *Biochim. Biophys. Acta* **175**, 136–141.
Kleinsmith, L. J., and Kish, V. M. (1974). *In* "Methods in Enzymology" (in press).
Kleinsmith, L. J., Allfrey, V. G., and Mirsky, A. E. (1966a). *Proc. Nat. Acad. Sci. U.S.* **55**, 1182–1189.
Kleinsmith, L. J., Allfrey, V. G., and Mirsky, A. E. (1966b). *Science* **154**, 780–781.
Kleinsmith, L. J., Heidema, J., and Carroll, A. (1970). *Nature (London)* **226**, 1025–1026.
Kostraba, N. C., and Wang, T. Y. (1972a). *Biochim. Biophys. Acta* **262**, 169–180.
Kostraba, N. C., and Wang, T. Y. (1972b). *Cancer Res.* **32**, 2348–2352.
Krebs, E. G. (1972). *Curr. Top. Cell. Regul.* **5**, 99–133.
Langan, T. A. (1967). *In* "Regulation of Nucleic Acid and Protein Biosynthesis" (V. V. Koningsberger and L. Bosch, eds.), pp. 233–242. Elsevier, Amsterdam.
Langan, T. A. (1968a). *In* "Regulatory Mechanisms for Protein Synthesis in Mammalian Cells" (A. San Pietro, M. R. Lamborg, and F. T. Kenney, eds.), pp. 101–118. Academic Press, New York.
Langan, T. A. (1968b). *Science* **162**, 579–580.
LeStourgeon, W. M., and Rusch, H. P. (1971). *Science* **174**, 1233–1236.
LeStourgeon, W. M., and Rusch, H. P. (1973). *Arch. Biochem. Biophys.* **155**, 144–158.
MacGillivray, A. J., and Rickwood, D. (1973). *Biochem. Soc. Trans.* **1**, 72–76.
Martelo, O. J., Woo, S. L. C., Reimann, E. M., and Davie, E. W. (1970). *Biochemistry* **9**, 4807–4813.
Meisler, M. H., and Langan, T. A. (1969). *J. Biol. Chem.* **244**, 4961–4968.
Moss, B., and Rosenblum, E. N. (1972). *J. Biol. Chem.* **247**, 5194–5198.
Ord, M. G., and Stocken, L. A. (1966). *Biochem. J.* **98**, 888–897.
Platz, R. D. (1972). Ph.D. Thesis, University of Michigan, Ann Arbor.
Platz, R. D., and Hnilica, L. S. (1973). *Biochem. Biophys. Res. Commun.* **54**, 222–227.
Platz, R. D., Kish, V. M., and Kleinsmith, L. J. (1970). *FEBS Lett.* **12**, 38–40.
Platz, R. D., Stein, G. S., and Kleinsmith, L. J. (1973). *Biochem. Biophys. Res. Commun.* **51**, 735–740.
Pratje, E., and Heilmeyer, L. M. G., Jr. (1972). *FEBS Lett.* **27**, 89–93.
Riches, P. G., Harrad, K. R., Sellwood, S. M., Rickwood, D., and MacGillivray, A. J. (1973). *Biochem. Soc. Trans.* **1**, 70–72.
Richter, K. H., and Sekeris, C. E. (1972). *Arch. Biochem. Biophys.* **148**, 44–53.
Rickwood, D., Threlfall, G., MacGillivray, A. J., and Paul, J. (1972). *Biochem. J.* **129**, 50P–51P.
Rickwood, D., Riches, P. G., and MacGillivray, A. J. (1973). *Biochim. Biophys. Acta* **299**, 162–171.
Roses, A. D., and Appel, S. H. (1973). *J. Biol. Chem.* **248**, 1408–1411.
Ruddon, R. W., and Anderson, S. L. (1972). *Biochem. Biophys. Res. Commun.* **46**, 1499–1508.
Schiltz, E., and Sekeris, C. E. (1969). *Hoppe-Seyler's Z. Physiol. Chem.* **350**, 317–328.
Schiltz, E., and Sekeris, C. E. (1971). *Experientia* **27**, 30–33.
Segal, H. L. (1973). *Science* **180**, 25–32.
Shea, M., and Kleinsmith, L. J. (1973). *Biochem. Biophys. Res. Commun.* **50**, 473–477.

Shepherd, G. R., Noland, B. J., and Hardin, J. M. (1971). *Arch. Biochem. Biophys.* **142,** 299–302.
Sherod, D., Johnson, G., and Chalkley, R. (1970). *Biochemistry* **9,** 4611–4615.
Soriano, R. Z., and Love, R. (1971). *Exp. Cell Res.* **65,** 467–470.
Takeda, M., Yamamura, H., and Ohga, Y. (1971). *Biochem. Biophys. Res. Commun.* **42,** 103–110.
Teng, C. S., Teng, C. T., and Allfrey, V. G. (1971). *J. Biol. Chem.* **246,** 3597–3609.
Teng, C. T., Teng, C. S., and Allfrey, V. G. (1970). *Biochem. Biophys. Res. Commun.* **41,** 690–696.
Tomkins, G. M., and Martin, D. W., Jr. (1970). *Annu. Rev. Genet.* **4,** 91–106.
Turkington, R. W., and Riddle, M. (1969). *J. Biol. Chem.* **244,** 6040–6046.
Vidali, G., Boffa, L. C., and Allfrey, V. G. (1972). *J. Biol. Chem.* **247,** 7365–7373.
Vidali, G., Boffa, L. C., Littau, V. C., Allfrey, K. M., and Allfrey, V. G. (1973). *J. Biol. Chem.* **248,** 4065–4068.
Whitfield, J. F., and Perris, A. D. (1968). *Exp. Cell Res.* **49,** 359–372.

5

Characterization of Nuclear Phosphoproteins in *Physarum polycephalum*

BRUCE E. MAGUN

I. Introduction	137
II. Phosphate Content of Phosphoproteins	139
A. Correspondence of ^{3}H and ^{32}P Incorporation following Continuous Double-Labeling	139
B. Calculation of Phosphates per Polypeptide	141
C. Calculation of Copies per Nucleus	142
D. Demonstration of the Protein-Bound Nature of Phosphate	145
III. Pulse Labeling in ^{32}P$_i$	145
IV. Labeling of Nuclei *in Vitro* with [γ-^{32}P]ATP	147
V. Kinetics of Phosphate Turnover	151
VI. Phosphorylation during Starvation	154
VII. Summary	157
References	158

I. INTRODUCTION

The presence of high concentrations of protein-bound phosphate in acidic nuclear protein fractions has resulted in the classification of those fractions as "phosphoproteins." Few investigators would argue that all nuclear proteins in a specific fraction are in fact phosphorylated. But the number of polypeptide species which are phosphorylated, as well as the degree of their phosphorylation, remain largely unknown (for a complete discussion of nuclear phosphoproteins see Kleinsmith, Chapter 4).

Although measurement of total phosphate content of protein fractions is easily obtained by standard colorimetric procedures, problems are encountered when phosphate content of individual proteins is sought after purification and resolution. Because most methods of acidic nuclear protein resolution depend on electrophoresis of microgram quantities of proteins on acrylamide gels, demonstration of phosphorus content and turnover are usually made by the use of ^{32}P and ^{33}P isotopes. Exposure of whole organisms *in vivo* (Johnson and Allfrey, 1972; Teng *et al.*, 1970; Teng *et al.*, 1971; Jungmann and Schweppe, 1972) or of cultured cells *in vitro* (Gershey and Kleinsmith, 1969; Kleinsmith *et al.*, 1966a; Platz and Hnilica, 1973; Platz *et al.*, 1973; Turkington and Riddle, 1969) to isotopes for short periods of time is the most common means of phosphoprotein labeling. Alternatively, *in vitro* labeling of isolated nuclei or chromatin fractions with [γ-^{32}P]ATP has been used (Ahmed and Ishida, 1971; Chiu *et al.*, 1973; Ishida and Ahmed, 1973; Johnson and Allfrey, 1972; Kleinsmith and Allfrey, 1969; Platz *et al.*, 1970). Continuous labeling in ^{32}P$_i$ for several generations has not yet been used to test the correlation between those proteins which have the highest phosphorus content and those which are metabolizing phosphate most rapidly. The presumption apparently has been that rapid phosphate turnover (Kleinsmith *et al.*, 1966b) necessarily results in speedy equilibration of the phosphoproteins with the intracellular phosphate pools.

If each of the isotopic labeling procedures mentioned above (i.e., continuous, pulse, or ATP) does in fact label the same phosphoproteins to the same degree, it must be so demonstrated before valid conclusions can be made and extrapolated to other experiments employing different labeling procedures. If, on the other hand, it is found that different classes of phosphoproteins are selectively labeled by a specific labeling procedure, adequate precaution should be exercised by those who wish to generalize regarding changes in "phosphorylation" during various functional cellular states.

The acellular slime mold *Physarum polycephalum* is well suited for investigations of the type described herein (for review, see Rusch, 1970). The organism is a syncytium of synchronously dividing nuclei from which large quantities of nuclear proteins can be easily obtained. The phenol-soluble acidic nuclear proteins (PSANP) have been separated into at least 50 detectable bands on SDS–polyacrylamide gels, and some proteins have been found to undergo quantitative changes during the processes of starvation and differentiation (LeStourgeon and Rusch, 1973). Differentiation is readily induced by starvation, which results in spherulation in microplasmodia or sporulation following illumination of surface-grown plasmodia.

In the experiments described below,* an attempt has been made to answer the following questions:

a. Of the numerous PSANP protein bands separable on gels, which ones are phosphorylated, and to what degree are they phosphorylated?
b. Do differences exist between phosphoproteins labeled *in vivo* and *in vitro*?
c. Do kinetic differences exist with respect to phosphate turnover among the detectable phosphoproteins?
d. What changes occur, if any, in the phosphoproteins during the process of differentiation (starvation)?

Physarum was grown in the presence of $^{32}P_i$ and ^{3}H-amino acids for several generations to allow for complete equilibration of the cellular phosphate and protein with the precursor pools. Following isolation and electrophoresis of the PSANP fraction (LeStourgeon and Rusch, 1973), ^{32}P and ^{3}H in individual gel slices were expressed as moles phosphate and moles protein. In addition to scintillation spectrometry, autoradiograms of longitudinal gel slices were used to maximize resolution of individual phosphoprotein species.

II. PHOSPHATE CONTENT OF PHOSPHOPROTEINS

A. Correspondence of ^{3}H and ^{32}P Incorporation following Continuous Double-Labeling

In order to determine molar ratios of phosphate to protein, experiments were designed as follows. Plasmodia were grown in ^{3}H-amino acids (leucine, arginine, isoleucine, and proline) and in $^{32}P_i$ for several generations. From isolated G_2 nuclei the PSANP fraction was extracted as described (LeStourgeon and Rusch, 1973), an aliquot of which was analyzed for Lowry protein and ^{3}H dpm in order to obtain a value for dpm ^{3}H per μg protein. Similarly, dpm ^{32}P per mole phosphate were calculated by colorimetric determination of phosphate (Ames and Dubin, 1960) in a plasmodial fraction (DNA) followed by scintillation counting of that fraction. It was found that the specific activity of plasmodial phosphate was identical to the specific activity of the culture medium with respect to *total* phosphate (organic plus inorganic). Following elec-

* Experiments described in this chapter will be published in full detail elsewhere (Magun et al., 1974).

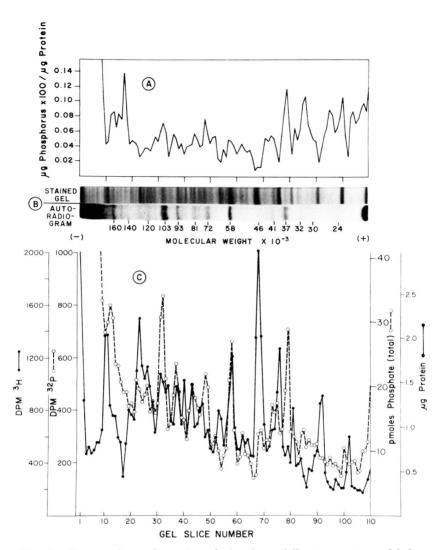

Fig. 1. Correspondence of protein and phosphorus following continuous labeling. Microplasmodia were grown for 48 hours in medium (Mohberg and Rusch, 1969) containing 10 µCi/ml ³H-amino acids (see text) and 40 µCi/ml KH₂³²PO₄, with a final phosphate concentration of 1.8 mM. These cultures were then grown as surface plasmodia on the same medium (fresh) for an additional 40 hours, at which time they were harvested 8 hours after mitosis III. The PSANP were isolated and electrophoresed on gels (LeStourgeon and Rusch, 1973) which were immediately frozen on dry ice and sliced into 1-mm transverse sections for scintillation spectrometry (C). A duplicate gel was fixed-stained for 12 hours in 0.045% Coomassie blue in 40% methanol–10% acetic acid and destained for 48 hours. The gel was then photographed (B, stained gel) and then sliced longitudinally into a 1-mm thick slice which was dried for autoradiography on Kodak No Screen x-ray film (B, autoradiogram). ³H and ³²P activity were converted to moles phosphate and protein, respectively, as described in text. Phosphorus/protein ratios (A) were calculated from scintillation data (C).

trophoresis, gels were immediately frozen, sliced into 1-mm transverse slices, dissolved in 0.5 ml hydrogen peroxide and counted for both ^{3}H and ^{32}P activity by scintillation spectrometry. Since SDS–acrylamide gel electrophoresis is able to separate proteins on a molecular weight basis (Shapiro et al., 1967), molecular weights of individual bands were calculated using marker proteins, and the protein content per gel slice was thereby able to be expressed as moles.

Results of the experiment (Fig. 1) demonstrated that few of the approximately 50 detectable stained bands represented phosphoproteins. Autoradiograms of the longitudinal gel slice revealed approximately 15 distinct bands corresponding to accumulations of ^{32}P (Fig. 1B). Most of the heavily stained bands, corresponding to ^{3}H peaks, did not correspond to maxima of ^{32}P activity. A protein band of molecular weight 58,000 was a notable exception. Computation of phosphate/protein ratios (Fig. 1A) demonstrated the marked fluctuations between phosphate and protein along the gel, the ratios varying from 0.02 to 0.12% phosphorus by weight.

B. Calculation of Phosphates per Polypeptide

From data obtained in the experiment shown in Fig. 1, the phosphate/protein ratios were obtained in transverse gel slices corresponding to the center slice through peaks of ^{32}P activity (Table I). The phosphate/protein ratios, ranging up to 4 phosphate groups per polypeptide,

TABLE I Molar Phosphate Content

$MW^a \times 10^{-3}$	gm phosphorus^b × 100 per gm protein^c	Moles phosphate per mole protein
160	0.085	4.22
103	0.072	2.31
93	0.052	1.51
81	0.044	1.10
72	0.080	1.79
58	0.048	0.86
41	0.040	0.51
37	0.012	1.36
30	0.007	0.67

[a] Calculated from electrophoretic mobilities of phosphorylase (94,000), bovine serum albumin (68,000), pyruvate kinase (57,000), ovalbumin (43,000), and chymotrypsinogen (26,000).

[b] Calculations based on 1046 DPM ^{32}P per pg phosphorus.

[c] Calculations based on 670 DPM ^{3}H per μg protein.

represent a *minimum* value, depending on the (unknown) proportion of protein in the gel slice corresponding to the specific phosphoprotein. As the contribution of specific phosphoprotein to total protein in a gel slice decreases, the molar phosphate/protein ratios correspondingly increase.

Of the total ^{32}P present in the extensively dialyzed PSANP sample prior to electrophoresis, at least 90% was recoverable on gels. Following electrophoresis, about half the ^{32}P appeared near or at the solvent front (i.e., ahead of the low molecular weight protein front) and was unassociated with protein. Measurement of total phosphate content of acidic nuclear protein fractions, even after exhaustive dialysis, may therefore be in excess of the real protein-bound phosphorus because of contamination by residual low molecular weight phosphorus compounds including, perhaps, orthophosphate. Alkali lability of ^{32}P activity might therefore not be diagnostic of protein-bound phosphate.

C. Calculation of Copies per Nucleus

Because the phosphoproteins have been implicated in regulation of gene activity, it is of interest to know the number of copies per nucleus of each of the phosphoprotein species. Are the phosphoproteins present in small enough quantities to be candidates for regulation of single genes or gene groups, or are they present in such large quantities that a pleiotropic function might be inferred?

Calculations were made using the G_2 value of 1.2 pg DNA per nucleus (Mohberg and Rusch, 1971) and the PSANP/DNA ratio of 0.67 (Le Stourgeon and Rusch, 1973). The molar phosphate ratio from Table I was used to calculate the protein corresponding to the total phosphate (measured planimetrically) in the major ^{32}P-containing bands. The number of copies per nucleus, shown in Table II, therefore represents a maximum value, since the molar phosphate ratio is a minimal value based on phosphoprotein purity through a center slice. The values obtained, on the order of 200,000 copies per nucleus, appear to be in excess of what one might expect if the phosphoproteins were regulators of individual genes, as has been shown with the *lac* repressor in *Escherichia coli* (10 copies per genome) (Gilbert and Müller-Hill, 1966). It has been estimated that a mammalian nucleus would require 10^3–10^4 molecules of a *lac*-type repressor for it to be effective (A. Riggs, personal communication, cited in Elgin *et al.*, 1974). However, since so little is known about control of transcription in eukaryotes, the relationship between the magnitude of protein redundancy and individual gene control is only conjectural.

5. Nuclear Phosphoproteins in P. polycephalum

TABLE II Copies per Nucleus of Phosphoproteins

MW × 10^{-3}	A Phosphate (pmoles)[a]	B Protein (pmoles)[b]	C Avogadro's number × B	D Copies per nucleus[c]
103	103.0	44.6	2.69×10^{13}	1.39×10^5
93	97.5	64.6	3.89×10^{13}	2.02×10^5
58	57.3	66.6	4.01×10^{13}	2.08×10^5
37	64.6	47.5	2.86×10^{13}	1.48×10^5

[a] Calculated planimetrically from gel slice scintillation data in Fig. 1.
[b] Calculated as A/molar phosphate ratio (Table I).
[c] Calculated as C/nuclear number. Nuclear number (1.93×10^8) obtained from DNA/phenol-protein ratio of 1.48 with 156 protein applied to gel.

The number of copies per nucleus of some of the major nonphosphorylated protein bands were also determined planimetrically from the data in Fig. 1, and the results are presented in Table III. In the gel band containing the most protein (molecular weight 46,000), protein was present at a concentration of about 8 μg per gel or a half-million copies per nucleus. Using bovine serum albumin as a standard, the limit of sensitivity of protein detection in the dye-binding reaction was determined to be about 0.5 μg protein per band. A protein of 50,000 daltons would therefore have to be present in excess of 30,000 copies per nucleus to be detectable as a stained band in the gel shown in Fig. 1. The detectable acidic nuclear proteins therefore would probably have to be discarded as candidates for individual gene regulation, and it is not surprising that nuclear protein profiles remain remarkably constant for cells progressing through cell cycles (Karn *et al.*, 1974; Jeter and Cameron, Chapter 8). Detection of such putative regulatory proteins,

TABLE III Copies per Nucleus of Nonphosphorylated Proteins

MW × 10^{-3}	Protein (μg)[a]	Protein (pmoles)	Avogadro's number × B	Copies per nucleus[b]
46	7.89	171.5	1.03×10^{14}	5.35×10^5
35	2.23	63.7	3.84×10^{13}	1.99×10^5
28	3.43	123.0	7.4×10^{13}	3.84×10^5
23	1.74	75.7	4.56×10^{13}	2.36×10^5

[a] Calculated planimetrically from gel slice scintillation data in Fig. 1.
[b] See Table II, footnote c for details of calculation.

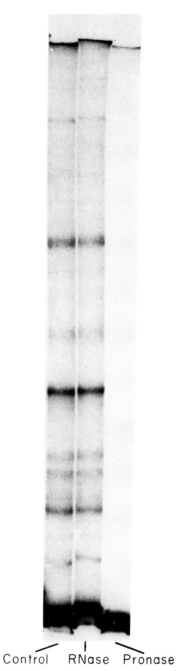

Fig. 2. Autoradiograms of PSANP digested in RNase and Pronase. Microplasmodia were grown for 48 hours in medium containing 20 μCi/ml $^{32}P_i$. The PSANP were extracted and dialyzed against 0.1% SDS and divided into three 200 μl portions, to each of which was added 100 μl 0.01% phosphate buffer (pH 7.5) plus 50

if they exist at all, would depend on increasing the sensitivity as well as the resolution of protein detection methods.

D. Demonstration of the Protein Bound Nature of Phosphate

Following dialysis of the PSANP fraction against 0.1% SDS buffer, the sample was digested in Pronase or ribonuclease prior to electrophoresis. It was determined that greater than 99% of the ^{3}H activity was rendered TCA-soluble by the pronase digestion, confirming that protein was the only ^{3}H-containing species in the sample. Autoradiograms of gels containing the control and digested samples (Fig. 2) demonstrated that pronase digestion prevented appearance of ^{32}P bands in the gel and that ribonuclease digestion did not diminish the intensity of ^{32}P activity in individual bands. The ^{32}P bands observable in gels therefore represent protein components to which phosphate groups are bound.

III. PULSE LABELING IN ^{32}P$_i$

In the previous section, experiments were described using continuous labeling in ^{32}P$_i$ to determine total phosphate content of phosphoprotein bands. In order to preferentially label those proteins which are most rapidly metabolizing phosphate, cultures were pulse labeled in medium containing a high specific activity of ^{32}P$_i$ (1 mCi/ml). Figure 3 demonstrates the results of an experiment in which plasmodia were continuously labeled in ^{33}P$_i$ and labeled for one hour in G_2 in ^{32}P$_i$. Marked ratio differences between the two phosphorus isotopes were found along the course of the gel (Fig. 3A). The pulse-labeling profile was found to be independent of time in the cell cycle. In the gel autoradiogram (Fig. 3B), note that phosphoproteins of molecular weights 103,000 and 58,000, which contain the greatest amount of total phosphate, are only slightly labeled with ^{32}P during pulse labeling. Phosphoproteins in the molecular weight regions 120,000–150,000, 91,000, and 83,000 are most heavily

---

µl aqueous solution containing water (control), Pronase (100 µg), or RNase (175 µg). Samples were incubated at 37°C for 2 hours. 100 µl 10% SDS–5% mercaptoethanol were added to each sample which was heated to boiling for 2 minutes and electrophoresed on gels. Following electrophoresis, gels were stained and sliced longitudinally for autoradiography. (The dye-binding capacity of the Pronase-digested gel was completely abolished also, indicating the proteinaceous nature of all stained material.)

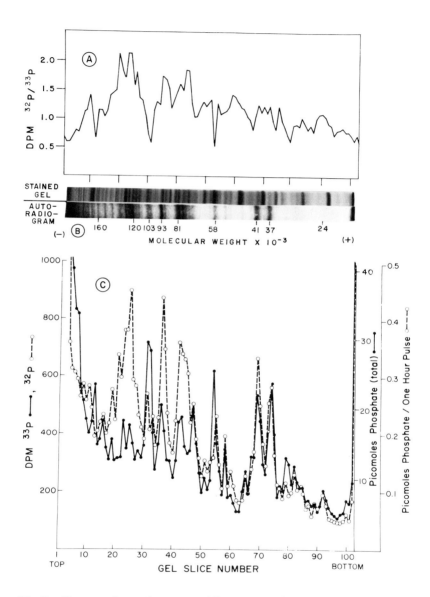

Fig. 3. Correspondence of protein and ^{32}P activity following pulse labeling. Experiment was identical to the one in Fig. 1, except that cultures were grown in medium without ^{32}PO$_4$, containing 10 μCi/ml ^{3}H-amino acids. One hour before harvesting, plasmodia were blotted dry and placed for one hour on 3 ml fresh medium containing 1 mCi/ml KH$_2^{32}$PO$_4$. PSANP were then processed as outlined as outlined in Fig. 1.

labeled following pulse labeling but are only faintly visible in autoradiograms of continuously labeled cultures (Fig. 1B). Phosphoproteins of molecular weights 41,000 and 37,000 rapidly incorporate ^{32}P and also accumulate ^{33}P during continuous labeling.

The marked disparities between pulse-labeled and continuously labeled proteins (Fig. 3A) indicate that only the most highly phosphorylated proteins are detectable following continuous labeling and that some of the less phosphorylated species apparently are actively turning over phosphate. These phosphoproteins become visible only following pulse labeling. Since the quantities of the protein components corresponding to the ^{32}P peaks are unknown in the gel system, the term "less phosphorylated" implies either a low phosphate/protein ratio or relatively few copies per nucleus. Because of the already low phosphate/protein ratios obtained for the continuously labeled phosphoproteins, the latter possibility appears to be the more tenable explanation. If true, then pulse labeling in ^{32}P$_i$ allows visualization of some phosphoproteins which are present in fewer copies per nucleus than the other resolvable species of acidic nuclear proteins. Extrapolating further, it is reasonable to expect that many other species of nonphosphorylated proteins are present in small quantities, but are masked in gels by the presence of major protein bands.

IV. LABELING OF NUCLEI in Vitro WITH [γ-^{32}P]ATP

Phosphorylation of nuclear proteins *in vitro*, which utilizes endogenous protein kinases and exogenous [γ-^{32}P]ATP, has been a means of phosphoprotein labeling adopted by many investigators. Johnson and Allfrey (1972) noted discrepancies in the ^{32}P gel profiles between phenol soluble proteins labeled *in vivo* with ^{32}P$_i$ and *in vitro* with [γ-^{32}P]ATP. Experiments were performed in this laboratory to compare ^{32}P profiles of PSANP labeled in isolated nuclei *in vitro* with [γ-^{32}P]ATP with those obtained by pulse and continuous labeling methods *in vivo*.

Plasmodia were labeled continuously in ^{33}P$_i$ prior to isolation of nuclei and incubation in a Mg^{2+}-rich medium containing [γ-^{32}P]ATP. The PSANP fraction was then isolated, electrophoresed, and gels were sliced for scintillation counting and autoradiography. The results (Fig. 4) demonstrate that relatively few bands are phosphorylated *in vitro* by ATP and that these do not correspond to major protein bands in the stained gel. There is also little correspondence between phosphoproteins labeled by [γ-^{32}P]ATP *in vitro* and ^{33}P$_i$ *in vivo*. Apparently *in vitro* phosphorylation labels some proteins (molecular weights 170,000 and

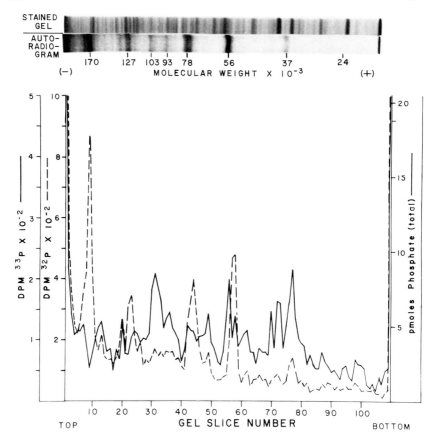

Fig. 4. Correspondence of ^{33}P *in vivo* with ^{32}P *in vitro*. Microplasmodia were grown for 24 hours in medium containing 20 μCi/ml KH$_2^{33}$PO$_4$ and then fused into surface plasmodia which were grown in the same medium for an additional 48 hours. Nuclei were isolated and extracted 3 times in 0.14 M NaCl and twice in buffer A (0.03 M Tris-HCl, pH 7.5, 0.115 mM NaCl, 5 mM MgCl$_2$, 0.1% Triton-X 100, 40 mM α-glycerophosphate, 12 mM NaF). Nuclei were then incubated for 30 seconds at 25°C in 1.0 ml buffer A containing 150 μCi γ-^{32}P-ATP (New England Nuclear). Nuclei were then extracted in 0.25 N HCl and processed routinely for PSANP extraction and electrophoresis.

56,000) which do not appear to be labeled *in vivo* by either continuous or pulse labeling (Fig. 5). Conversely, many other phosphoproteins are labeled *in vivo*, but do not appear to be labeled *in vitro*.

The inability to label *in vitro* all the phosphoproteins demonstrable in the *in vivo* situation is not surprising. Nonoptimal conditions, removal of essential proteins or cofactors, and phosphate saturation of phospho-

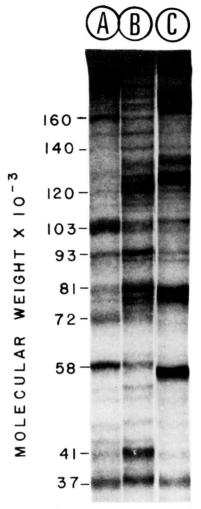

Fig. 5. Comparison of autoradiograms of continuous, pulse, and *in vitro* labeling. Differences between the ^{32}P profiles are demonstrated by comparison of autoradiograms of gels obtained after continuous labeling in ^{32}P$_i$ (A, from experiment in Fig. 1), pulse labeling in ^{32}P$_i$ (B, from experiment in Fig. 3), and *in vitro* labeling of nuclei in [γ-^{32}P]ATP (C, from experiment in Fig. 4).

proteins are all factors to be considered. It is unexpected, however, that some of the phosphoproteins should be intensely labeled *in vitro*, while these same proteins are apparently unlabeled *in vivo*. The labeling may be artifactual and unrelated to *in vivo* phosphorylation or it may represent *in vivo* phosphorylation which is normally repressed by factors which are present in the cell. Whatever the explanation, it would be

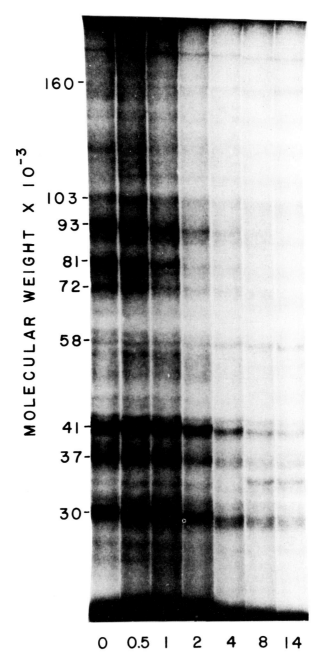

inappropriate to extrapolate *a priori* from the *in vitro* to the *in vivo* situation.

V. KINETICS OF PHOSPHATE TURNOVER

Phosphorylation of proteins has been associated with rapidly changing intracellular states in which there is an amplification of regulatory signals (Holzer and Duntze, 1971; Krebs, 1972). Proteins with high rates of phosphate turnover may be important as regulators of gene activity by virtue of their ability to respond quickly to internal signals. The discrepancies already noted between continuously labeled and pulse-labeled phosphoprotein profiles indicate that differences in phosphorylation kinetics do in fact exist. In an attempt to quantify those differences, cold–chase experiments were undertaken in which turnover rates of the phosphate groups were determined following pulse or continuous labeling of microplasmodial cultures. Nonsynchronous microplasmodia were used in order to eliminate cell cycle-related changes in phosphate turnover. Cultures were labeled continuously for several generations in ^{3}H-amino acids throughout all phases of the experiments. Gels containing the labeled proteins were sliced and the ^{32}P/^{3}H ratios in midpeak slices were determined as a measure of ^{32}P specific activity in individual phosphoproteins.

Gel autoradiograms from the pulse-chase experiment show a rapid loss of ^{32}P activity beginning at 2 hours in all bands except those of molecular weights 58,000 and 103,000, in which the decrease in ^{32}P activity appears to be less complete (Fig. 6). The half-lives of the ^{32}P activity were found to range from 1.5 to 3 hours.

The use of a cold chase following continuous labeling in ^{32}P$_i$ allows one to observe the decay in ^{32}P specific activity of those proteins which accumulate phosphate slowly (Figs. 7, 8). This group of proteins (molecular weights 160,000, 103,000, 58,000, and 190,000) lose their phosphate groups at a slower rate than other phosphoproteins of a lower molecular

Fig. 6. Cold chase following pulse labeling in ^{32}P$_i$ (autoradiograms). Microplasmodia were grown for 96 hours in medium containing 5 μCi/ml ^{3}H-amino acids. Cultures were centrifuged, resuspended, and incubated for 1 hour in fresh medium containing ^{3}H-amino acids (5 μCi/ml) and KH$_2$^{32}PO$_4$ (1 mCi/ml). Microplasmodia were then centrifuged and resuspended twice in fresh medium, followed by incubation for varying times in fresh medium containing 5 μCi/ml ^{3}H-amino acids (cold-chase medium). Microplasmodia were harvested at 0 time and at later points up to 14 hours, and PSANP were extracted as described. Each time point represents the autoradiographic profile of a gel containing 100 μg protein.

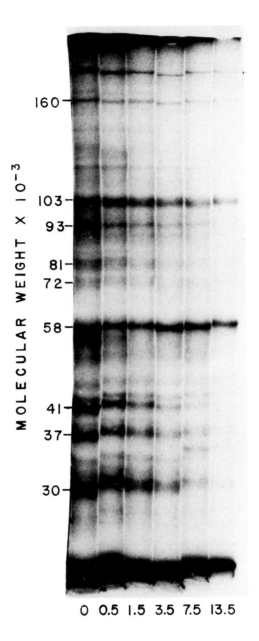

Fig. 7. Cold chase following pulse labeling in $^{32}P_i$ (autoradiograms). Microplasmodia were grown for 96 hours in medium containing 5 µCi/ml ^{3}H-amino acids and 20 µCi/ml $^{32}P_i$. Cultures were centrifuged, resuspended twice in fresh medium and incubated for varying times in fresh medium containing 5 µCi/ml ^{3}H-amino acids (cold-chase medium). Microplasmodia were taken at 0 time and at other points up to 13.5 hours, and PSANP were extracted as described. Each time point represents the autoradiographic profile of a gel containing 100 µg protein.

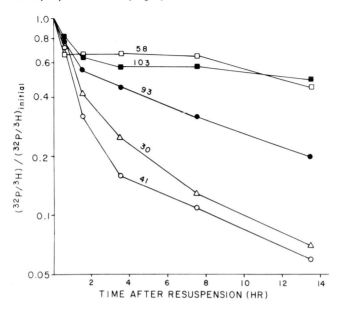

Fig. 8. Decrease in ^{32}P specific activity in experiment described in Fig. 7. Gels from various time points were sliced transversely and analyzed by scintillation spectrometry for ^{3}H and ^{32}P activity ^{32}P/^{3}H ratios were calculated for gel slices through midpeak regions of phosphoproteins of molecular weights 30,000, 41,000, 58,000, 93,000, and 103,000 daltons. The curves represent the decrease in ^{32}P specific activities during the time course of the cold chase.

weight (30,000 and 41,000). Except for an initial decline (thought to be caused by centrifugation and resuspension) the specific activity of the phosphoprotein at molecular weight 58,000 apparently does not decrease over the chase period. Since it was determined that the microplasmodial mass more than doubled during the chase period, one would expect that the polypeptide of molecular weight 58,000 would also double. Since the ^{32}P/^{3}H ratio did not decline over the chase period (in the presence of ^{3}H-amino acids), it might be concluded that either phosphorylation of the polypeptide occurred from a pool whose ^{32}P specific activity was unaffected by the chase (e.g., polyphosphate) or that the protein was phosphorylated before application of the chase. If the latter explanation is the correct one, it can be concluded that: (a) phosphorylation of the polypeptide is conservative, and (b) there exists a lag period of several hours from the time of phosphorylation for the polypeptide to appear in the PSANP fraction.

It should be noted, importantly, that the half-lives of the phosphorylated polypeptides are not known and cannot be obtained until resolved from other proteins. The apparent decay of phosphate from the chase experiments may represent turnover of whole polypeptides as well as, or exclusive of, phosphate groups. The problem of interpretation becomes even more difficult if it is considered that polypeptides which have multiple phosphate groups may have different turnover rates for each phosphate group.

The important conclusions which should be drawn from both types of chase experiments are (a) a range of phosphorylated PSANP proteins exists in which phosphate is rapidly turning over in some species and is apparently stable in others; (b) a polypeptide exists in *Physarum* of molecular weight 58,000 which apparently is conservatively phosphorylated and may exhibit a time lag of hours between the time of phosphorylation and its appearance on the chromosome; and (c) pulse labeling of other organisms or cells with $^{32}P_i$ may also demonstrate only phosphoproteins which are rapidly metabolizing phosphate, and may not reflect at all the relative phosphate content of the phosphoproteins.

VI. PHOSPHORYLATION DURING STARVATION

Physarum is an excellent model system for the study of the process of differentiation, which takes the form of spherulation in microplasmodia or sporulation in surface-grown plasmodia (Sauer *et al.*, 1969). Starvation in a nonnutrient medium, which is a precondition for both spherulation and sporulation, is accompanied by changes in some of the PSANP proteins (LeStourgeon and Rusch, 1973). It was of interest to examine changes in phosphorylation of the PSANP fraction during the process of starvation, when gene activity is changing from vegetative to differentiated functions.

Cultures were continuously labeled in $^{32}P_i$ for several generations prior to the induction of starvation, which was effected by transferring logarithmically growing surface cultures to distilled water for varying times. Cultures were labeled for one hour in $^{33}P_i$ before harvesting. In Fig. 9 it can be seen that the phosphate content (DPM $^{32}P/\mu g$ protein) of the PSANP drops to about half during the course of starvation. The ability to incorporate exogenous phosphate ($^{33}P_i$) decreases immediately after induction of starvation and is negligible at 32 hours. It is not known, however, to what extent the $^{33}P_i$ is incorporated into the plasmodial precursor pools throughout the course of starvation. The possibility exists, therefore, that the organism is not taking up exogenously added phosphate during the course of starvation.

In Fig. 10 the autoradiographic profile over the course of starvation

5. Nuclear Phosphoproteins in P. polycephalum

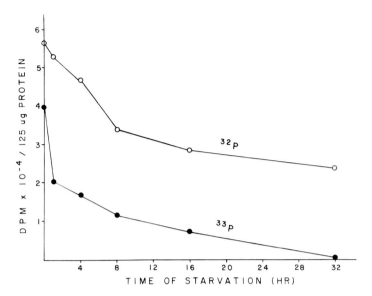

Fig. 9. Phosphorylation of PSANP during starvation. Microplasmodia were grown for 48 hours in medium containing 20 μCi/ml KH$_2^{32}$PO$_4$ and then fused into surface plasmodia which were grown for an additional 20 hours on the same (fresh) medium. At that time all plasmodia were transferred to distilled for varying times (starvation). One hour before time of harvesting, plasmodium was blotted dry and placed on 0.5 ml KH$_2^{33}$PO$_4$ (1 mCi/ml) for 1 hour. Following extraction of PSANP from nuclei, 125 μg PSANP were applied to each gel for electrophoresis. Gels were sliced transversely and analyzed for ^{33}P and ^{32}P simultaneously. DPM ^{32}P in Fig. 9 represent the sum of all isotope from top of gel to protein front, inclusive.

demonstrates a generalized decrease in phosphate content in all demonstrable phosphoproteins. Differential changes in the phosphate content of specific phosphoproteins do not appear nor do new species of phosphoproteins become apparent.

Increases in phosphorylation have been described in many systems stimulated to a higher degree of gene activity (see Kleinsmith, Chapter 4). During the process of spherulation in *Physarum*, synthesis of new RNA species has been reported (Chet and Rusch, 1970), but a general decrease in total RNA synthesis occurs (Chet and Rusch, 1969). It appears, then, that activation of new portions of the *Physarum* genome is not accompanied by differential phosphorylation of PSANP species. That the level of phosphorylation parallels the decline in RNA synthesis may suggest a generalized role of phosphorylation as a means of regulation of the *quantity* rather than the *character* of transcribed RNA. It

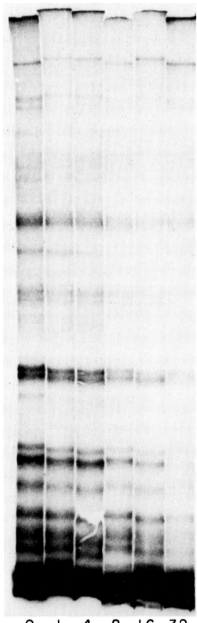

is also possible that increase in phosphorylation is the effect, rather than the cause, of increased gene activity.

The present methods of phosphoprotein demonstration may lack the sensitivity and resolution to detect phosphorylation changes in phosphoproteins present in low quantities. Only better methodology will be able to provide answers to many of the questions raised in this chapter.

VII. SUMMARY

Continuous labeling of *Physarum polycephalum* in $^{32}P_i$ has permitted accurate determination of phosphate content of the phenol-soluble proteins. Of the apparently few phosphorylated proteins resolvable on gels, most do not correspond to major protein bands. Based on calculations of minimal phosphate content, there is no reason to assume *a priori* that these nuclear phosphoproteins are highly phosphorylated (more than 4 phosphates per polypeptide). Although most of the resolvable phenol-soluble proteins are present (per nucleus) in quantities which appear to exclude them from consideration as regulatory proteins such as the *lac* repressor, the phosphorylated proteins may be present in small enough numbers to qualify as regulators.

The phosphorylated proteins of the phenol-soluble protein fraction have markedly different rates of phosphorus turnover. Some phosphoproteins appear to be conservatively phosphorylated, and these can be detected only following long-term labeling in $^{32}P_i$. Other phosphoproteins which are rapidly metabolizing phosphorus can be detected only after pulse labeling in a medium of higher $^{32}P_i$ specific activity. Of the detectable phosphoproteins, the latter group is probably present in small quantities per nucleus.

During the process of differentiation in *Physarum*, there is a decrease in both phosphorus content and rate of phosphorus incorporation. New species of phosphoproteins do not appear. Phosphorylated acidic nuclear proteins may therefore act as regulators of the quantity rather than the character of transcribed RNA.

Fig. 10. PSANP during starvation (autoradiograms). 74 μg PSANP from experiment described in Fig. 9 were applied to each gel, which was sliced longitudinally for autoradiography after electrophoresis. Saran wrap was placed between dried gels and x-ray film to eliminate possibility of film exposure due to ^{33}P. Autoradiograms therefore represent only ^{32}P activity, which decreases markedly during the course of starvation. Notice that high concentrations of ^{32}P appear ahead of the protein front (at bottom of gels) at all time points. This high concentration of ^{32}P activity is not associated with protein and remains in gel even after days of washing in 10% acetic acid.

REFERENCES

Ahmed, I., and Ishida, H. (1971). *Mol. Pharmacol.* **7**, 323.
Ames, B. H., and Dubin, D. T. (1960). *J. Biol. Chem.* **235**, 769.
Chet, I., and Rusch, H. P. (1969). *J. Bacteriol.* **100**, 673.
Chet, I., and Rusch, H. P. (1970). *Biochim. Biophys. Acta* **209**, 559.
Chiu, J. F., Craddock, C., Getz, S., and Hnilica, L. (1973). *FEBS Lett.* **33**, 247.
Elgin, S. C. R., Boyd, J. B., Hood, L. E., Wray, W., and Wu, F. C. (1974). *Cold Spring Harbor Symp. Quant. Biol.*, **38**, 821.
Gershey, E. L., and Kleinsmith, L. J. (1969). *Biochim. Biophys. Acta* **194**, 519.
Gilbert, W., and Müller-Hill, B. (1966). *Proc. Nat. Acad. Sci. U.S.* **56**, 1891.
Holzer, H., and Duntze, W. (1971). *Annu. Rev. Biochem.* **40**, 345.
Ishida, H., and Ahmed, K. (1972). *Exp. Cell. Res.* **78**, 31.
Johnson, E. M., and Allfrey, V. G. (1972). *Arch. Biochem. Biophys.* **152**, 786.
Jungmann, R. A., and Schweppe, J. S. (1972). *J. Biol. Chem.* **247**, 5535.
Karn, J., Johnson, E. M., Vidali, G., and Allfrey, V. G. (1974). *J. Biol. Chem.* **249**, 667.
Kleinsmith, L. J., and Allfrey, V. G. (1969). *Biochim. Biophys. Acta* **175**, 136.
Kleinsmith, L. J., Allfrey, V. G., and Mirsky, A. E. (1966). *Science* **154**, 780.
Kleinsmith, L. J., Allfrey, V. G., and Mirsky, A. E. (1966b). *Proc. Nat. Acad. Sci. U.S.* **55**, 1182.
Krebs, E. G. (1972). *Curr. Top. Cell. Regul.* **5**, 99.
LeStourgeon, W. M., and Rusch, H. P. (1973). *Arch. Biochem. Biophys.* **155**, 144.
Magun, B. E., Burgess, R. R., and Rusch, H. P. (1974). In preparation.
Mohberg, J., and Rusch, H. P. (1969). *J. Bacteriol.* **97**, 1411.
Mohberg, J., and Rusch, H. P. (1971). *Exp. Cell Res.* **66**, 305.
Platz, R., and Hnilica, L. S. (1973). *Biochem. Biophys. Res. Commun.* **54**, 222.
Platz, R. D., Kish, V. M., and Kleinsmith, L. J. (1970). *FEBS Lett.* **12**, 38.
Platz, R. D., Stein, G. S., and Kleinsmith, L. J. (1973). *Biochem. Biophys. Res. Commun.* **51**, 735.
Rusch, H. P. (1970). *Advan. Cell Biol.* **1**, 297–327.
Sauer, H. W., Babcock, K. L., and Rusch, H. P. (1969). *Exp. Cell Res.* **57**, 319.
Shapiro, A. L., Viñuela, E., and Maizel, J. V. (1967). *Biochem. Biophys. Res. Commun.* **28**, 815.
Teng, C. S., Teng, C. T., and Allfrey, V. G. (1971). *J. Biol. Chem.* **246**, 3597.
Teng, C. T., Teng, C. S., and Allfrey, V. G. (1970). *Biochem. Biophys. Res. Commun.* **41**, 690.
Turkington, R. W., and Riddle, M. (1969). *J. Biol. Chem.* **244**, 6040.

6

The Nuclear Acidic Proteins in Cell Proliferation and Differentiation

WALLACE M. LESTOURGEON, ROGER TOTTEN,
AND ARTHUR FORER

I. Introduction 159
II. The Heterogeneity of the Nuclear Acidic Proteins and Cell
   Proliferation 161
III. The Major Acidic Nuclear Proteins and Specific Gene Regulation 171
IV. The Contractile Proteins of Isolated Chromatin and Considerations
    of Their Possible Role in the Regulation of Cell Proliferation ... 174
   A. Recognition That Contractile Proteins Were Major Components of Isolated Nuclei and Chromatin 174
   B. Preliminary Evidence That Contractile Proteins May Be True Nuclear Components 178
   C. Functional Aspects of Intranuclear Contractile Proteins 181
References 187

I. INTRODUCTION

The present body of information concerning the means of cell differentiation favors the hypothesis that differentiation is the result of programmed sequential activation, inactivation, or otherwise quantitative regulation of specific genes. By definition, then, efforts to elucidate the mechanisms of genetic regulation should be facilitated by investigations on differentiating cells. Operationally, in the eukaryotes most of these studies are focused on dissecting the nuclear macromolecules in order to determine their respective influence on the genetic material. These studies are largely still in a descriptive phase since critical methods have not been developed for identifying all the functional constituents of the genetic complex. However, it has been through this approach that a promising class of acidic proteins has been identified as true components of the genetic complex and, like the basic nuclear proteins and nucleic acids, the acidic proteins have been isolated and examined with respect to their biological activities. Among the more important discoveries concerning the acidic chromatin proteins have been the observations that during differentiative periods these proteins show considerable activity with respect to their rate of synthesis, changing intranuclear concentrations, and chemical modifications.

Because cell differentiation may be considered to be the result of any intrinsically regulated change in the metabolic capabilities of a cell, all but the most descriptive published observations concerning the nuclear acidic proteins relate to differentiative phenomena. Emphasis in this chapter will be placed on the activities of the nuclear acidic proteins during periods when actively growing cells differentiate into nonproliferative forms and when nonproliferative cells "dedifferentiate" into actively growing cells. This, the most fundamental form of cell differentiation, conveniently serves as an ideal starting point to consider some of the functional roles of the residual acidic proteins. Based on the information which is now available, together with the known identity of several of these proteins, a mechanism whereby certain of these proteins may regulate cell proliferation and quiescence will be presented. In support of the conclusions to be drawn in this chapter several general "interpretations" will be established both from the literature and from studies conducted in this laboratory. In summary these are as follows:

1. When metabolically active, proliferating cells differentiate into nondividing forms there is a decrease in the heterogeneity of the residual acidic proteins and, in reverse, an increase in protein heterogeneity is associated with the establishment of proliferative cells states. The increased or decreased heterogeneity occurs primarily in the high molecular weight (50,000–250,000) proteins and these proteins also constitute a "fingerprint" region in that the greatest number of tissue-specific proteins are of high molecular weight.

2. Most of the 100 or more acidic proteins, which are present in sufficient intranuclear concentrations to be visualized through high resolution electrophoresis, do not function as specific gene regulators but rather have enzymatic, structural, and kinetic roles.

3. The major proteins of isolated nuclei, actin, myosin, and the regulatory proteins for condensation, tropomyosin and troponin, may regulate heterochromatization and thus cell proliferation. Mechanistically these proteins may be responsible for the generally observed phenomena that

nuclei from quiescent nonproliferative cells are usually condensed and show increased heterochromatization as compared to the swollen and metabolically active nuclei from proliferative cells.

Since many of the preliminary observations suggesting that the nuclear acidic proteins may possess specific gene regulatory activity have been reviewed elsewhere (Stellwagen and Cole, 1969; Baserga and Stein, 1971; Stein and Baserga, 1971; Elgin *et al.*, 1971; McClure and Hnilica, 1972; Spelsberg *et al.*, 1972) and in other chapters in this book, the literature discussed here will be in support of the above statements.

II. THE HETEROGENEITY OF THE NUCLEAR ACIDIC PROTEINS AND CELL PROLIFERATION

A now important but otherwise overlooked discovery, reported simultaneously with the first demonstration that chromatin contains significant amounts of acidic protein (Stedman and Stedman, 1943), is the observation that the chromatin from proliferative tissue is rich in acidic protein compared to nondividing cells. In these preliminary studies the ratio of acidic protein to histone was found to correlate with the presence or absence of proliferative cell states. For example, in the extreme cases reported, the chromatin from Walker rat carcinoma contained as much as 72% acidic protein and only traces of histone, while the highly differentiated cod sperm chromatin contained as much as 23% histone and about 60% acidic protein. The correlation between the amount of residual acidic protein and metabolic activity is also apparent in the studies of Mirsky and Ris (1951), in which the chromosomes of carp erythrocyte nuclei were found to contain only 4% residual protein while calf liver contained as much as 39% acidic protein.

In a more detailed study Dingman and Sporn (1964) analyzed the chromatin from nine embryonic and adult tissues of the chicken and found that proliferating embryonic tissues generally contain more residual nonhistone protein than mature and quiescent cells. More specifically, the highest levels of nonhistone protein occurred in whole embryos from eggs incubated 1.5 days and the lowest levels occurred in the chromatin from mature erythrocytes. Associated with the observation that the ratio of nonhistone protein to DNA is related to the level of genetic activity is the demonstration that RNA polymerase activity could be recovered from embryonic brain chromatin but not from mature erythrocytes. Similarly, in duck erythrocytes the immature and metabolically active cell nuclei contain about 3 times more residual acidic protein than the mature and highly heterochromatinized forms (Gershey and Kleinsmith, 1969). Along the same lines, Marushige and Dixon (1969) observed that during spermatogenesis and heterochromatization in trout testes there is a manyfold decrease in the content of nonhistone protein, and a significant loss in the ability of isolated chromatin to support RNA synthesis is associated with the loss of these proteins. Native chromatin from mature sperm did not support RNA synthesis. Obviously related to these findings is the report of Frenster (1965) that isolated genetically active euchromatin from calf thymus lymphocytes (Frenster et al., 1963; Littau et al., 1964) contains almost twice as much residual nonhistone protein as the condensed and repressed heterochromatin yet the levels of histone in both types of chromatin is essentially identical.

In further studies designed to investigate whether chemicals or hormones can invoke changes in the composition of rat liver chromatin (Sporn and Dingman, 1966), it was observed that tumors induced with chemical carcinogens as well as normal liver cells stimulated with pituitary hormones show increased amounts of nonhistone protein. In other studies where the composition of chromatin from normal and neoplastic cells was compared, Grunicke et al. (1970) showed that rapidly growing Novikoff hepatoma cells in culture contain increased amounts of residual nuclear protein as compared to normal rat liver.

These reports have suggested that, as a rule, metabolically active or proliferating cells contain quantitatively more acidic protein than nondividing and metabolically quiescent cells. However, while these observations are perhaps true in cases where actively dividing cells are compared to highly differentiated and quiescent forms, they cannot be applied to cells of all tissues. This seems especially clear since Steele and Busch (1963) found no significant quantitative difference in the chromatin composition between rat Walker tumor and normal liver. Also Grunicke et al. (1970) did not observe significant quantitative differences between the acidic chromatin proteins of normal rat liver and Morris 5123 C hepatoma. In this regard, however, it should be pointed out that mitosis in Walker tumor and Morris 5123 C hepatoma (vs. normal liver) is not grossly increased as is mitosis in Novikoff hepatoma cells in culture.

Since in none of the above studies were the residual acidic proteins resolved into individual components little could be said as to the nature of the quantitative difference. In studies designed to gain further information on the naturally occurring metabolic activities of the nuclear acidic proteins during growth, differentiation, and the establishment of nonproliferative states, LeStourgeon and Rusch (1971) demonstrated electrophoretically that the quantitative differences observed by others

6. Cell Proliferation and Differentiation

between proliferative and nonproliferative cell states is actually the result of a changed population of acidic chromatin proteins. In further studies on the acidic nuclear proteins of the lower eukaryote *Physarum polycephalum*, LeStourgeon and Rusch (1973) separated the residual acidic proteins into 2 fractions and used an electrophoretic procedure (see Chapter 3) which acutely resolved the acidic proteins into more than 100 protein bands. In these studies it was determined that in response to natural stimuli which induce quiescence (starvation-induced encystment or spore formation), most of the specific changes which occur are confined to the protein components of dispersed extranucleolar chromatin. More specifically during the establishment of nonproliferative states in *Physarum* there is a loss of many of the high molecular weight proteins (52,000 to 150,000) (Fig. 1). Also disappearing from the nucleus is a major protein of lower molecular weight (band 36, Fig. 1). The electrophoretic profiles pictured in Fig. 1 show the changes which result after prolonged starvation (4 days) leading to haploid spore formation. Except for the increase in bands 1 and 2 (see gel 2, Fig. 1), the same changes develop during the rapid formation (16 hours) of diploid cysts (Fig. 2). As will be discussed in detail later these changes also occur in response to other stimuli unfavorable for growth (i.e., refrigeration, high plasmodial density). While the changes which occur during differentiation into nonproliferative states can be characterized as a generalized loss of most of the high molecular weight proteins a severalfold increase always occurs in a single protein of molecular weight 46,000 (band 32, Figs. 1 and 2). In *Physarum* then the net result of normal differentiation into nonproliferative cell states is a significant loss of protein heterogeneity.

The decrease in protein heterogeneity during the establishment of nonproliferative states is not however restricted to *Physarum*. Essentially the same protein changes develop in HeLa S_3 chromatin in response to the stress of high cell density and nutritional depletion (LeStourgeon et al., 1973b) (Fig. 3). As in *Physarum*, there is a generalized loss of many of the high molecular weight proteins, the specific loss of a major low molecular protein (band 36, Fig. 3), and a significant increase in a major protein (band 32) with an identical molecular weight to the major protein of *Physarum* which also increased on the induction of nonproliferative states. In a separate study (Vidali et al., 1973) where high resolution acrylamide gel electrophoresis was also used to characterize the acidic proteins of avian erythrocyte chromatin, a decrease in protein heterogeneity was found to be associated with heterochromatization during erythrocyte maturation. As in *Physarum* and HeLa cells,

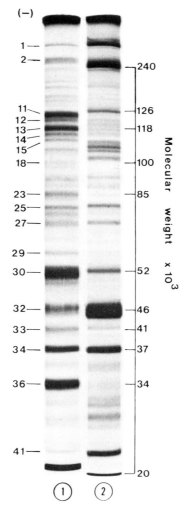

Fig. 1. Electrophoretic banding patterns of the nuclear acidic proteins of *Physarum polycephalum* isolated from actively growing plasmodia (gel 1) and after the nonproliferative state was induced by starvation (gel 2). Acidic proteins extracted and separated electrophoretically as described in Chapter 3. Note the generalized loss of many of the high molecular weight proteins as well as the major proteins 30 and 36. Note also the significant increase in band 32. The approximate molecular weights listed above for bands 30 and 36 differ slightly from previously published values (see LeStourgeon and Rusch, 1973 and LeStourgeon *et al.*, 1973b). The above values are based on molecular weight studies with the purified proteins. Courtesy of Academic Press Inc.

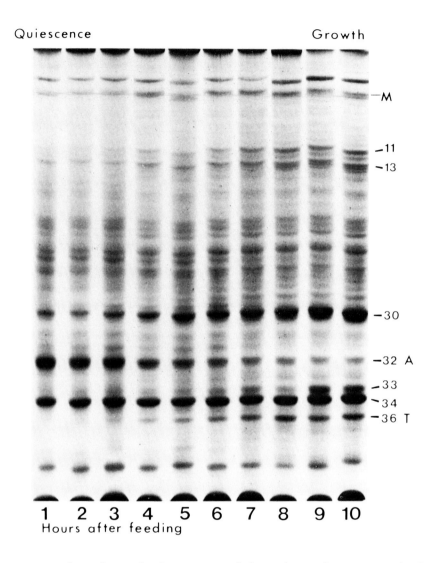

Fig. 2. Electrophoretic banding patterns of the nuclear acidic proteins isolated hourly after refeeding quiescent *Physarum* microplasmodia. Gel 1 shows the complement of acidic nuclear proteins present in nonproliferative states and gel 10 shows the protein complement just preceding the first postfeeding mitosis. Note the decrease in the major protein at band 32 and the concomitant reappearance of protein 36. Note also the increasing heterogeneity of the high molecular weight proteins during this period of dedifferentiation. The sequential synthesis and intranuclear accumulation of these proteins are described in Chapter 3. Myosin (M), actin (A), tropomyosin (T). From LeStougeon *et al.*, 1973c. Courtesy Academic Press, Inc.

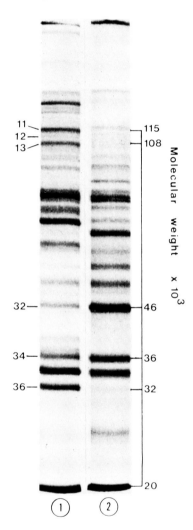

Fig. 3. Electrophoretic banding patterns of the nuclear acidic proteins of HeLa S$_3$ cells extracted from logarithmically growing cells (gel 1) and from cells allowed to remain in culture 48 hours beyond peak exponential growth (gel 2). Note that the changes which occur in HeLa cell nuclei are similar to those which occur in the lower eukaryote *Physarum polycephalum* and in mouse embryo fibroblasts (Figs. 1 and 6), and note also that band 32 is of identical molecular weight in all 3 cells. From LeStourgeon *et al.*, 1973b. Courtesy Academic Press, Inc.

Vidali et al. demonstrated a generalized decrease in most of the high molecular weight acidic proteins and also the loss of at least one protein again of lower molecular weight (28,000). The correlation between the amount and heterogeneity of the high molecular weight acidic proteins and the presence or absence of proliferative cell states was also demonstrated by Weisenthal and Ruddon (1972). in their studies the acidic nuclear proteins from chronic human lymphocytic leukemia cells were found to be mostly of low molecular weight, yet within 69 minutes after PHA stimulation the protein complement was altered such that almost all of the proteins present were of high molecular weight. A point of considerable importance in these studies was the observation that as a result of PHA-induced dedifferentiation the new complement of acidic proteins was essentially identical to that of highly proliferative Burkitt lymphoma cells in culture. That decreased acidic protein heterogeneity is also associated with the establishment of quiescent nuclear states in plants was demonstrated by Pipkin and Larson (1973). In their studies a general loss of many specific proteins was associated with the differentiation of vegetative pollen nuclei into quiescent generative nuclei in *Hippeastrum belladonna*.

Perhaps the clearest demonstration that *increased* protein heterogeneity is associated with cell "dedifferentiation" has come from experiments where nondividing cells are stimulated to proliferate (LeStourgeon *et al.*, 1973c; Weisenthal and Ruddon, 1972). For example, if starved *Physarum* microplasmodia are harvested just prior to cyst wall formation and transferred into fresh nutrient media, a 12-hour period of "dedifferentiation" leading to DNA synthesis, mitosis, and the reestablishment of plasmodial proliferation is induced (LeStourgeon *et al.*, 1973c). During this period of dedifferentiation and chromatin reactivation, the high molecular weight proteins, lost during encystment, are resynthesized and reaccumulate in the nucleus (Fig. 2). Likewise the major low molecular weight protein (band 36) is also resynthesized and reappears in the nucleus. During this period of chromatin reactivation the protein of molecular weight 46,000 (band 32), which increased severalfold in intranuclear concentration during establishment of the nonproliferative state, decreases to its concentration unique to exponential growth (Fig. 2).

The correlation between increased heterogeneity of the high molecular weight acidic proteins and euchromatization has been further substantiated through the findings of Simpson and Reeck (1973), MacGillivray *et al.* (1972), and Wu *et al.* (1973). Simpson and Reeck isolated condensed and dispersed rabbit liver chromatin and found that the dispersed, low-melting chromatin fractions contained not only increased amounts of nonhistone protein but showed increased heterogeneity of the high molecular weight proteins as well. The condensed, high-melting chromatin fractions, while containing fewer high molecular weight protein species in general, was especially rich in a single nonhistone protein apparently of middle molecular weight (i.e., about 50,000). This protein may be functionally analogous to protein 32 (46,000) of *Physarum*, HeLa cells, and mouse embryo fibroblasts in that in these cells this protein increases severalfold during periods of chromatin condensation and inactivation (LeStourgeon and Rusch, 1973; LeStourgeon *et al.*, 1973b, 1974). MacGillivray *et al.* (1972) and Wu *et al.* (1973) found, respectively, in mouse and rat brain chromatin increased heterogeneity of the high molecular weight acidic proteins as compared to other tissues. As pointed out by Wu *et al.*, this occurrence may account for the findings of Grouse *et al.* (1972) that *in vivo* transcription of unique DNA sequences in different mouse tissues is restricted to about 4% of the DNA in most tissues but may be as high as 11% in brain. That the high molecular weight regions in SDS–acrylamide gels contain the greatest number of tissue-specific proteins can be clearly observed in Fig. 4.

The resynthesis and intranuclear accumulation of the acidic proteins during periods of chromatin reactivation in *Physarum* and HeLa cells perhaps explains the frequently observed phenomena that stimuli which induce proliferative states in other cells, invoke a period of increased acidic protein synthesis. Examples can be found in the studies on WI-38 human diploid fibroblasts and 3T6 mouse fibroblasts (Rovera and Baserga, 1971; Tsuboi and Baserga, 1972) where, in response to refeeding, an increase occurs in the synthesis of the acidic proteins before the initiation of DNA synthesis and mitosis. While electrophoretic profiles of stained gels were not shown in these studies, the amino acid isotope incorporation profiles presented in the study of Tsuboi and Baserga show an increased synthesis of a protein (or proteins) in the lower molecular weight region of the gels beginning about 3 hours after refeeding. It seems possible that this increased isotope incorporation could reflect the synthesis of a major protein perhaps similar to the lower molecular weight protein 36 of *Physarum* and HeLa cells (see Figs. 1–3) which is resynthesized during the induction of proliferative cell states (LeStourgeon *et al.*, 1973c). Further support for this possibility comes from the demonstration (LeStourgeon *et al.*, 1974) that when C3H 10 T½ CL8 mouse embryo fibroblasts in culture respond to the stress of contact inhibition, a major low molecular weight protein (band 36, Fig. 5) is lost from the nucleus yet is resynthesized in response to refeeding. The mouse embryo fibroblasts, again like *Physarum* and HeLa cells, not only lose the lower molecular weight protein on the

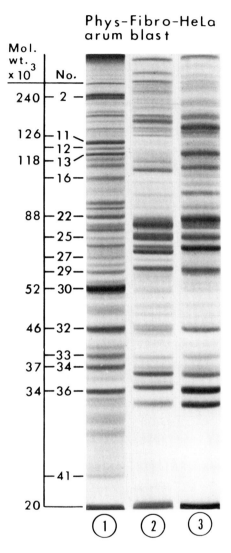

Fig. 4. Electrophoretic profiles of the phenol-soluble acidic nuclear proteins from actively growing *Physarum* (gel 1); mouse embryo fibroblasts (gel 2); and from HeLa S₃ cells (gel 3). Note that protein band 32 is of identical molecular weight (46,000) in all these cell types and that the proteins from HeLa cells and mouse embryo fibroblasts are very similar with cell-specific differences limited to the high molecular weight proteins.

establishment of nonproliferative states but also show a severalfold increase in the 46,000 molecular weight component (Fig. 5).

From the now rather numerous reports that increased synthesis of

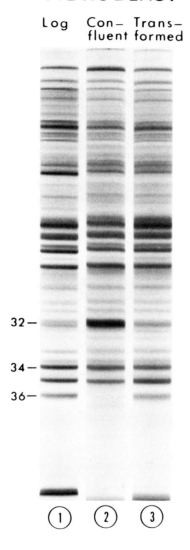

Fig. 5. Electrophoretic banding patterns of the acidic nuclear proteins from C3H 10 T½ CL8 mouse embryo fibroblasts extracted from actively growing normal cells (gel 1), from nongrowing 2-day confluent cultures (gel 2), and from oncogenically transformed cells at densities greater than necessary to inhibit normal cell growth (gel 3). Numbered bands correspond to similarly numbered bands in Figs. 1–3.

6. Cell Proliferation and Differentiation 171

the residual acidic proteins is associated with the induction of cell proliferation or increased metabolic activity, several apparently consistent observations can be summarized as follows:

1. Stimuli which induce quiescent cells to proliferate invoke a generalized increase in the synthesis of essentially all of the acidic proteins (Levy et al., 1973; LeStourgeon et al., 1973c; Stein and Baserga, 1970).

2. Stimuli which induce increased metabolic activity rather than cell proliferation usually invoke an increased synthesis in only a few specific acidic chromatin proteins (Shelton and Allfrey, 1970; Enea and Allfrey, 1973; Jungmann and Schweppe, 1972; Ruddon and Rainey, 1970; Ledinko, 1973).

3. The increased protein synthesis induced by stimuli for cell proliferation is not necessarily coupled to, but usually precedes, DNA synthesis and mitosis (Rovera and Baserga, 1971; Rovera et al., 1971; Tsuboi and Baserga, 1972; Teng and Hamilton, 1969; Stein and Baserga, 1970; Levy et al., 1973; LeStourgeon et al., 1973c), although the maximum intranuclear accumulation of the newly synthesized proteins occurs just preceding and during DNA synthesis (LeStourgeon et al., 1973c).

4. The increased synthesis of the residual acidic proteins occurs in individual proteins in a *temporal* fashion during the prereplicative period leading to DNA synthesis and mitosis (Levy et al., 1973; LeStourgeon et al., 1973c) and results in an increased heterogeneity of the acidic chromatin proteins.

From the foregoing discussion and summarizing statements it now seems clear that the loss in acidic nuclear protein heterogeneity associated with the establishment of nonproliferative states explains the early observations that differentiated and nonproliferative cells contain less residual acidic protein than proliferative cell nuclei. Also, the increased synthesis associated with induced "dedifferentiation" seems to be an important prerequisite for reestablishing the increased protein heterogeneity associated with euchromatin and proliferative cell states. The possible function of many of the acidic proteins associated with dispersed chromatin and proliferative cell states will be considered below.

III. THE MAJOR ACIDIC NUCLEAR PROTEINS AND SPECIFIC GENE REGULATION

The demonstration that many specific nuclear acidic proteins disappear from the nucleus when cells differentiate into quiescent forms and that these proteins are specifically resynthesized and reaccumulate in the nucleus during periods of dedifferentiation has suggested that these proteins may play an important role in the regulation of cell proliferation. While a mechanism whereby certain known residual acidic proteins may regulate cell proliferation will be discussed in the next section of this chapter, it should be constructive to first consider whether or not these proteins act as specific gene regulatory agents. Certainly most of the publications concerning the residual acidic proteins allude to the evidence suggesting a gene regulatory role for these proteins and it seems clear that regulatory macromolecules are in fact present in many residual protein fractions. However, several lines of evidence argue that most of the proteins which can be visualized through high resolution acrylamide gel electrophoresis do not function in the capacity of specific gene regulatory agents, as follows.

If genetic regulation in the eukaryotes is accomplished through mechanisms similar to those operating in prokaryotic cells (i.e., the *lac* operon) it may be postulated, based simply on differences in DNA content [4.25×10^{-15} gm DNA/*E. coli* genome (Helmstetter and Cooper, 1968) vs. 1.2×10^{-12} gm DNA/*Physarum* nucleus (Mohberg and Rusch, 1971)], that the nuclei of *Physarum* for example would contain 3500 times more regulatory protein than *E. coli*. Thus, if 10 copies of *lac* repressor are present per *E. coli* genome then 35,000 copies of a regulatory protein might be expected to be present in an interphase *Physarum* nucleus. While this no doubt is an impractical and perhaps invalid analogy, it will serve to demonstrate that even if regulatory proteins were present in these concentrations in eukaryotes, they would still not be detected through currently used procedures for separating and identifying the residual acidic proteins. This statement is based on the estimations of LeStourgeon *et al.* (1973c) and LeStourgeon *et al.* (1974) that even the most minor protein components detectable in SDS–acrylamide gels are present in intranuclear concentrations in the range of 60,000 to 80,000 copies per nucleus. As many as one million molecules of the major proteins (bands 30, 32, 34, 36, Fig. 5) are present in the interphase nuclei of *Physarum*. Furthermore, only about 100–200 individual proteins can be detected in the residual acidic protein fractions and one would suspect a requirement for far more protein species if each gene is under the regulation of specific proteins. Since several of the individual protein bands have been isolated, purified, identified, and found to be single polypeptide species (LeStourgeon *et al.*, 1974), the argument that most of the protein bands detected through SDS–acrylamide gel electrophoresis are composed of many polypeptides all with identical molecular weights is invalid, although in some gel systems still used this no doubt is possible.

6. Cell Proliferation and Differentiation

The rather high intranuclear concentrations of the residual acidic proteins which can be visualized in SDS–acrylamide gels might suggest that many of these proteins are in fact enzymes or components of enzyme systems. In preliminary studies (A. Burgess and W. M. LeStourgeon, unpublished) bands 11, 12, 13, and 14 in the phenol-soluble acidic protein fraction of *Physarum* (Figs. 1 and 2) coelectrophoresed with the 2 major and 2 minor polypeptide bands present in purified *Physarum* RNA polymerase fractions. The approximate molecular weights determined for these proteins are also in the range expected for the components of eukaryotic RNA polymerase. As might be expected, these proteins disappear from the nucleus upon the establishment of quiescent cell states in *Physarum* and HeLa cells (see Figs. 1–3), but are among the proteins preferentially synthesized during periods of chromatin reactivation and dedifferentiation (LeStourgeon et al., 1973c). Further evidence for the presence of numerous enzymes (i.e., protein phosphokinase, methylase, RNA and DNA polymerase, deoxyribonuclease, NAD nucleosidase, glutamate dehydrogenase, glutamic-oxalacetic transaminase, lactic dehydrogenase, malate dehydrogenase, and ATPase) in residual nuclear protein fractions has been reviewed by Elgin et al. (1971).

The rather high intranuclear protein concentrations of many of the residual acidic proteins may also suggest that some of these proteins may possess kinetic and even structural roles. For example protein band 30 (Figs. 1 and 2) of *Physarum* is of similar molecular weight to the microtubular components of higher eukaryotes (52,000) (Bibring and Baxandall, 1971) and the amino acid composition of this protein (W. M. LeStourgeon, unpublished) is also similar to tubulin from other sources. The decrease in the intranuclear concentration of this protein (Figs. 1 and 2) during the establishment of nonproliferative states would seem to be consistent with the associated cessation of mitosis. Like the proteins thought to be components of the RNA-polymerase complex, the tubulin-like protein is also preferentially synthesized in response to stimuli which induce cell proliferation (LeStourgeon et al., 1973c).

Similarly, the major protein bands 2, 32, and 36 of *Physarum* (Figs. 1, 2, and 5) have been isolated and identified as the contractile proteins myosin, actin, and tropomyosin, respectively (LeStourgeon et al., 1974), and as described in the next section of this chapter, these proteins are apparently major components of the residual proteins of isolated HeLa cell and mouse embryo fibroblast nuclei. Thus, with the demonstration that major quantitative changes do not occur in the residual acidic proteins during the cell cycle (LeStourgeon and Rusch, 1971, 1973), with the tentative evidence that some of the residual acidic proteins are enzymes and structural proteins involved in the metabolic activities of chromatin, and with the demonstration that at least three contractile proteins are components of the residual protein complex, it seems unlikely that all of the other proteins present in similar intranuclear concentrations act as specific gene regulatory agents. Since most of the published experiments (e.g., Kleinsmith et al., 1966; Gilmour and Paul, 1969; Teng and Hamilton, 1969; Kamiyama and Dastugue, 1971; Shelton and Allfrey, 1970; Lemborska and Georgiev, 1972; Teng et al., 1971; O'Malley et al., 1972; Kostraba and Wang, 1973; Shea and Kleinsmith, 1973; Axel et al., 1973) which have suggested a specific regulatory role for some of the acidic proteins have relied on more sensitive techniques (i.e., isotope incorporation, hybridization, effects on in vitro RNA synthesis, etc.) to detect protein changes or the results thereof, the above arguments apply only to those proteins which can be visualized after staining acrylamide gels. The most probable exception would be the experiments on the proteins associated with the polytene chromosomes of Drosophila (Helmsing and Berendes, 1971; Berendes, 1972; Helmsing, 1972), where numerous gene copies may require higher levels of regulatory macromolecules.

IV. THE CONTRACTILE PROTEINS OF ISOLATED CHROMATIN AND CONSIDERATIONS OF THEIR POSSIBLE ROLE IN THE REGULATION OF CELL PROLIFERATION

A. Recognition That Contractile Proteins Were Major Components of Isolated Nuclei and Chromatin

From the observations described in the first two sections of this chapter it seems clear that the increased acidic protein heterogeneity associated with euchromatin and proliferative cell states results to a large extent from the presence of many functional components involved in maintaining chromatin activity. Since it has not been clearly established that the loss of these anionic macromolecules can lead to a histone-mediated heterochromatization of native chromatin and since the contractile proteins actin, myosin, and tropomyosin have been identified (LeStourgeon et al., 1974) as major components of isolated nuclei, a reexamination of the phenomenon of dispersed and condensed chromatin seems relevant. This becomes even more apparent with the observations of Nations et al. (1974) that during periods of chromatin inactivation the concentrations of the contractile proteins in isolated nuclei undergo rapid and significant quantitative changes before decreases in the high molecular weight proteins can be detected.

While it is beyond the scope of this chapter to describe in detail all the experimental evidence leading to the identification of actin, myosin, and native tropomyosin-like proteins in isolated nuclei of *Physarum*, HeLa cells, and mouse embryo fibroblasts, a brief summary of the evidence will be presented.

As previously reported (LeStourgeon and Rusch, 1971; 1973; LeStourgeon et al., 1973b,c, 1974; Nations et al., 1974) and summarized above, the nuclei of *Physarum*, HeLa cells, and C3H 10 T$\frac{1}{2}$ CL8 mouse embryo fibroblasts lose or show a significant decrease in a major low molecular weight acidic protein (band 36, Figs. 1–4) on the establishment of nonproliferative states. Occurring concomitantly with the decrease in protein 36 is a severalfold increase in the intranuclear concentration of a second major protein (band 32, Figs. 1–4). Since these changes were associated with increased chromatin condensation or dispersion, depending on the direction of cell differentiation, a strong correlation was established between the presence or absence of these proteins and the presence or absence of proliferative cell states. In studying these major proteins further (LeStourgeon et al., 1974), it was observed that extracting isolated *Physarum* nuclei with 1.0 M KCl, pH 7.0, would solubilize both the major proliferation-associated protein (band 36) and nonproliferation-associated protein (band 32) as well as numerous other nonhistone proteins (see gel 2, Fig. 6).

It was subsequently observed that if the 1.0 M KCl extracts were dialyzed into low salt (0.05 M KCl, containing 5 mM Mg^{2+}), an insoluble complex formed, containing the major proteins associated with "proliferation" and "nonproliferation" (Fig. 6), as well as two other proteins. Through high resolution electrophoresis it was determined that the four proteins in the insoluble complex had molecular weights of 240,000, 46,000, 34,000, and 26,000, and as determined by SDS–acrylamide gel electrophoresis were the same proteins as bands 2, 32, 36, and 40 in the residual acidic protein fraction soluble in phenol (Fig. 6). Subsequent observations revealed that the very high molecular weight component (band 2, Fig. 6) was the only protein which when isolated and purified to electrophoretic homogeneity was insoluble in 0.05 M KCl, 5 mM Mg^{2+}, and that there was a rigid stoichiometry among the components in the complex regardless of the amount in the extract. Thus, the high molecular weight component caused stoichiometric precipitation of the other components, which suggested a specific interaction between the proteins in the insoluble complex. The similarity of molecular weights between the components of the insoluble complex of *Physarum* nuclei

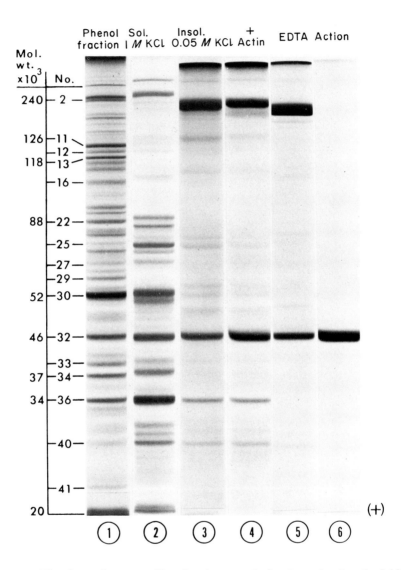

Fig. 6. The electrophoretic profiles of various protein fractions: the phenol-soluble acidic nuclear proteins of *Physarum* (gel 1), the nuclear proteins soluble in 1.0 M KCl, pH 7.0 (gel 2), the 1.0 M KCl-soluble nuclear proteins which form an insoluble complex on dialysis into 0.05 M KCl containing 5 mM Mg^{2+} (gel 3), the same protein fraction to which 3 µg of rabbit muscle actin was added (gel 4), the 1.0 M KCl-soluble proteins which form an insoluble complex in 0.05 M KCl containing 4 mM EDTA (gel 5), and purified rabbit muscle actin (gel 6). Note that 1.0 M KCl solubilizes only certain specific nuclear acidic proteins and especially few of the high molecular weight proteins.

6. Cell Proliferation and Differentiation

and the contractile proteins myosin, actin, tropomyosin, and troponin of muscle (Perry, 1967; Spudich and Watt, 1971; Weber and Murray, 1973) suggested that these proteins may function like their muscle counterparts and that chromatin may contain a contractile system.

That these proteins could in fact be components of a contractile system was suggested by several lines of evidence. First, myosin and actin had been isolated from whole plasmodia of *Physarum* (Hatano and Oosawa, 1966; Hatano and Takahashi, 1971; Hatano and Ohnuma, 1970; Adelman and Taylor, 1969; Tanaka and Hatano, 1972; Hatano and Tazawa, 1968; Nachmias *et al.*, 1970; Nachmias, 1972) and evidence for tropomyosin-like proteins in extracts had also been presented (Tanaka and Hatano, 1972). Second, Jockusch and co-workers (1970, 1971) had previously isolated an actin-like protein from isolated *Physarum* nuclei and had also identified a myosin-like protein in nuclear extracts and in electron micrographs of sectioned nuclei (Jockusch *et al.*, 1973); they considered the intranuclear "actomyosin" to be involved either in nuclear membrane cleavage or in chromosome movement (during *Physarum*'s intranuclear mitosis). Third, both actin and myosin had been isolated from calf thymus nuclei (Ohnishi *et al.*, 1963; Ohnishi *et al.*, 1964) and finally, the chromatin and nucleoli of many cells condense during inactivation, and actin–myosin involvement seemed possible since actin and myosin are universally involved in contractility in eukaryotes (Huxley, 1969; Pollard and Korn, 1971; Behnke *et al.*, 1971; Cohen and Cohen, 1972).

The classic procedures for identification and characterization of muscle proteins were used to confirm that the major protein associated with "nonproliferation" (band 32) is actin; that the major protein associated with "proliferation" (band 36) is like the regulatory protein for contraction, tropomyosin; and that the major component of the insoluble complex was myosin. Briefly the actin component (band 32) was found to have an identical molecular weight (46,000) to rabbit muscle actin and like actin from muscle and other sources contained the unusual amino acid N^{τ}-methyl histidine at a concentration of 1 mole per mole of protein. The nuclear actin formed filaments in 0.6 M KCl which, as examined with the electron microscope, were of the same dimensions as muscle actin. These filaments also reacted with heavy meromyosin (HMM) from rabbit muscle to form the typical "arrowhead" structures unique to these proteins, and, like actin from other sources, ATP removed the "arrowheads" from the actin filaments. Purified nuclear actin reacted with nuclear myosin (band 2) to form an actomyosin complex in the weight ratio of 4:1 myosin to actin and this complex showed Ca^{2+}-dependent ATPase activity with specific activities similar to muscle actomyosin. Also, the actomyosin complex formed after dialyzing the 1.0 M KCl nuclear extracts against 0.05 M KCl, 5 mM $MgCl_2$ "superprecipitated" (i.e., it contracted to about one-half its volume) on the addition of 5 mM ATP (pH 7.0).

The major component of the actomyosin complex formed in low salt (band 2, Fig. 6), like myosin from other sources, is of high molecular weight (approximately 240,000) and is split with trypsin into two major components analogous to heavy and light meromyosin of muscle. Purified nuclear myosin formed aggregates in 0.05 M KCl, 5 mM Mg^{2+} which, as determined with the electron microscope, display both "tail-to-tail" and "head-to-head" filaments as previously described by Nachmias (1972). The purified nuclear myosin reacted with nuclear actin as described above.

Evidence that bands 36 and 40 are like the native tropomyosin–troponin proteins of muscle was seen in their ability to bind stoichiometrically to the actomyosin complex if divalent cations were present (Fig. 6), by their similarity of molecular weights (34,000 and 26,000) to several of the native tropomyosin components of muscle, by their inhibition of the Ca^{2+}-stimulated ATPase activity of nuclear actomyosin, and by the fact that when these proteins were present in the actomyosin complex the ATP-mediated contraction of these proteins specifically required Ca^{2+}. Thus, from the observations summarized above and from other observations described in detail by LeStourgeon *et al.* (1974), it has been conclusively established that bands 2 and 32 of the residual acidic proteins of isolated *Physarum* nuclei are myosin and actin, respectively, and that proteins 36 and 40 have tropomyosin-like activities.

B. Preliminary Evidence That Contractile Proteins May Be True Nuclear Components

Since the contractile proteins present in isolated nuclei of *Physarum* and other cell nuclei are also present in the cytoplasm of the same cells, it is not as yet possible to ascribe a functional role for these proteins in the regulation of chromatin activity. This seems apparent since during nuclear isolation these proteins may penetrate the nuclear membrane and bind to the chromatin complex. As pointed out by LeStourgeon *et al.* (1974), however, several lines of evidence suggest that these proteins may be true nuclear components. For example, in *Physarum* the plasmodium is a true syncytium and, because of this, complete dispersion of plasmodial organelles and cytoplasm is achieved within 2–3 seconds of homogenization. More specifically, if for example 3 gm of wet weight plasmodium is homogenized on a blender in 200 ml of nuclear isolation solution, the plasmodial constituents are homogeneously dispersed essentially at the instant blending is initiated. In an initial effort to determine the extent to which cytoplasmic actin, myosin, and tropomyosin may contaminate nuclei under these conditions, an excess of soluble, purified, and highly radioactive *Physarum* actomyosin was added to a nonradioactive plasmodium at the instant of plasmodial dispersion and the homogenate was allowed to stand 10 minutes for binding to occur. Having known the approximate amount of total actomyosin present in the control plasmodium as well as the amount of each radioactive protein added, it was possible to estimate from the specific activity of the contractile proteins recovered and from the dilution constants for each protein in the homogenate, the extent of nuclear protein contamination by soluble cytoplasmic constituents. These preliminary experiments demonstrated that only about 6–10% of the nuclear actin and tropomyosin-like proteins could be derived from the cytoplasm. In these experiments, however, considerably more nuclear contamination by myosin could be detected although radioactive myosin was added to at least a fivefold excess over the nonradioactive myosin in the homogenate.

Since plasmodia are bound by membrane only, cytoplasmic organelles and membrane could be separated by centrifugation (5×10^4 g for 30 minutes at 4°C) leaving a clear yellow and viscous cytoplasm above the solid residue. Nuclei (0.1 ml wet packed volume, approximately 10^8 nuclei) isolated from actively growing microplasmodia were suspended in 4.0 ml of cytoplasmic preparations from quiescent plasmodia and incubated at 4°C for 10 min prior to isolating the nuclear proteins. In these experiments the electrophoretic profiles of the residual acidic proteins from incubated and control samples were identical. It should be pointed out, however, that only actin could be detected in the soluble cytoplasm of *Physarum*.

To determine if actin-like filaments were attached to nuclear membrane following nuclear isolation, nuclei were placed in a glycerol-standard salt solution, incubated with heavy meromyosin from rabbit skeletal muscle, and observed with the electron microscope. Neither "arrowhead" complexes (which are diagnostic for actin) nor other filaments were seen adhering to the nuclei. Furthermore, in a separate study (J. R. Jeter, personal communication), it has been demonstrated by removing the nuclear membrane with Triton that the contractile proteins are not components of the nuclear membrane.

In further experiments the possibility that traces of cytoplasmic and membrane material could contribute all of the contractile proteins recovered was tested by comparing the proteins from highly purified nuclei with nuclear preparations in which two filtrations and one centrifugation procedure were omitted. As judged empirically by phase contrast microscopy, the impure preparations were about 80% nuclei and the rest of the material could be identified as membrane and trapped cytoplasmic residue. Thus, if minute traces of extranuclear material in the purified nuclear preparations were the source of all the contractile proteins, then a manyfold increase in actin, myosin, and tropomyosin should be observed in the contaminated preparations. However, the highly contaminated nuclei showed no increase in the tropomyosin-like component and only moderately elevated actin concentrations when compared to pure nuclear preparations.

Added support for the *in vivo* intranuclear occurrence of myosin and actin can be found in the studies of Jockusch *et al.* (1973) and Ryser (1970) who have observed myosin and actin-like filaments, respectively, in nuclei fixed *in situ*. In the studies of Jockusch *et al.*, whole *Physarum* plasmodia were incubated in a 1:1 mixture of glycerol and standard salts, fixed, embedded, sectioned, and observed with the electron microscope. Thick filaments (80–120 Å) with dimensions like those of muscle myosin were observed dispersed *within* the nucleoplasm. Furthermore, in the presence of 5mM pyrophosphate still larger filaments could be observed. In the studies of Ryser filaments with the same dimensions as F-actin were observed in the nucleoplasm of fixed and sectioned *Physarum* plasmodia.

Other evidence that the contractile proteins may be true nuclear components can be seen in the experiments of Nations *et al.* (1974). In these experiments it was observed that some of the nuclear protein changes which occur when microplasmodia are starved for 16 hours can be induced within 45 minutes if microplasmodia growing in shake culture are gently pelleted by centrifugation and allowed to stand at room temperature for 45 minutes before isolating the nuclei and the nuclear proteins. It was demonstrated that in response to this treatment the intranuclear concentration of actin increased by 110% and the tropomyosin-like component (band 36) decreased in intranuclear concentration by 51%. While it is not clear as yet whether the stimulus is anoxia or other effects induced by high plasmodial density, it was demonstrated that the great majority of actin which entered the nucleus was not synthesized during the 45-minute period; cycloheximide (50 μg/ml) inhibited incorporation of labeled amino acids into pelleted plasmodial protein by 86%, yet the intranuclear actin concentration increased by 82%. Since the specific activities of the intranuclear actin in prelabeled control and prelabeled pelleted plasmodia were essentially identical (286 dpm/μg vs. 288 dpm/μg, respectively), it does not appear that the increase in nuclear actin reflects simply a 110% net increase in cytoplasmic actin and subsequent nuclear contamination, but rather a specific transfer of preformed protein.

6. Cell Proliferation and Differentiation 181

In summary, the evidence so far available suggests that the contractile proteins of isolated nuclei may in fact be nuclear in origin and that they are not components of the nuclear membrane. Since the contractile proteins are not quantitatively removed when the weakly associated nucleoplasmic proteins are briefly extracted with dilute saline (0.14 M NaCl) (LeStourgeon and Rusch, 1973), it appears that these proteins are true components of the chromatin material.

C. Functional Aspects of Intranuclear Contractile Proteins

With the identification of actin, myosin, and tropomyosin-like proteins in isolated nuclei of *Physarum*, and with the apparent presence of these proteins in isolated nuclei of HeLa cells and C3H 10 T$\frac{1}{2}$ CL8 mouse embryo fibroblasts, it seems of interest to consider the possibility of a chromatin-specific function for these proteins. Because of the well-documented role of these proteins in the contraction of muscle and the apparent universal presence of these proteins in other contractile systems, it seems possible that the intranuclear actomyosin–tropomyosin complex has some contractile function. This seems even more likely since there is an apparent increase in the intranuclear concentration of actin and decrease in the tropomyosin-like protein during periods when the nuclei and chromatin of *Physarum*, HeLa cells, and fibroblasts condense. As discussed above the increase in nuclear actin, the decrease in the major tropomyosin-like protein, and the condensation of chromatin are all associated with the establishment of nonproliferative cell states. Similarly, it is widely known that the cells of metabolically active proliferating tissues usually contain large swollen nuclei whereas quiescent nonproliferative cells possess condensed and relatively inactive nuclei (i.e., neoplastic cells vs. normal differentiated cells; spermatocytes vs. sperm; immature avian erythrocytes vs. mature erythrocytes; proliferative vs. nonproliferative *Physarum*, *Acanthamoeba*, etc.).

Based on the information now available, together with relevant observations from the literature, a summary of the changes in intranuclear concentration of the contractile proteins is presented in Fig. 7. This figure outlines some of the nuclear events which occur when cells respond to conditions unfavorable for growth, the result of these events perhaps being a generalized attenuation of nuclear activity and cessation of cell proliferation. This inactivation of nuclear activity through condensation may not however be limited to a response to environmental stress, since cells which normally differentiate into nonproliferative forms (i.e., avian erythrocytes, sperm, spores, etc.) might also utilize similar mechanisms. Before discussing possible mechanisms whereby contractile proteins may

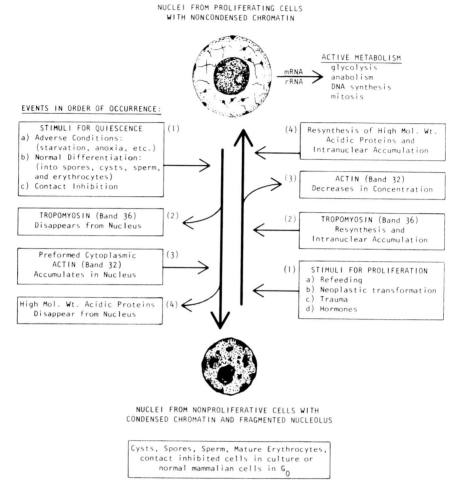

Fig. 7. A schematic summary and suggested results of the acidic nuclear protein changes associated with cell differentiation into proliferative and nonproliferative states. Discussion in text.

regulate cell proliferation, other pertinent information must be considered.

1. The Chemistry of Contractile Systems

In order that the reader can better understand the data presented in this discussion, a brief review of contractile protein chemistry will be presented. For more detailed reviews, see Perry (1967), Spudich and Watt (1971), Weber and Murray (1973), Huxley (1969), Szent-

Györgyi (1951), Taylor (1972), and Bendall (1969). In physiological ionic environments purified skeletal muscle myosin has a low level of Ca^{2+}-ATPase activity (EC 3.6.1.3). Actin interacts with myosin stoichometrically and stimulates a Mg^{2+}-ATPase by nearly a factor of 20, such that *pure* skeletal muscle actomyosin has high Mg^{2+}- and low Ca^{2+}-ATPase activity. If, however, tropomyosin and troponin, which are native components of contractile systems, are added to the actomyosin preparations, the Mg^{2+}-ATPase is inhibited *unless* a low level of Ca^{2+} (10^{-6} M) is present. The presence of tropomyosin-troponin, thus, confers a Ca^{2+} requirement for ATPase activity with the result that in the presence of Ca^{2+}, actomyosin contracts on the addition of ATP and relaxes when Ca^{2+} is removed. That actomyosin may contract or relax depending on pH, ionic strength, Mg^{2+}, Ca^{2+}, ATP level, and the presence or absence of the regulatory proteins (tropomyosin and troponin) is quite relevant to the suggested role of actomyosin in nuclei, especially since Ca^{2+} and Mg^{2+} have profound effects on nuclear size and stability and on chromatin condensation.

2. Nuclear vs. Chromatin Condensation

Since *Physarum* nuclei contract to about half their size during the change of state from "proliferation" to "nonproliferation" (Guttes and Guttes, 1969), it might be concluded that the intranuclear actomyosin complex is involved simply in causing this change in nuclear volume. While some of the nuclear actomyosin in *Physarum* may have this function, it seems more likely that most of the actomyosin is involved in causing chromatin to condense, for the following reasons. As determined by phase contrast microscopy, nuclei isolated from density inhibited (confluent) fibroblasts were not markedly different in size from the nuclei of proliferating fibroblasts; however, the chromatin appeared more granular and the nucleoli were fragmented into numerous small bodies. Also, nuclei isolated from starved (nonproliferative) HeLa cells were not significantly different in size from those of proliferative cells, although the chromatin again appeared condensed and granular. Similarly, when tadpole liver cells in tissue culture are cooled, the chromatin condenses, the nucleoli fragment, and there is no change in nuclear volume (Kriegstein and Bennett, 1973). Therefore, it seems most likely that the general effect is chromatin condensation and that nuclear volume is reduced in only some cases where cell volume is also reduced (such as in spores or cysts).

The chromatin of isolated nuclei may be dispersed or condensed depending on pH, ionic strength, and the presence or absence of divalent cations in the medium (Anderson and Wilbur, 1952; Anderson, 1956;

Philpot and Stanier, 1956; Robbins et al., 1970; Olins and Olins, 1972). Specifically, divalent cations such as Mg^{2+} or Ca^{2+} produce granularity while removal of divalent cation produces dispersion. Directly related to the presence of contractile proteins in chromatin are the results of Olins and Olins (1972) who compared chromatin granulation and dispersion in isolated rat liver and chick erythrocyte nuclei in response to experimentally altered pH and divalent cation concentration. Based on the changes which occur in the cells so far examined, rat liver nuclei (from metabolically active cells) should contain little actin but should be relatively rich in tropomyosin, whereas the mature erythrocyte nuclei should be rich in actin and contain no tropomyosin [from the work of Vidali et al. (1973) this may be true since erythrocyte chromatin contains a major 46,000 molecular weight component]. The loss of tropomyosin from the avian erythrocyte nucleus should have the effect of lowering the Mg^{2+} requirement for chromatin condensation. This is indeed what Olins and Olins found, for they report "an interesting difference" in the granulating and dispersion characteristics between the two nuclear types. More specifically very low levels of Mg^{2+} cause erythrocyte chromatin to condense while the rat liver chromatin remained dispersed.

In the context of these experiments on isolated nuclei it is relevant to point out that contractile threads have been made from DNA–myosin complexes (Vorobyev, 1957), and the contractile properties of these threads are consistent with the chromosomal and nuclear behavior so far observed; namely, the threads will *contract* upon the addition of ATP if divalent cations are present, but will *elongate* in the absence of divalent cation.

3. General Results

As diagrammed in Fig. 7, nuclei from proliferative cells are swollen and metabolically active. The information summarized in this chapter suggests the possibility that in response to stimuli which inhibit growth, or as a result of normal differentiative events, the intranuclear concentration of actin increases and the tropomyosin-like protein disappears from the nucleus. As pointed out, the intranuclear accumulation of actin appears to result from a flux of actin into the nucleus from preformed cytoplasmic pools (Nations et al., 1974), and this can occur in a very short period of time. For example, as pointed out previously the intranuclear concentration of actin in *Physarum* can increase by 110% within 45 minutes of pelleting plasmodia. From contractile protein chemistry, the loss of tropomyosin can be expected to result in a *sensitization* of intranuclear actomyosin and contraction is thus stimulated. Since tropomyosin is the actual regulatory molecule, the intranuclear actin concentration need not change for condensation to occur.

Inactive nuclei or chromatin complexes in nonproliferative cells are condensed. In response to stimuli which induce dedifferentiation and cell proliferation, the intranuclear concentration of actin apparently decreases and there is a parallel increase in the inhibitor protein for contraction, tropomyosin. Previous studies (LeStourgeon et al., 1973c) have shown that the reappearance of tropomyosin in *Physarum* results from new protein synthesis and intranuclear accumulation. In the same report it was also demonstrated that while the intranuclear concentration of actin decreases during induction of proliferation, the rate of cytoplasmic actin synthesis increases sharply. Thus, there appears to be a competitive or otherwise regulated flux of actin into the nuclei of proliferative cells. Again from contractile protein chemistry, the reappearance of tropomyosin would have the effect of inhibiting intranuclear actomyosin contraction (in the absence of Ca^{2+}), resulting in nuclear swelling.

As summarized in Fig. 7, it is presumed that the regulation of cell proliferation by the contractile proteins of nuclei may be a very general method whereby cells either proliferate or do not. However, there is also preliminary evidence for intermediate states. When *Physarum* is cultured under the best known conditions, mitosis occurs synchronously approximately every 8 hours at 26°C. Under these conditions there is always more of the major tropomyosin-like protein than actin in the nucleus. However, if plasmodia are cultured at 18°C, mitosis occurs only once every 12-14 hours and growth is inhibited accordingly. The nuclei from plasmodia cultured at 18°C contain approximately half as much tropomyosin as nuclei from 26°C plasmodia, and more actin than tropomyosin is present in the nucleus. Thus the absolute rate of growth perhaps may be correlated with differing amounts of actin and tropomyosin in the nucleus. Since the mechanisms of actomyosin regulation, by tropomyosin, troponin, divalent cations, and perhaps other proteins, are not precisely known, it should be pointed out that other undefined parameters may also influence chromatin activity. Thus, regulatory mechanisms, where the intranuclear concentrations of actin, myosin, tropomyosin, and troponin are not altered precisely as outlined here, may also be operative.

4. Implications

Heterochromatin is the condensed and genetically inactive portion of interphase chromatin, while euchromatin is the dispersed and genetically active form. If the condensation of heterochromatin occurs by the same actomyosin-mediated mechanism by which entire nuclei may condense, one might expect heterochromatin to be relatively rich in actin, with little or no tropomyosin. In this regard it is relevant to point out that isolated proliferative interphase nuclei always contain some actin, just as they always contain *some* heterochromatin.

In other cells with completely inactive nuclei, such as mature avian red blood cells or epidermal cells before stimulation with growth factor (Cohen, 1965; Savage and Cohen, 1973), one might predict that if the chromatin in these nuclei is turned off by an actomyosin-mediated mechanism then these nuclei should be relatively rich in actin and relatively poor in tropomyosin as compared to proliferating cells. Indeed, mature avian erythrocyte nuclei are rich in a 46,000 molecular weight component (Vidali *et al.*, 1973). The cell fusion experiments of Harris *et al.* (1969), Ringertz *et al.* (1972), and Carlsson *et al.* (1973) are relevant in this regard: quiescent chick erythrocyte nuclei become activated after fusion with "proliferating" tissue culture cells, upon which they swell (decondense) markedly, and take up proteins from both the cytoplasm and nucleus of the tissue culture cells. If the changes during this nuclear activation are similar to those which occur upon refeeding *Physarum* and HeLa cells, it may be predicted that the tropomyosin-like protein is one of the proteins entering the chick nucleus.

5. Mechanistic Considerations and Concluding Implications

Generally, it has been assumed that the regulation of chromatin activity is a highly sophisticated and complex process, in that each gene is under specific regulation through various inducer-repressor phenomena. As discussed here, however, the possibility exists that many genes such as those which must be expressed for cell proliferation may be inactivated in a rather nonspecific fashion. The ability of cells to respond within minutes to stimuli unfavorable to growth and grossly redirect metabolic activities and energy utilization is not without obvious evolutionary and survival implications. Several mechanisms whereby contractile proteins may interact with chromatin and invoke an attenuation or shutdown of nuclear activity can be considered.

As discussed previously, there is evidence that myosin can interact with DNA and form contractile threads, apparently without altering the actin-binding sites of myosin. Since relatively little myosin is present in the chromatin complex, compared to DNA, the possibility exists that myosin may normally be complexed with portions of the DNA at specific places along the chromosomes. Thus, when actin enters the nucleus during condensation it may complex and link the myosin sites. This possibility may be supported by the fact that there is very little myosin in

6. Cell Proliferation and Differentiation

isolated nuclei as compared to actin and tropomyosin and by the fact that the intranuclear concentration of myosin does not change significantly during rapid condensation phenomena in *Physarum*.

Another possibility, however, is that actin may interact directly with DNA. As pointed out in the first section of this chapter, during periods of chromatin condensation many of the high molecular weight acidic nuclear proteins decrease in intranuclear concentration. The possibility exists that action may displace these proteins during condensation. In support of this possibility it can be pointed out that F-actin filaments contain tightly bound ADP (Oosawa et al., 1965) and the monomers which make up the F-actin filaments are arranged in a helix with a longitudinal repeat of around 374 Å (Moore et al., 1970; Spudich et al., 1972). The actin repeat corresponds to 11 repeats of the DNA helix and it is conceivable that the repeating bound nucleotides of the actin could interact with repeating DNA units. As another possibility, one might consider that some of the numerous high molecular weight acidic proteins, which increase as actin decreases during nuclear swelling and reactivation, may function as specific inducer agents, as has been previously suggested (Frenster, 1965; Teng et al., 1971; Paul, 1972). It might also be reasoned that the contractile proteins never interact with DNA but interact with other chromosomal components.

While it is interesting to consider the mechanisms whereby contractile proteins could interact with specific nuclear components to induce nonproliferative states, it is still not clear whether these proteins have a functional role in the nucleus. However, the possibility that contractile proteins may possess a role in the regulation of cell proliferation and quiescence cannot be overlooked since it has been shown (LeStourgeon et al., 1974) that oncogenically transformed mouse embryo fibroblasts fail to accumulate actin or lose the tropomyosin-like protein in response to stimuli which induce nonproliferative states.

Note Added in Proof

Continuing studies on the *Physarum* nuclear protein with tropomyosin-like activities (band 36) have suggested that this protein is a component of or is the major constituent of the 30 S ribonucleoprotein (RNP) complex (also known as "informofers"). Band 36 protein is the same molecular weight (34,000) as the major constituent of RNP particles from mammalian cells (T. Martin et al., *Cold Spring Harbor Symp. Quant. Biol.* **38**, 1973). Like RNP particle protein from mammalian cells, band 36 protein contains an unidentified basic residue (about 3% total residue) which elutes from PA-35 resin at approximately the same position as described for the unknown residue in rat liver RNP protein (see Table

I, this volume, page 73 and A. Sarasin, *FEBS Lett.* **4**, 327, 1969). All other residues are present in similar amounts including glycine at 16% (T. Martin *et al.*, as above). Also, like RNP particle protein from other sources, band 36 protein forms 200-Å spherical particles in low ionic environments which look like the structures formed by RNP particle protein from mammalian cells. These findings may suggest that the tropomyosin-like activities of band 36 protein are nonspecific phenomena or that this protein may function to maintain a euchromatin state as well as complex DNA-like RNA.

ACKNOWLEDGMENTS

Portions of the work described in this chapter was supported in part by NIH Grants CA-5002 and CA-07175.

REFERENCES

Adelman, M. R., and Taylor, E. W. (1969). *Biochemistry* **8**, 4976–4988.
Anderson, N. G. (1956). *Quart. Rev. Biol.* **31**, 169.
Anderson, N. G., and Wilbur, K. M. (1952). *J. Gen. Physiol.* **35**, 781–796.
Axel, R., Cedar, H., and Felsenfeld, G. (1973). *Proc. Nat. Acad. Sci. U.S.* **70**, 2029–2032.
Baserga, R., and Stein, G. (1971). *Fed. Proc., Fed. Amer. Soc. Exp. Biol.* **30**, 1752–1759.
Behnke, O., Kristensen, B. I., and Nielsen, L. E. (1971). *J. Ultrastruct. Res.* **37**, 351–369.
Bendall, J. R. (1969). "Muscles, Molecules, and Movement." Heinemann, London.
Berendes, H. D. (1972). *In* "Developmental Studies on Giant Chromosomes" (W. Beermann, ed.), pp. 181–207. Springer-Verlag, Berlin and New York.
Bibring, T., and Baxandall, J. (1971). *J. Cell Biol.* **48**, 324–339.
Carlsson, S. A., Moore, G. P. M., and Ringertz, N. R. (1973). *Exp. Cell. Res.* **76**, 234–241.
Cohen, I., and Cohen, C. (1972). *J. Mol. Biol.* **68**, 383–387.
Cohen, S. (1965). *Develop. Biol.* **12**, 394–407.
Dingman, C. W., and Sporn, M. B. (1964). *J. Biol. Chem.* **239**, 3483–3492.
Elgin, S. C. R., Froehner, S. C., Smart. J. E., and Bonner, J. (1971). *Advan. Cell Mol. Biol.* **1**, 1–57.
Enea, V., and Allfrey, V. G. (1973). *Nature (London)* **242**, 265–267.
Frenster, J. H. (1965). *Nature (London)* **206**, 680–683.
Frenster, J. H., Allfrey, V. G., and Mirsky, A. E. (1963). *Proc. Nat. Acad. Sci. U.S.* **50**, 1026–1032.
Gershey, E. L., and Kleinsmith, L. J. (1969). *Biochim. Biophys. Acta* **194**, 519–525.
Gilmour, R. S., and Paul, J. (1969). *J. Mol. Biol.* **40**, 137–139.
Grouse, L., Chilton, M. D., and McCarthy, B. J. (1972). *Biochemistry* **11**, 798–805.
Grunicke, H., Potter, V. R., and Morris, H. P. (1970). *Cancer Res.* **30**, 776–787.
Guttes, E., and Guttes, S. (1969). *Experientia* **25**, 1168–1170.

Harris, H., Sidebottom, E., Grace, D. M., and Bramwell, M. E. (1969). *J. Cell Sci.* **4**, 499–525.
Hatano, S., and Ohnuma, J. (1970). *Biochim. Biophys. Acta* **205**, 110–120.
Hatano, S., and Oosawa, F. (1966). *Biochim. Biophys. Acta* **127**, 488–498.
Hatano, S., and Takahashi, K. (1971). *Mechanochem. Cell Motility* **1**, 7.
Hatano, S., and Tazawa, M. (1968). *Biochim. Biophys. Acta* **154**, 507–519.
Helmsing, P. J. (1972). *Cell Differentiation* **1**, 19–24.
Helmsing, P. J., and Berendes, H. D. (1971). *J. Cell Biol.* **50**, 893–896.
Helmstetter, C. E., and Cooper, S. (1968). *J. Mol. Biol.* **31**, 507–518.
Huxley, H. E. (1969). *Science* **164**, 1356–1366.
Jockusch, B. M., Brown, D. F., and Rusch, H. P. (1970). *Biochem. Biophys. Res. Commun.* **38**, 279–283.
Jockusch, B. M., Brown, D. F., and Rusch, H. P. (1971). *J. Bacteriol.* **108**, 705–714.
Jockusch, B. M., Ryser, U., and Behnke, O. (1973). *Exp. Cell Res.* **76**, 464–466.
Jungmann, R. A., and Schweppe, J. S. (1972). *J. Biol. Chem.* **247**, 5535–5542.
Kamiyama, M., and Dastugue, B. (1971). *Biochim. Biophys. Res. Commun.* **44**, 29–36.
Kleinsmith, L. K., Allfrey, V. G., and Mirsky, A. E. (1966). *Science* **154**, 780–781.
Kostraba, N. C., and Wang, T. Y. (1973). *Exp. Cell Res.* **80**, 291–296.
Kriegstein, H., and Bennett, T. P. (1973). *Exp. Cell Res.* **80**, 152–158.
Ledinko, N. (1973). *Virology* **54**, 294–298.
Lemborska, S. A., and Georgiev, G. P. (1972). *Cell Differentiation* **1**, 245–251.
LeStourgeon, W. M., and Rusch, H. P. (1971). *Science* **174**, 1233–1236.
LeStourgeon, W. M., and Rusch, H. P. (1973). *Arch. Biochem. Biophys.* **155**, 144–158.
LeStourgeon, W. M., Goodman, E. M., and Rusch, H. P. (1973a). *Biochim. Biophys. Acta* **317**, 524–528.
LeStourgeon, W. M., Wray, W., and Rusch, H. P. (1973b). *Exp. Cell Res.* **79**, 487–490.
LeStourgeon, W. M., Nations, C., and Rusch, H. P. (1973c). *Arch. Biochem. Biophys.* **159**, 861–872.
LeStourgeon, W. M., Forer, A., Bertram, J. S., Yang, Y., and Rusch, H. P. (1974). Submitted for publication.
Levy, R., Levy, S., Rosenberg, S. A., and Simpson, R. T. (1973). *Biochemistry* **12**, 224–228.
Littau, V. C., Allfrey, V. G., Frenster, J. H., and Mirsky, A. E. (1964). *Proc. Nat. Acad. Sci. U.S.* **52**, 93–100.
McClure, M. E., and Hnilica, L. S. (1972). *Sub-Cell. Biochem.* **1**, 311–332.
MacGillivray, A. J., Cameron, A., Krauze, R. J., Rickwood, R. J., and Paul, J. (1972). *Biochim. Biophys. Acta* **277**, 384–402.
Marushige, K., and Dixon, G. H. (1969). *Develop. Biol.* **19**, 397–414.
Mirsky, A. E., and Ris, H. (1951). *J. Gen. Physiol.* **34**, 475–493.
Mohberg, J., and Rusch, H. P. (1971). *Exp. Cell Res.* **66**, 305–316.
Moore, P. B., Huxley, H. E., and DeRosier, D. J. (1970). *J. Mol. Biol.* **50**, 279–295.
Nachmias, V. T. (1972). *Cold Spring Harbor Symp. Quant. Biol.* **37**, 607–612.
Nachmias, V. T., Huxley, H. E., and Kessler, D. (1970). *J. Mol. Biol.* **50**, 83–90.
Nations, C., LeStourgeon, W. M., Magun, B. E., and Rusch, H. P. (1974). *Exp. Cell Res.* (in press).
Ohnishi, T., Kawamura, H., and Yamamoto, T. (1963). *J. Biochem. (Tokyo)* **54**, 298–300.

Ohnishi, T., Kawamura, H., and Tanaka, Y. (1964). *J. Biochem. (Tokyo)* **56**, 6–15.
Olins, D. E., and Olins, A. L. (1972). *J. Cell Biol.* **53**, 715–736.
O'Malley, B. W., Spelsberg, T. C., Schroeder, W. T., Chytil, F., and Steggles, A. W. (1972). *Nature (London)* **235**, 141–144.
Oosawa, F., Asakura, S., Higashi, S., Kasai, M., Kobayashi, S., Nakano, E., Ohnishi, T., and Taniguchi, M. (1965). In Molecular Biology of Muscular Contraction (S. Ebashi *et al.*, eds.), pp. 56–76. Elsevier, Amsterdam.
Paul, J. (1972). *Nature (London)* **238**, 444–446.
Perry, S. V. (1967). *Progr. Biophys.* **17**, 327–381.
Philpot, J. St. L., and Stanier, J. E. (1956). *Biochem. J.* **63**, 214–223.
Pipkin, J. L., Jr., and Larson, D. A. (1972). *Exp. Cell Res.* **79**, 28–42.
Pollard, T. D., and Korn, E. D. (1971). *J. Cell Biol.* **48**, 216–219.
Ringertz, N. R., Carlsson, S. A., and Savage, R. E. (1972). *Advan. Biosci.* **1**, 219.
Robbins, E., Pederson, T., and Klein, P. (1970). *J. Cell Biol.* **44**, 400–416.
Rovera, G., and Baserga, R. (1971). *J. Cell. Physiol.* **77**, 201–212.
Rovera, G., Farber, J., and Baserga, R. (1971). *Proc. Nat. Acad. Sci. U.S.* **68**, 1725–1729.
Ruddon, R. W., and Rainey, C. H. (1970). *Biochem. Biophys. Res. Commun.* **40**, 152–160.
Ryser, U. (1970). *Z. Zellforsch. Mikrosk. Anat.* **110**, 108–130.
Savage, C. R., Jr., and Cohen, S. (1973). *Exp. Eye Res.* **15**, 361–366.
Shea, M., and Kleinsmith, L. J. (1973). *Biochem. Biophys. Res. Commun.* **50**, 473.
Shelton, K. R., and Allfrey, V. G. (1970). *Nature (London)* **228**, 132–134.
Simpson, R. T., and Reeck, G. R. (1973). *Biochemistry* **12**, 3853–3858.
Spelsberg, T. C., Wilhelm, J. A., and Hnilica, L. S. (1972). *Sub-Cell. Biochem.* **1**, 107–145.
Sporn, M. B., and Dingman, C. W. (1966). *Cancer Res.* **26**, 2488–2495.
Spudich, J. A., and Watt, S. (1971). *J. Biol. Chem.* **246**, 4866–4871.
Spudich, J. A., Huxley, H. E., and Finch, J. T. (1972). *J. Mol. Biol.* **72**, 619–632.
Stedman, E., and Stedman, E. (1943). *Nature (London)* **152**, 267–269.
Steele, W. J., and Busch, H. (1963). *Cancer Res.* **23**, 1153–1163.
Stein, G. S., and Baserga, R. (1970). *J. Biol. Chem.* **245**, 6097–6105.
Stein, G. S., and Baserga, R. (1971). *Advan. Cancer Res.* **15**, 287–330.
Stellwagen, R. H., and Cole, R. D. (1969). *J. Biol. Chem.* **244**, 4878–4887.
Szent-Györgyi, A. (1951). "Chemistry of Muscular Contraction," 2nd rev. ed. Academic Press, New York.
Tanaka, H., and Hatano, S. (1972). *Biochim. Biophys. Acta* **257**, 445–451.
Taylor, E. W. (1972). *Annu. Rev. Biochem.* **41**, 577–611.
Teng, C. S., and Hamilton, T. H. (1969). *Proc. Nat. Acad. Sci. U.S.* **63**, 465–472.
Teng, C. S., Teng, C. T., and Allfrey, V. G. (1971). *J. Biol. Chem.* **246**, 3597–3609.
Tsuboi, A., and Baserga, R. (1972). *J. Cell. Physiol.* **80**, 107–118.
Vidali, G., Boffa, L. C., Littau, V. C., Allfrey, K. M., and Allfrey, V. G. (1973). *J. Biol. Chem.* **248**, 4065–4068.
Vorobyev, V. I. (1957). *Biochemistry (USSR)* **22**, 555–564.
Weber, A., and Murray, J. M. (1973). *Physiol. Rev.* **53**, 612–673.
Weisenthal, L. M., and Ruddon, R. W. (1972). *Cancer Res.* **32**, 1009–1017.
Wu, F. C., Elgin, S. C. R., and Hood, L. E. (1973). *Biochemistry* **12**, 2792–2797.

7

Nonhistone Proteins of Dipteran Polytene Nuclei

H. D. BERENDES AND P. J. HELMSING

I. Introduction 191
II. Cytology and Cytochemistry of Genome Activity 192
      A. Puffs: Morphological Manifestation of Gene Activity ... 192
      B. Puff Cytochemistry 194
III. Chemical Modification of Polytene Chromosome Proteins 199
      A. Acetylation 199
      B. Phosphorylation 200
      C. Methylation and Ethylation 201
IV. Qualitative and Quantitative Changes in Protein during Gene Activation 201
V. The Possible Role of Proteins in Gene Activation 206
References 209

I. INTRODUCTION

Nowadays, nuclear nonhistone proteins are generally regarded to fulfill essential roles in the regulation and coordination of nuclear activity in eukaryote cells (Stellwagen and Cole, 1969; Baserga and Stein, 1971; MacGillivray et al., 1972). A major fraction of nonhistone protein species commonly present in chromatin extracts of various cell types of an organism (Elgin and Bonner, 1970, 1972; Shaw and Huang, 1970; MacGillivray et al., 1971; Barrett and Gould, 1973) may in conjunction with the histones participate in the constitution and maintenance of the structural organization of the chromosome complement, as well as in the control of basic genome activities common to a wide variety of cell types. Nonhistone protein species restricted in their occurrence to a particular tissue or cell type may be involved in the control of cell-specific genome activity. These protein species may have a specific function in the intranuclear transport of gene-activating stimuli (e.g., steroid hormone molecules), in the initiation of transcription, in the quantitative regulation of transcription, in the processing of primary gene products, or in the intranuclear transport of mature gene products. So far, however, detailed information as to the actual role and the mechanism of action of tissue-specific nonhistone protein fractions isolated from chromatin preparations is scarce. Moreover, the wide variety of procedures employed in the isolation and characterization of nuclear nonhistone proteins as well as the different approaches to elucidate their function hamper the classification and precise definition of the function of those common and tissue-specific nuclear nonhistone protein species that have been described for a wide range of eukaryotes. Although essentially similar problems are encountered in comparing the nonhistone protein patterns of nuclei or chromatin from polytene cell types of dipteran insects, the giant polytene chromosomes do offer a unique opportunity to elucidate the function of some nonhistone protein species in the process of local chromosome activity. In particular, the elegant technique for dissection of defined regions of the giant chromosomes by micromanipulation, as developed by Pelling (1970) and Edström and co-workers (Edström and Daneholt, 1967; Daneholt et al., 1969a,b), provides a powerful tool for the elucidation of the possible relationships between certain nonhistone protein species and the initiation and regulation of local transcription as well as the preparation of the ultimate, locus-specific, ribonucleoprotein product.

II. CYTOLOGY AND CYTOCHEMISTRY OF GENOME ACTIVITY

A. Puffs: Morphological Manifestations of Gene Activity

In a classic paper, Beermann (1952) described the variations in local modifications of the structure of polytene chromosomes during larval development of the midge *Chironomus tentans*. These modifications, puffs and Balbiani rings (giant pufflike structures), were interpreted by him as morphological expressions of local genome activity. Since then a considerable quantity of data has accumulated in support of his view. Detailed analysis of puffing patterns, in which location and size of puffs were compared in the chromosome complement of a tissue at subsequent stages in development, revealed, in addition to fluctuations in size of the majority of the puffs, characteristic changes at certain moments in development. A comparison of puffing patterns of different tissues of the same larva indicated the presence of tissue-specific puffs. These, however, constitute only a minor fraction as compared to the number of puffs commonly present in the tissues investigated. These data, which have been obtained from studies of several *Drosophila* and *Chironomus* species (for reviews, see Clever, 1963; Kroeger and Lezzi, 1966; Pelling, 1966; Berendes and Beermann, 1969; Ashburner, 1970b, 1972; Panitz, 1972), suggest differential gene activity during development and in functionally different cell types. It could be suggested that the differences in the puffing patterns of different tissues are to some extent reflected in the pattern of nuclear nonhistone proteins. A comparison of nonhistone protein patterns of nuclei from midgut and salivary gland cells of *Drosophila hydei*, prepared by the same procedure, revealed consistent differences between the two cell types (Helmsing and van Eupen, 1973). The significance of this finding in relation to the differences in the puffing patterns of the two tissues remains to be elucidated, however.

It has been firmly established now that puffs and Balbiani rings are chromosome sites undergoing active transcription. Using tritiated RNA precursors, Pelling (1959) demonstrated an enhanced precursor incorporation in puffed chromosome loci. Puff size reflected, in many instances, the rate of precursor incorporation (Pelling, 1964). The kinetics of local RNA synthesis has been analyzed in great detail for some active chromosome loci (e.g., Balbiani ring (BR) 2 in *Chironomus tentans* salivary gland chromosome IV). Following pulse labeling of the salivary glands, the Balbiani ring (and other parts of the chromosome complement) was isolated by microdissection, and the RNA extracted and analyzed electrophoretically (Daneholt et al., 1969b, 1970; Pelling, 1970). *In situ* hybridization experiments in which RNA extracted from BR 2 was applied to a salivary gland squash preparation revealed exclusive hybrid formation with the (repeated) DNA sequences present in the Balbiani ring 2 (Lambert, 1972).

It was further shown that a high molecular weight fraction isolated from the nuclear sap and hybridized *in situ* to denatured salivary gland chromosomes annealed specifically to Balbiani ring 2 DNA (Lambert et al., 1972, 1973). Recently, a particular high molecular weight RNA species (75S), synthesized in Balbiani ring 2, was found in the nuclear sap and in the cytoplasm of the salivary gland cells (Daneholt and Hosick, 1973). On the basis of these data, it was speculated that this high molecular weight RNA species may have a messenger function, a suggestion which is compatible with cytogenetic evidence indicating a relationship between the presence of a certain Balbiani ring and the occurrence of a specific polypeptide in the secretory product of *Chironomus* salivary gland cells (Grossbach, 1969). This and other observations indicating a correlation between the activity of certain puffs and the occurrence of particular cell products (Beermann, 1961; Berendes, 1965; Baudisch and Panitz, 1968; Poels, 1972; Leenders and Beckers, 1972; Berendes *et al.*, 1973) provide strong support for the view that puffs and Balbiani rings are indeed morphological manifestations of transcription of a genetic entity, the product of which is meaningful to the metabolism of the cell.

Some insight into the processes at the chromosomal level, e.g., local changes in chromosome structure and macromolecular composition, involved in the puffing process, has been gained by the application of particular agents or treatments which cause specific modifications in the chromosomal puffing pattern (induction of new puffs and/or regression of already present puffs). Although a wide variety of such agents has been described (for review, see Ashburner, 1970b; Berendes, 1972), the steroid hormone ecdysone and derivatives and treatments affecting the cellular respiratory metabolism have been most frequently applied to investigate puff formation under experimentally controlled conditions (Clever and Karlson, 1960; Ritossa, 1962, 1964; Berendes, 1967a, 1968; Ashburner, 1970a, 1971; Panitz *et al.*, 1972; Leenders and Berendes, 1972; Leenders et al., 1973).

B. Puff Cytochemistry

On the basis of their nucleic acid metabolism two categories of puffs have been distinguished. As already outlined above, most puffs and the Balbiani rings are characterized by intensive RNA synthesis, the local DNA content remaining proportional to the rest of the chromosome complement throughout the puffing process. Some loci in the polytene chromosomes of *Sciara* and *Rhynchosciara* species, however, display local DNA amplification accompanying the puffing process (Pavan and Breuer, 1955; Rudkin and Corlette, 1957; Crouse and Keyl, 1968; Gabrusewycz-Garcia, 1964, 1971). In addition to DNA amplification brought about by disproportionate DNA synthesis and leading to a strictly geometric increase in local DNA quantity, these puffs ("DNA-puffs") display intensive RNA synthesis.

Most puffs can be easily identified by performing staining reactions with metachromatic dyes discriminating between deoxyribo- and ribonucleic acids such as azure B (Swift, 1964) or toluidine blue (Pelling, 1964). Also the methyl green-pyronin method has been employed in the detection of RNA-containing regions (Breuer and Pavan, 1955; Kress, 1972). Application of these procedures reveals, in the studies reported so far, many more RNA-containing loci than were identified on the basis of morphological criteria attributed to the detection of puffs (Kress, 1972; see also Berendes, 1972).

RNA and DNA puffs can also be identified following staining procedures involving dye binding to proteins (Fig. 1). In particular, the acidic fast green and the light green-orange G methods have been frequently applied for this purpose (Swift, 1962; Clever, 1962; Berendes, 1968). These staining procedures reveal a pattern of staining regions along the chromosome complement which closely matches the pattern observed after staining for RNA according to one of the procedures mentioned above.

For certain experimentally induced puffs, a comparative study has been made on the staining reaction following the application of a variety of protein-staining procedures (Holt, 1970). The results of this study, summarized in Table I, suggest that the protein-staining reaction of the puffs is based on the presence of nonhistone protein(s) which, after the fixation and processing procedures applied in this study, fail to react to staining reactions specific for SH groups or free α-amino groups.

Microspectrophotometric analysis of the relative dye-binding capacity (naphthol yellow S) of two puffs of *Drosophila hydei* (2-48C and 4-81B) at successive stages of their development following a puff-inducing treatment revealed a significant increase in naphthol yellow S binding coinciding with puff formation (Holt, 1970) (Fig. 2). It was suggested that this increase is a consequence of a local increase in quantity of nonhistone protein(s). This suggestion agrees with results of previous studies revealing a higher UV absorbancy (at 231 nm) in puffed as compared to nonpuffed regions in *Drosophila melanogaster* chromosomes (Rudkin, 1962, 1964). On account of the densitometric studies on puffed regions following staining of the chromosomes for histones according to the method of Alfert and Geschwind (1953), it may be concluded that the histone to DNA ratio remains essentially constant during puff formation (Swift, 1964; Gorovski and Woodard, 1966, 1967). Moreover, extraction of histones with 0.2 N HCl (4 hours at 20°C) does not alter the staining properties of puffs when a "total protein" stain (e.g., fast green at low pH) is applied (Swift, 1964).

Further support for a local increase in protein quantity during puff formation came from interferometric studies (Holt, 1971). These studies indicated a gradual local increase in solid material during puff formation in slides which were extensively treated with ribonuclease to remove the RNA from the puffs (Fig. 2). In contrast to these findings are the

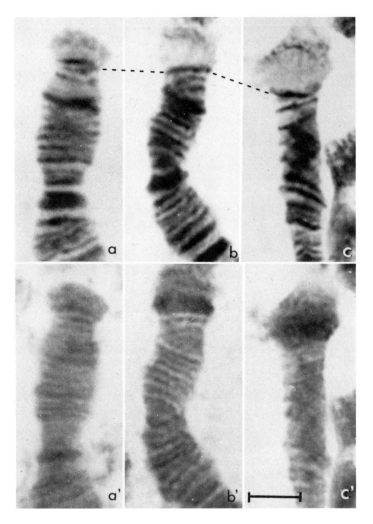

Fig. 1. Acidic fast green staining reaction of puff 2-48C in *Drosophila hydei* salivary gland chromosomes at subsequent stages after the onset of a puff-inducing treatment (temperature treatment). The DNA is stained by the Feulgen procedure. (a, a') nonpuffed control; (b, b') puff size at 6 minutes after onset of puff induction; (c, c') puff size at 60 minutes after onset of puff induction; (a), (b), and (c) photographed with a green filter (635 nm); (a'), (b'), and (c') photographed with a red filter (505 nm). The series (a'), (b'), (c') demonstrates the increase in fast green dye binding during puff formation. Bar = 5 μm.

data of Paul and Mateyko (1970) reporting similar refractive indices, 1.3598 and 1.3594, respectively, for puffs and interband regions in *Drosophila melanogaster* as determined by interference microscopy.

It seems well established now that the local increase in nonhistone

TABLE I
Staining Properties of Experimentally Induced Puffs[a] in *Drosophila hydei* Salivary Gland Chromosomes after Fixation of the Glands in 7.5% Neutral Formaldehyde

Staining procedure	Response of the puffs 2-48C, 2-36A, and 4-81[a]
a. Mercuric bromphenol blue on basic and acidic proteins	Intense blue puffs
b. Naphthol yellow S at pH 2.4 on free cationic groups of proteins	Yellow puffs
c. N-(1-naphthyl)ethylenediamine dihydrochloride on indole groups of tryptophan	Blue-purple puffs
d. Fast green at pH 2.4 on free cationic groups of proteins	Intense green puffs
e. As d following hot TCA extraction (5% TCA, 90°C, 20 minutes)	Intense green puffs
f. As d following RNase digestion (1 mg/ml, 37°C, 4 hours)	Intense green puffs
g. As d following pronase digestion (1 mg/ml, 37°C, 1 hour)	No staining reaction
h. Fast green at pH 8.0 after removal of nucleic acids (histone staining)	Very pale green puffs
i. Mercury orange on SH groups	No staining reaction
j. Ninhydrin-Schiff reagent on free α-amino groups	No staining reaction

[a] These puffs occur as a consequence of a temperature treatment in which larvae are transferred from 25° to 37°C and kept in a moist chamber at 37°C for 15 minutes.
[b] From Holt, 1970, and unpublished data.

protein during puff formation occurs independently of protein synthesis. Experimental puff induction in the presence of inhibitors of protein synthesis (e.g., puromycin or cycloheximide) at concentrations inhibiting over 95% of the cytoplasmic protein synthesis results in the appearance of the puffs specific to the treatment applied. These puffs display their characteristic protein-staining properties (Clever, 1964a, 1967; Clever and Romball, 1966; Ashburner, 1970a; Poels, 1972). Furthermore, it appears from autoradiographic studies on the incorporation of tritiated amino acids that neither puffs which arise during normal development nor experimentally induced puffs incorporate amino acids to a detectably higher degree than nonpuffed chromosome regions (Pettit and Rasch, 1966; Berendes, 1967b; Holt, 1970). Altogether, these data strongly suggest that during gene activation, expressed morphologically as puff formation, preexisting nonhistone protein(s) accumulates within the puff-forming chromosome locus.

Before discussing some aspects of chemical modifications of polytene chromosome protein components, the cytochemistry of puffed loci at the submicroscopic level may be briefly considered. Since the first description of the occurrence of discrete spherical and elongated particles

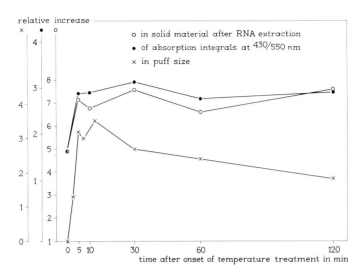

Fig. 2. Changes in dye-binding capacity and total solid material (after extraction of RNA) in relation to changes in puff size of region 2-48C. Puff size is represented as a ratio between the measured diameters of the puff and a particular nonpuffed band. Dye binding was determined microspectrophotometrically on chromosomes stained with Feulgen and naphthol yellow S. The ratio between the absorbancies at 430 and 550 nm of the puff region was compared with the ratio for a particular nonpuffed band. Total solid material, retained after RNase digestion, was measured under conditions of interference contrast at 546 nm. The values represented are ratios between the total solid material in the puff region and that in a nonpuffed reference area of the same chromosome. Because the relative quantities of DNA in the puff and the reference area in nonreplicating cells remain the same throughout the puffing process and the RNA has been digested, the relative increase in solid material as shown by the graph should be derived mainly from an increase in protein (see also Holt, 1971).

300 Å in diameter in a Balbiani ring of Chironomus (Beermann and Bahr, 1954), the occurrence of such particles, varying in size from 250 to 650 Å, has been reported for a large number of puffs and Balbiani rings in *Drosophila* and *Chironomus* (Swift, 1962, 1964, 1965; Sorsa, 1969; Berendes, 1969, 1972; Yamamoto, 1970; Leenders et al., 1973) (Fig. 3). In some puffs of *Drosophila* salivary gland chromosomes, particles of 180–230 Å may associate with a protein core to form large 0.1–0.3 μm ribonucleoprotein complexes (Swift, 1962, 1965; Derksen et al., 1973) (Fig. 3b). The 250–650 Å particles, which have been found in the nuclear sap as well as associated with the nuclear membrane (sometimes even in the perinuclear cytoplasm, are generally considered as the ultimate puff products by means of which the newly synthesized (puff-specific) RNA is transported to the cytoplasm. The ribonucleoprotein character

7. Nonhistone Proteins of Dipteran Polytene Nuclei

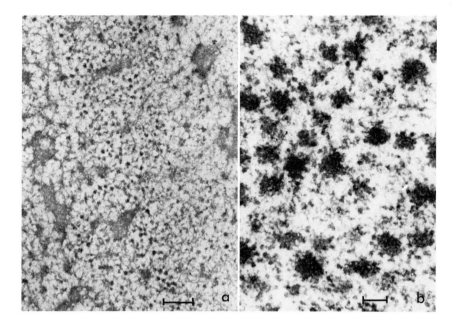

Fig. 3. Electron micrographs of two different puffs of *Drosophila hydei* salivary gland chromosomes showing different types of RNP particles. (a) Unknown puff region with 300–350 Å particles. (b) Puff region 2-48C displaying large 0.1–0.3 μm RNP complexes composed of a protein core with 180–230 Å particles attached to it. Bar = 0.2 μm.

of these particles has been demonstrated by a variety of techniques including alkaline hydrolysis by which the RNA is extracted from the particles (Derksen et al., 1973), a uranyl-EDTA–lead staining procedure indicating the presence of RNA in these structures (Bernard, 1969; Vazquez-Nin and Bernard, 1971) and enzyme digestion with RNase and proteolytic enzymes (Stevens and Swift, 1966).

III. CHEMICAL MODIFICATION OF POLYTENE CHROMOSOME PROTEINS

A. Acetylation

It has been proposed that enzymatic acetylation of ϵ-amino groups of the histones F_{2a1}, F_{2a2}, and F_3 (Vidali et al., 1968; Gershey et al., 1968; Pogo et al., 1968) and of σ-seryl groups of histone F_3 (Pogo et al., 1968) may be involved in the regulation of transcription. A number of correlations between histone acetylation and an increase in template activity in intact cells as well as in in vitro assays using fractionated chromatins provided support for this suggestion (for review, see Stein and Baserga, 1972; MacGillivray et al., 1972).

However, autoradiographic studies on the incorporation of ^{3}H-sodium acetate into giant chromosomes of Chironomus and Drosophila have failed to provide clear evidence for a correlation between local gene activity (puffing) and acetate incorporation (Clever, 1967; Ellgaard, 1967; Allfrey et al., 1968; Holt, 1970). This failure, however, could be merely due to the procedure used to prepare chromosome squashes. This procedure regularly involves the use of acetic acid which extracts arginine-rich histones (F_{2a1} and F_3). However, also after fixation of the tissue with formaldehyde, glutaraldehyde, or picric acid, procedures which largely prevent histone extraction, no specific labeling could be detected in puffs of Drosophila melanogaster and Chironomus thummi induced experimentally (by temperature treatment) in the presence of ^{3}H-acetate (Clever and Ellgaard, 1970; H. D. Berendes, unpublished). In order to investigate whether or not the enzymatic acetylation might have been impaired by the use of ^{3}H-acetate, the incorporation of acetyl groups derived from ^{3}H-acetyl coenzyme A into experimentally induced puffs in Drosophila hydei was studied. To assess exclusively for postsynthetic acetylation, isolated salivary glands, kept in a complex medium (Poels, 1972) supplied with 10 μg/ml cycloheximide and 0.5 μCi ^{3}H-acetyl coenzyme A, were submitted to a temperature treatment (10 minutes at 37°C) and subsequently fixed in neutral formaldehyde. In the autoradiographs of the squashed chromosomes none of the induced puffs displayed specific labeling (J. Limpens, L. Berlowitz, and H. D. Berendes, unpublished). It, therefore, seems that gene activation in insect chromosomes does not involve acetylation of histones or nonhistone proteins to an extent which can be detected by the procedures employed.

B. Phosphorylation

Postsynthetic phosphorylation of serine and threonine residues of histone as well as nonhistone proteins has also been suggested to bear a relationship to transcriptional regulation (Langan, 1968; Turkington and Riddle, 1969; Teng et al., 1971).

With regard to histone phosphorylation, it has been reported that this process can be enhanced by cAMP-stimulated protein kinase activity (Langan, 1968; Ruddon and Anderson, 1972). Although histone phosphorylation has so far not been studied in polytene chromosome systems, it could be assumed that local histone phosphorylation is involved in puff formation. A study on the effect of cAMP in combination with the steroid hormone ecdysone revealed that, though cAMP alone does not elicit puff formation at any particular locus, it does increase the size of the puffs induced by the hormone (Leenders et al., 1970).

Phosphorylation of nonhistone proteins in the polytene chromosomes of *Sciara* has been suggested on the basis of an autoradiographic study of [γ-^{32}P]ATP incorporation in chromosomes from which histones and nucleic acids were extracted. However, the incorporation was not found to be specifically associated with puff regions (Benjamin and Goodman, 1969).

C. Methylation and Ethylation

Recently, autoradiographic studies on the incorporation of ^{3}H-methyl-L-methionine, ^{14}C-(carboxyl)-methionine, and S-adenosyl-^{3}H-methyl methionine into the polytene chromosomes of *Sciara coprophila* revealed postsynthetic methylation of nonhistone proteins in these chromosomes. These data based upon the labeling pattern obtained after incubation of the salivary glands with ^{3}H-methyl methionine in the presence of either cycloheximide or streptovitacin and subsequent extraction of histones with 0.2 N HCl further revealed that the pattern of protein methylation was correlated with the pattern of DNA synthesis but not with the pattern of RNA synthesis along the chromosome (Goodman and Benjamin, 1972). The latter finding is in agreement with the observations of Holt (1970) on *Drosophila hydei* chromosomes in which puffs induced in the presence of ^{3}H-methyl-L-methionine did not display specific high labeling intensities as compared to the rest of the chromosome.

Autoradiographic studies on the incorporation of ^{3}H-ethyl ethionine, though resulting in diffuse labeling of the polytene chromosomes of *Drosophila hydei* (Holt, 1970) and *Sciara coprophila* (Goodman and Benjamin, 1972), did not provide any indication for a specific relationship of postsynthetic ethylation of nonhistone protein with genetic activity.

The data so far presented suggest that gene activation (the initiation of local transcription) in dipteran polytene chromosomes does not involve a local modification of a chromosomal protein component by either acetylation, phosphorylation, methylation, or ethylation.

IV. QUALITATIVE AND QUANTITATIVE CHANGES IN PROTEIN DURING GENE ACTIVATION

Studies on qualitative and quantitative changes in nuclear protein have been performed following the experimental induction of changes in the chromosomal puffing pattern by treatments affecting the cellular respiratory metabolism and by the hormone ecdysone. A variety of treatments interfering with cellular respiration (see Leenders and Berendes, 1972) causes the appearance of at least 4 new puffs (2-32A, 2-36A, 2-48C, and 4-81B) in the polytene chromosomes of Drosophila hydei. Two of these treatments, a temperature treatment described before and recovery from anaerobiosis, have been used in studies on nuclear proteins.

Ecdysone applied *in vivo* or *in vitro* induces a complex pattern of changes in the chromosomal puffing pattern, entirely different from that resulting from a temperature treatment. This pattern of changes includes the appearance of new puffs, an increase in size of existing puffs, and the disappearance of puffs, all within 60 minutes after the administration of the hormone (Berendes, 1967a). All newly appearing puffs as well as those increasing in size display a strong staining reaction with total protein-staining dyes. Moreover, interferometric studies on two regions developing a puffed appearance following a temperature treatment revealed a 150% (region 2-48C) and a 210% (region 4-81B) increase in solid material during puff formation. Since RNA has been digested and the DNA content does not increase disporportionally, the increase in solid material should be mainly due to a local accumulation of protein (Holt, 1971; Fig. 2).

In order to investigate whether experimentally induced changes in the chromosomal puffing pattern, involving local accumulation of nonhistone protein(s) at several loci, are accompanied by qualitative changes in the overall pattern of nuclear nonhistone proteins, salivary gland nuclei were isolated following an *in vivo* temperature treatment or an *in vitro* ecdysone treatment. The phenol–SDS soluble nuclear proteins were fractionated on a molecular weight basis by SDS–polyacrylamide electrophoresis. The electrophoretic pattern obtained from nuclei with either temperature-induced or ecdysone-induced puffs was compared with that obtained from nuclei of nontreated glands. Nuclei with temperature-induced puffs displayed, in addition to 10 major and 15 minor bands also present in the protein extracts of nontreated nuclei, the presence of a minor new band in the 23,000 MW region (Fig. 4). Also nuclei displaying ecdysone-induced puffs contained a minor new band. This band, however, was situated in the 42,000 MW region of the gels. These data not only indicated that a new protein fraction appears within the nucleus concurrent with the appearance of new puffs, they also suggest a specific relationship between the protein fraction and the group of chromosome loci which is activated. A relationship between the 23,000 MW protein fraction and the occurrence of the puffs 2-32A, 2-36A, 2-48C,

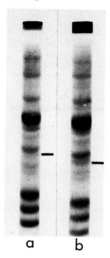

Fig. 4. Electrophoretic separation of nonhistone proteins extracted from isolated *Drosophila hydei* salivary gland nuclei in SDS–polyacrylamide gels. A comparison of nonhistone polypeptides from nuclei displaying a group of at least 4 puffs induced by a temperature treatment (b) with those from nuclei of control glands (a) revealed the presence of one extra polypeptide band in the pattern of the nuclei displaying the newly induced puffs (markers).

and 4-81B (temperature-induced) was further substantiated by the finding of the same protein fraction in salivary gland nuclei which were isolated from larvae which had just recovered from an anaerobic treatment causing the appearance of the same group of puffs as found after a temperature treatment. Furthermore, it was established that the 42,000 MW protein fraction was also present in salivary gland nuclei of larvae at a developmental stage shortly before puparium formation. These nuclei have developed ecdysone-specific puffs as a consequence of a natural increase in hormone titer in the hemolymph occurring 8–10 hours before puparium formation (Helmsing and Berendes, 1971).

Neither the 23,000 MW nor the 42,000 MW nonhistone protein fraction was restricted in its occurrence to the salivary gland nuclei. Both protein fractions also appeared in midgut nuclei following the appropriate puff-inducing treatment (Helmsing, 1972). In order to resolve the question of whether or not the particular nonhistone protein fractions result from a *de novo* synthesis following the puff-inducing treatment, an ecdysone treatment was carried out in the presence of tritiated amino acids and the radioactivity distribution in the gels was compared with the pattern of protein bands. No indication was obtained for a *de novo* synthesis of the 42,000 MW fraction occurring within the nucleus after a 45-minute ecdysone treatment (Helmsing and Berendes, 1971).

These data suggest that the development of a new group of puffs as a consequence of a particular natural or experimental stimulus probably involves the migration of a specific preexisting cytoplasmic protein fraction into the nucleus. As yet, however, it is not excluded that this new nuclear protein fraction results from cleavage of a preexisting nuclear protein, even though a comparison of the electrophoretic profiles of the nonhistone proteins from nuclei with and without specific puffs does not support this idea.

It is tempting to assume that the specific protein fraction is among the nonhistone protein(s) accumulating at the chromosome loci responding to the appropiate stimulus with puff formation. The question of whether or not such a relationship does exist is presently approached by extraction and characterization of nonhistone proteins from defined, experimentally induced puffs, isolated by microdissection according to the procedures described by Edström and co-workers. In this procedure nuclei and, subsequently, subnuclear components are dissected from salivary glands fixed with ethanol–acetic acid (3:1 v/v) (see Fig. 5). Because this fixative, known to extract some histones, could produce a loss of nonhistone proteins, the electrophoretic profiles of these proteins, extracted from equal aliquots of freshly prepared and ethanol–acetic acid-fixed salivary gland nuclei, were compared. This comparison revealed that qualitatively the pattern of nuclear nonhistone proteins separated by SDS–polyacrylamide electrophoresis is essentially identical in the two nuclear preparations. It was obvious, however, that the ethanol–acetic acid fixation did produce a significant quantitative reduction of several nonhistone protein fractions (Fig. 6). In spite of the latter disadvantage, which has as a consequence that more material has to be used, the method as such seems promising with regard to the possibility of identifying qualitative differences in nonhistone protein patterns from different chromosome puffs resulting either from a different or from the same stimulus.

Recently, it has been reported that in response to steroid hormone stimulation a specific nonhistone protein fraction, synthesized *de novo*, accumulates within the nuclei of target cells. The accumulation of a new 41,000 MW protein fraction in rat liver nuclei began approximately 2 hours after the administration of cortisol and reached its maximum about 6 hours later (Shelton and Allfrey, 1970). In uterus nuclei of ovariectomized rats treated with estradiol-17β, the synthesis of a new nuclear nonhistone protein reached a peak at 12 hours after the administration of the steroid (Teng and Hamilton, 1970). Also in *Drosophila hydei* salivary gland nuclei a newly synthesized nonhistone protein fraction, migrating into the 40,000–45,000 MW range, can be identified

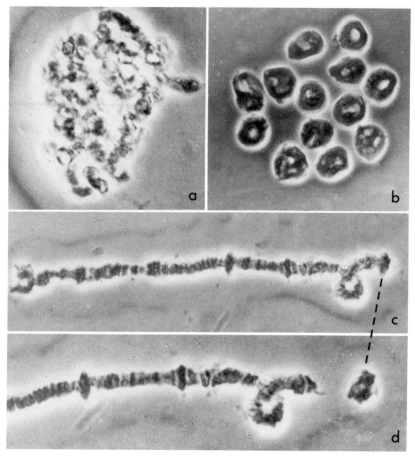

Fig. 5. Components of salivary gland nuclei from *Drosophila hydei* isolated by microdissection from ethanol–acetic acid-fixed glands. (a) Complete nucleus; (b) nucleoli; (c) chromosome 2 with puff 48C present; and (d) as (c) after dissection of puff 48C (courtesy Dr. T. Brady).

following coelectrophoresis of nuclear proteins extracted from glands incubated for 16 hours with ecdysone in the presence of ^{14}C-labeled amino acids and from glands incubated for the same period of time in medium containing ^{3}H-labeled amino acids but no steroid (P. J. Helmsing, unpublished).

So far, only speculations can be made as to the significance of these newly synthesized protein fractions for the nuclear metabolism. The newly synthesized protein fraction occurring within the *Drosophila* salivary gland nuclei after long-term ecdysone treatment could be a protein which is translated from a message transcribed in a chromosome locus

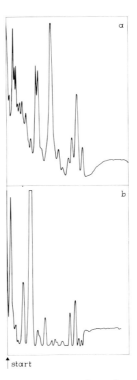

Fig. 6. Densitometric analyses of SDS–polyacrylamide gels following electrophoresis of total nuclear nonhistone protein (phenol–SDS-soluble fraction) extracted from (a) freshly prepared salivary gland nuclei and (b) ethanol–acetic acid (3:1) fixed salivary gland nuclei of *Drosophila hydei*.

activated immediately after the ecdysone administration ("early ecdysone puffs"). As has been suggested recently, such an "early" protein fraction could be responsible for the activation of a new series of chromosome loci, which in *Drosophila* become active at 4–6 hours after the onset of hormone treatment ("late ecdysone puffs"), as well as for the repression of the activity of the "early ecdysone puffs" (Ashburner, 1973; see also Clever, 1964b, 1966).

V. THE POSSIBLE ROLE OF PROTEINS IN GENE ACTIVATION

The data presented above are compatible with the idea that the activation of various genome loci, all reacting to the same stimulus, is accompanied by the appearance in the nucleus of only one particular nonhistone protein which is specifically related to the particular genome response observed. However, because the identification of the specific protein fractions has, so far, been based exclusively on their migration in SDS–polyacrylamide gels which separate the proteins according to their molecular weights, it is feasible that the specific protein fractions which were identified consist of more than one protein species. These protein species could differ from one another in their amino acid sequences as well as in their amino acid composition. Furthermore, the detection of any particular protein species by the procedure applied, depends largely on its quantity in the nuclear extract.

In speculating about the role of nonhistone proteins in the process of puffing, Swift (1964) suggested three possible functions. The nonhistone protein could be, in the case of ecdysone-induced gene activity, a protein responsible for the transfer of steroid molecules from the cytoplasm into the nucleus. Following its entry in the nucleus the protein could initiate transcription at specific genome sites responsive to the hormone. It is not unlikely that a similar mechanism could be involved in the induction of specific puffs by treatments affecting the cellular respiratory metabolism (e.g., temperature treatment, recovery from anaerobiosis). Following these treatments (see also Leenders and Berendes, 1972), a particular intracellular stimulus may be released from the mitochondria, bound to a specific nonhistone protein, and transported into the nucleus.

As a second possibility, Swift suggested that the nonhistone protein could be an enzyme involved in transcription (one of the RNA polymerases) or in the processing of newly synthesized RNA. The third suggestion implied a role of the nonhistone protein in the transfer of the newly synthesized RNA from its site of synthesis to the cytoplasm.

The first suggestion finds support from autoradiographic studies indicating a transfer of ^{3}H-labeled ecdysone from cytoplasm to the nucleus in *Drosophila* salivary gland cells (Emmerich, 1969; Claycomb et al., 1971). Moreover, it was reported recently that specific ecdysone-binding proteins are present in the cytoplasm, the nucleus, and in chromatin of *Drosophila hydei* salivary gland cells (Emmerich, 1972). These findings and the results of microinjection experiments which revealed that the genome response following injection of ecdysone directly into the cell nucleus was significantly smaller than after injection of the same hormone concentration into the cytoplasm (Berendes, 1973) indicate that the cytoplasmic ecdysone-binding protein may be a requirement for the transport of the hormone into the nucleus. However, the hormone–protein complexes isolated from nuclei and cytoplasm differ with respect to their sedimentation properties (Emmerich, 1972). Thus, it remains to be demonstrated that the nuclear ecdysone-binding protein is actually a modified cytoplasmic ecdysone-binding protein, rather than a specific protein species which is bound to the nucleus. If it is assumed that the nuclear steroid-binding protein is derived from the cytoplasm, it could well be identical with the protein fraction migrating into the nucleus as a consequence of a hormone treatment. This protein or the protein–steroid complex could then be involved in the activation of the hormone specific genome sites. In that case, it could be suggested that the various genome loci responding to the hormone treatment possess identical recognition sites for the protein or steroid–protein complex, a suggestion which is compatible with the theoretical model for the regulation of genome activity in eukaryotes as proposed by Britten and Davidson (1969). Such a mechanism would require only *one* specific protein to elicit the adequate genome response.

With regard to the response of the *Drosophila* genome to various treatments affecting the cellular respiratory metabolism, a similar mechanism could be assumed. However, some of the loci responding in common to a temperature treatment can be activated selectively or in different combinations (Leenders *et al.*, 1973). These loci, therefore, should possess additional recognition sites, different from the one serving a simultaneous action of the group of genome loci responding to a temperature treatment.

As to the second suggestion, proposing an enzyme function for the protein accumulating at a chromosome locus undergoing puff formation, it could be supposed that at least part of the accumulating protein is RNA polymerase or another enzyme involved in the processing of RNA. Neither the staining reactions nor the interferometric measurements discriminate between enzyme molecules and nonhistone proteins with other functions. On the other hand, it is very unlikely that different stimuli which provoke essentially the same phenomenon, gene activation, should induce the migration of different enzyme entities into the nucleus.

As to the third possible function proposed by Swift, it seems likely that a major fraction of the protein(s) accumulating at the puff-forming chromosome loci is indeed involved in the transport of the newly synthesized RNA from the puff to the cytoplasm. The protein-staining reactions at subsequent stages of puff development revealed the presence of a thin protein-staining band in presumptive puff regions and a successive increase in width of this band when the puff grows in size (Fig. 1). The increase in nonhistone protein quantity as well as the expansion of its distribution during the development of a puff are correlated with and possibly reflect the distribution of RNP–complexes within the puff (Berendes, 1972).

Whereas some chromosome regions activated by the same stimulus revealed the presence of morphologically different RNP complexes, some puff regions activated by different stimuli display morphologically similar RNP particles. Because, so far, no information about the protein composition of RNP complexes from polytene nuclei has been presented, it remains to be elucidated whether or not morphologically different RNP particles have essentially the same polypeptide composition. Studies on the nuclear RNP complexes from a variety of vertebrate cell types have indicated a certain universality of the protein component which consists of only a few subunit polypeptides (Georgiev and Samarina, 1971; Martin and Swift, 1973). In this context, it seems extremely unlikely that the specific 42,000 MW polypeptide fraction appearing in the nucleus as a consequence of a hormone treatment and the 23,000 MW polypeptide found in the nucleus after a temperature treatment are polypeptides involved in the transport of newly synthesized RNA.

In summarizing the experimental data presented, it may be concluded that the activation of a particular set of genome sites by an endogenous or exogenous stimulus probably involves the migration of a specific nonhistone protein into the nucleus. This protein may be required for transmission of the stimulus from the cytoplasm to the nucleus and may cause the first detectable change in protein concentration at the genome site to be translated. It is not likely that this nonhistone protein has an enzyme function related with transcription nor that it is involved in the transfer of newly synthesized RNA from the activated genome loci.

The local accumulation of nonhistone protein in the activated chromosome regions displaying puff formation may reflect a local increase in the concentration of RNA polymerase and other enzymes involved in the RNA metabolism, in addition to an accumulation of polypeptides involved in the transport of newly synthesized RNA. The latter polypeptides, being present as free molecules or associated with RNA, should be the major protein components in the chromosome puffs.

The isolation and characterization of proteins extracted from isolated puffs and the analysis of particular RNP complexes from polytene nuclei will, in the near future, certainly provide new results and raise new questions relevant to the function of nuclear nonhistone proteins.

REFERENCES

Alfert, M., and Geschwind, I. I. (1953). *Proc. Nat. Acad. Sci. U.S.* 39, 991–999.
Allfrey, V. G., Pogo, B. G. T., Littau, V. C., Gershey, E. L., and Mirsky, A. E. (1968). *Science* 159, 314–316
Ashburner, M. (1970a). *Chromosoma* 31, 356–376.

Ashburner, M. (1970b). *Advan. Insect Physiol.* **7**, 1–95.
Ashburner, M. (1971). *Nature (London) New Biol.* **230**, 222–223.
Ashburner, M. (1972). In "Developmental Studies on Giant Chromosomes" (W. Beermann, ed.), Vol. 4, pp. 101–151. Springer-Verlag, Berlin and New York.
Ashburner, M. (1974). *Cold Spring Harbor Symp. Quant. Biol.* **38**, (in press).
Barrett, T., and Gould, H. J. (1973). *Biochim. Biophys. Acta* **294**, 165–170.
Baserga, R., and Stein, G. (1971). *Fed. Proc., Fed. Amer. Soc. Exp. Biol.* **30**, 1752–1759.
Baudisch, W., and Panitz, R. (1968). *Exp. Cell Res.* **49**, 470–476.
Beermann, W. (1952). *Chromosoma* **5**, 139–198.
Beermann, W. (1961). *Chromosoma* **12**, 1–25.
Beermann, W., and Bahr, G. F. (1954). *Exp. Cell Res.* **6**, 195–201.
Benjamin, W. B., and Goodman, R. H. (1969). *Science* **166**, 629–630.
Berendes, H. D. (1965). *Chromosoma* **17**, 35–77.
Berendes, H. D. (1967a). *Chromosoma* **22**, 274–293.
Berendes, H. D. (1967b). *Drosophila Inform. Serv.* **42**, 102.
Berendes, H. D. (1968). *Chromosoma* **24**, 418–437.
Berendes, H. D. (1969). *Ann. Embryol. Morphog.* **1**, Suppl., 153–164.
Berendes, H. D. (1972). In "Developmental Studies on Giant Chromosomes" (W. Beermann, ed.), Vol. 4, pp. 181–207. Springer-Verlag, Berlin and New York.
Berendes, H. D. (1973). *Proc. Int. Congr. Endocrinol., 4th, 1972* **273**, 311–314.
Berendes, H. D., and Beermann, W. (1969). In "Handbook of Molecular Cytology" (A. Lima-de-Faria, ed.), pp. 501–519. North-Holland Publ., Amsterdam.
Berendes, H. D., Alonso, C., Helmsing, P. J., Leenders, H. J., and Derksen, J. (1973). *Cold Spring Harbor Symp. Quant. Biol.* **38** (in press).
Bernard, W. (1969). *J. Ultrastruct. Res.* **27**, 250–265.
Breuer, M. E., and Pavan, C. (1955). *Chromosoma* **7**, 371–386.
Britten, R. J., and Davidson, E. H. (1969). *Science* **165**, 349–357.
Claycomb, W. C., LaFond, R. E., and Villee, C. A. (1971). *Nature (London)* **234**, 302–304.
Clever, U. (1962). *J. Insect Physiol.* **8**, 357–376.
Clever, U. (1963). In "Funktionelle und morphologische Organisation der Zelle," pp. 30–39. Springer-Verlag, Berlin and New York.
Clever, U. (1964a). In "The Nucleohistones" (J. Bonner and P. O. P. Ts'o, eds.), pp. 317–334. Holden-Day, San Francisco, California.
Clever, U. (1964b). *Science* **146**, 794–795.
Clever, U. (1966). *Develop. Biol.* **14**, 421–438.
Clever, U. (1967). In "The Control of Nuclear Activity" (L. Goldstein, ed.), pp. 161–186. Prentice-Hall, Englewood Cliffs, New Jersey.
Clever, U., and Ellgaard, E. G. (1970). *Science* **169**, 373–374.
Clever, U., and Karlson, P. (1960). *Exp. Cell Res.* **20**, 623–627.
Clever, U., and Romball, C. G. (1966). *Proc. Nat. Acad. Sci. U.S.* **56**, 1470–1476.
Crouse, H. V., and Keyl, H. G. (1968). *Chromosoma* **25**, 357–364.
Daneholt, B., and Hosick, H. (1973). *Proc. Nat. Acad. Sci. U.S.* **70**, 422–446.
Daneholt, B., Edström, J.-E., Egyházi, E., Lambert, B., and Ringborg, U. (1969a). *Chromosoma* **28**, 399–417.
Daneholt, B., Edström, J.-E., Egyházi, E., Lambert, B., and Ringborg, U. (1969b). *Chromosoma* **28**, 418–429.
Daneholt, B., Edström, J.-E., Egyházi, E., Lambert, B., and Ringborg, U. (1970). *Cold Spring Harbor Symp. Quant. Biol.* **35**, 513–519.

Derksen, J., Berendes. H. D., and Willart, E. (1973). *J. Cell Biol.* **59**, 661–668.
Edström, J.-E., and Daneholt, B. (1967). *J. Mol. Biol.* **28**, 331–343.
Elgin, S. C. R., and Bonner, J. (1970). *Biochemistry* **9**, 4440–4447.
Elgin, S. C. R., and Bonner, J. (1972). *Biochemistry* **11**, 772–781.
Ellgaard, E. G. (1967). *Science* **157**, 1070–1072.
Emmerich, H. (1969). *Exp. Cell Res.* **58**, 261–270.
Emmerich, H. (1972). *J. Gen. Comp. Endocrinol.* **19**, 543–551.
Gabrusewycz-Garcia, N. (1964). *Chromosoma* **15**, 312–344.
Gabrusewycz-Garcia, N. (1971). *Chromosoma* **33**, 421–435.
Georgiev, G. P., and Samarina, O. P. (1971). *Advan. Cell Biol.* **2**, 47–110.
Gershey, E. L., Vidali, G., and Allfrey, V. G. (1968). *J. Biol. Chem.* **243**, 5018–5022.
Goodman, R. M., and Benjamin, W. B. (1972). *Exp. Cell Res.* **77**, 63–72.
Gorovski, M. A., and Woodard, J. (1966). *J. Cell Biol.* **31**, 41A.
Gorovski, M. A., and Woodard, J. (1967). *J. Cell Biol.* **33**, 723–728.
Grossbach, U. (1969). *Chromosoma* **28**, 136–187.
Helmsing, P. J. (1972). *Cell Differentiation* **1**, 19–24.
Helmsing, P. J., and Berendes, H. D. (1971). *J. Cell Biol.* **50**, 893–896.
Helmsing, P. J., and van Eupen, O. (1973). *Biochim. Biophys. Acta* **308**, 154–160.
Holt, T. K. H. (1970). *Chromosoma* **32**, 64–78.
Holt, T. K. H. (1971). *Chromosoma* **32**, 428–435.
Kress, H. (1972). *Sitzungs ber. Bayer. Akad. Wiss., Math.-Naturwiss, Kl.* pp. 129–149.
Kroeger, H., and Lezzi, M. (1966). *Annu. Rev. Entomol.* **11**, 1–22.
Lambert, B. (1972). *J. Mol. Biol.* **72**, 65–75.
Lambert, B., Wieslander, L., Daneholt, B., Egyházi, E., and Ringborg, U. (1972). *J. Cell Biol.* **53**, 407–418.
Lambert, B., Daneholt, B., Edström J.-E., Egyházi, E., and Ringborg, U. (1973). *Exp. Cell Res.* **67**, 381–389.
Langan, T. A. (1968). *Science* **162**, 579–580.
Leenders, H. J., and Beckers, P. J. A. (1972). *J. Cell Biol.* **55**, 257–265.
Leenders, H. J., and Berendes, H. D. (1972). *Chromosoma* **37**, 433–444.
Leenders, H. J., Wullems, G. J., and Berendes, H. D. (1970). *Exp. Cell Res.* **63**, 159–164.
Leenders, H. J., Derksen, J., Maas, P. M. J. M., and Berendes, H. D. (1973). *Chromosoma* **41**, 447–460.
MacGillivray, A. J., Carroll, D., and Paul, J. (1971). *FEBS Lett.* **13**, 204–208.
MacGillivray, A. J., Paul, J., and Threlfall, G. (1972). *Advan. Cancer Res.* **15**, 93–162.
Martin, T. E., and Swift, H. (1974). *Cold Spring Harbor Symp. Quant. Biol.* **38** (in press).
Panitz, R. (1972). In "Developmental Studies on Giant Chromosomes" (W. Beermann, ed.), Vol. 4, pp. 209–227. Springer-Verlag, Berlin and New York.
Panitz, R., Wobus, U., and Serfling, E. (1972). *Exp. Cell Res.* **70**, 154–160.
Paul, J. S., and Mateyko, G. M. (1970). *Exp. Cell Res.* **59**, 227–236.
Pavan, C., and Breuer, M. E. (1955). In "Symposium on Cell Secretion" (G. Schreiber, ed.), pp. 90–99. Belo Horizonte, Brazil.
Pelling, C. (1959). *Nature (London)* **184**, 655–656.
Pelling, C. (1964). *Chromosoma* **15**, 71–122.
Pelling, C. (1966). *Proc. Roy. Soc., Ser. B* **164**, 279–289.
Pelling, C. (1970). *Cold Spring Harbor Symp. Quant. Biol.* **35**, 521–531.
Pettit, B. J., and Rasch, R. W. (1966). *J. Cell. Comp. Physiol.* **68**, 325–334.

Poels, C. L. M. (1972). *Cell Differentiation* **1**, 63–78.
Pogo, B. G. T., Pogo, A. O., Allfrey, V. G., and Mirsky, A. E. (1968). *Proc. Nat. Acad. Sci. U.S.* **59**, 1337–1344.
Ritossa, F. M. (1962). *Experientia* **18**, 571–572.
Ritossa, F. M. (1964). *Exp. Cell Res.* **35**, 601–607.
Ruddon, R. W., and Anderson, S. L. (1972). *Biochem. Biophys. Res. Commun.* **46**, 1499–1508.
Rudkin, G. T. (1962). *Ann. Histochim., Suppl.* **2**, 77–84.
Rudkin, G. T. (1964). *In* "The Nucleohistones" (J. Bonner and P. O. P. Ts'o, eds.), pp. 184–192. Holden-Day, San Francisco, California.
Rudkin, G. T., and Corlette, S. L. (1957). *Proc. Nat. Acad. Sci. U.S.* **43**, 964–968.
Shaw, L. M. J., and Huang, R. C. (1970). *Biochemistry* **9**, 4530–4542.
Shelton, K. R., and Allfrey, V. G. (1970). *Nature (London)* **228**, 132–134.
Sorsa, M. (1969). *Ann. Acad. Sci. Fenn., Ser. A* **150**, 1–21.
Stein, G., and Baserga, R. (1972). *Advan. Cancer Res.* **15**, 287–330.
Stellwagen, R. H., and Cole, R. D. (1969). *Annu. Rev. Biochem.* **38**, 951–990.
Stevens, B. J., and Swift, H. (1966). *J. Cell Biol.* **31**, 55–77.
Swift, H. (1962). *In* "The Molecular Control of Cellular Activity" (J. M. Allen, ed.), pp. 73–125. McGraw-Hill, New York.
Swift, H. (1964). *In* "The Nucleohistones" (J. Bonner and P. O. P. Ts'o, eds.), pp. 169–183. Holden-Day, San Francisco, California.
Swift, H. (1965). *In Vitro* **1**, 26–49.
Teng, C. S., and Hamilton, T. H. (1970). *Biochem. Biophys. Res. Commun.* **40**, 1231–1238.
Teng, C. S., Teng, C. T., and Allfrey, V. G. (1971). *J. Biol. Chem.* **246**, 3597–3609.
Turkington, R. W., and Riddle, M. (1969). *J. Biol. Chem.* **244**, 6040–6046.
Vazquez-Nin, G., and Bernard, W. (1971). *J. Ultrastruct. Res.* **36**, 842–860.
Vidali, G., Gershey, E. L., and Allfrey, V. G. (1968). *J. Biol. Chem.* **243**, 6361–6366.
Yamamoto, H. (1970). *Chromosoma* **32**, 171–190.

8

Acidic Nuclear Proteins and the Cell Cycle

JAMES R. JETER, JR. AND IVAN L. CAMERON

I. Introduction 213
II. Synchronous Cell Cycle Systems, a Comparative Analysis 215
III. Acidic Proteins of the Nucleus in Relation to the Cell Cycle 221
    A. General Synthesis of Acidic Nuclear Proteins 221
    B. Acidic Nuclear Proteins in Cells Stimulated to
       Proliferate 222
    C. Acidic Nuclear Proteins in More Highly Synchronized
       Cell Populations 227
    D. Phosphorylation of Acidic Nuclear Proteins 234
    E. Electron Microscopic Evidence for Gene Activation 235
IV. Summary and Conclusions 240
    References 242

I. INTRODUCTION

The cell cycle of a growing eukaryotic cell is thought of as an orderly progression of changing biosynthetic and morphological events. These events eventually culminate in cell division and the production of daughter cells in which the same cycle recurs. The systematic fluctuations in cellular activity occurring throughout the cell cycle are ultimately dependent on variations in transcriptional or gene activity which may occur as a response to changes in the cellular environment either at the supra- or subcellular level (Mueller, 1969; Mitchison, 1971; Stein and Baserga, 1972). Since the early 1940's when the Steadmans first suggested it, evidence has accumulated in support of the idea that two groups of chromosomal proteins, the histones and the acidic nuclear proteins, are part of the regulatory mechanism involved in the control of differential gene expression (Huang and Bonner, 1962; Allfrey et al., 1963; Hnilica, 1967; Wang, 1967; MacGillivray et al., 1972). The histones, a group of chromosomal basic proteins, have been intensively investigated in order to define their role as gene regulators. Accumulated evidence indicates that their role in gene regulation is a secondary one (Phillips, 1971; McClure and Hnilica, 1972; MacGillivray et al., 1972; Stein and Baserga, 1972).

Recently attention has turned toward the second group of chromosomal proteins, the acidic nuclear proteins. The preliminary studies of these proteins indicate that they may be the molecules primarily responsible for conferring specificity to the control of transcriptional activity. Unlike the histones the acidic nuclear proteins exhibit a specificity of distribution in regions of the chromatin and among various tissues and cell types such that the amounts of acidic protein found in chromatin varies not only between eu- and heterochromatin but with the tissue source. These variations may be directly correlated with the level of RNA synthesis found in that tissue (Frenster, 1965; Dingman and Sporn, 1964; Marushige and Ozaki, 1967; Marushige and Dixon, 1969; Marushige and Bonner, 1971). In addition these proteins have been shown to have a degree of tissue and species specificity not seen in the histones (Wang, 1967; Loeb and Cruzet, 1969; Chytyl and Spelsberg, 1971; LeStourgeon and Rusch, 1973). Though they exhibit a degree of tissue and species specificity, several laboratories have reported that a major fraction of the nuclear acidic proteins appear to be homologous between tissues and species (Benjamin and Gellhorn, 1968; Elgin and Bonner, 1970, 1972; MacGillivray et al., 1972). This suggests that many of the acidic nuclear proteins are probably proteins concerned with the basic metabolic and transcriptional activities common to all chromatins while the unique proteins are probably involved in cell-specific gene activity.

Further evidence implicating these proteins as having a major role in the control of transcriptional activity comes from experiments showing that they stimulate *in vitro* RNA synthesis (Wang, 1968; Gilmour and Paul, 1969; Kamiyama and Wang, 1971; Teng et al., 1971; Kostraba and Wang, 1972) and that *de novo* synthesis and/or accumulation of specific nuclear acidic proteins occurs in response to hormonal and other stimuli (Teng and Hamilton, 1969, 1970; Shelton and Allfrey, 1970; Enea and Allfrey, 1973; Helmsing and Berendes, 1971; Helmsing, 1972; LeStourgeon and Rusch, 1973; Nations et al., 1973).

The nuclear acidic proteins also demonstrate increased rates of synthesis in normally quiescent cells which are stimulated to proliferate (Stein and Baserga, 1970b; Rovera and Baserga, 1971) and are actively turning over throughout the cell cycle (Borun and Stein, 1972; Jeter, 1973). Many of the nuclear acidic proteins are phosphoproteins which show increased rates of phosphorylation at times of gene activation (Kleinsmith *et al.*, 1966; Teng *et al.*, 1970, 1971).

In view of the multiple facets of acidic nuclear proteins which implicate them as playing a major role in the regulation of gene expression, studies on the relationship of these proteins to the varying transcriptional activity occuring throughout the cell cycle (Mueller, 1969; Cameron *et al.*, 1971a; Rudick and Cameron, 1972; Stein and Matthews, 1973) may offer a particularly good approach for defining their biological activity.

Studies on the differential patterns of growth and synthesis throughout the cell cycle are generally approached in either of two ways. The temporal sequence of these cellular events may be measured in single cells or in large quantities of synchronized cells. Since single cells do not provide enough material for bulk biochemical analyses, experiments on the relationship of the acidic nuclear proteins to gene activity must, of necessity, employ masses of synchronized cells. Though synchrony allows us to examine large quantities of cells that are in the same phase of the cell cycle it, too, has inherent problems which may differ from one technique to another. In this chapter we will undertake two things; first, to describe and compare the advantages and disadvantages of various methods of synchronizing cells, both *in vivo* and *in vitro*, and second, to examine the relationship of the acidic nuclear proteins to differential gene activity.

II. SYNCHRONOUS CELL CYCLE SYSTEMS, A COMPARATIVE ANALYSIS

If one is studying changes in the cell cycle it is important to know the degree of synchrony of the system, as well as the assumptions and the perturbations which are involved in the synchrony procedure. For instance, if one is attributing importance to a small change which occurs at one point in the cell cycle it is important to know that the cells which were used in the analysis were all from the same part of the cell cycle and were not contaminated with cells from another part of the cell cycle. Biochemical analysis of events which take place during the cell cycle require masses of synchronized cells where the cycles of individual cells are in unison. Thus synchronous cultures of cells offers an opportunity to amplify the biochemical events which take place in individual cells, to the point where standard biochemical analysis can be used.

Synchronous cell cycle systems can be divided into three broad categories as summarized in Table I. *Induction synchrony* includes those systems in which a treatment is applied to an asynchronous culture

TABLE I
Cell Cycle Synchrony[a]

I. Induction synchrony
   A. Inhibitors of DNA synthesis and then release
      1. Excess thymidine and release (often used as a double block)
      2. Fluorodeoxyuridine (FUdR)
      3. Amethopterin (methotrexate) folic acid antagonist
      4. Others, hydroxyurea, deoxyadenosine
   B. Metaphase Inhibitors
      1. Colchicine or colcemid
      2. Vinblastine
   C. Selective kill of cells in S phase
      1. Hydroxyurea (in some cells)
      2. ^{3}H-thymidine
   D. Starvation–refeeding
      1. General
      2. Key component
   E. Single change in environment
      1. Light
      2. Temperature
      3. Other, pressure, radiation, etc.
   F. Multiple changes of environment
      1. Light
      2. Temperature
      3. Others
   G. Stimulated systems in mammals: proliferative responses to a treatment, hormone, or drug, i.e.,
      Estradiol—uterus
      Isoproterenol—parotid gland
      Poietins—blood cells
      Antigen or PHA—peripheral lymphocytes
      Partial hepatectomy—liver

II. Selection synchrony
   A. Wash off or shake off, swirl petri dish in low Ca^{2+} medium, collect dividers in vortex
   B. Sedimentation separation, differential or gradient sedimentation in sucrose, Ficoll or a protein gradient
   C. Filtration
   D. Membrane elution (grow off or baby factory)
   E. Mechanical electronic or sorting

III. Natural synchrony
   A. Early embryos of various animals (sea urchins) and plants (endosperm)
   B. Lily anthers in genus *Trillium* and *Lilium*
   C. *Physarum polycephalum* acellular slime mould

[a] Modified after James (1966); for specific references, see Zeuthen (1964), Cameron and Padilla (1966), Helmstetter (1969), Mitchison (1971), Nias and Fox (1971), and Stein and Baserga (1972).

8. Acidic Nuclear Proteins and the Cell Cycle

which forces the cells into synchrony. This category also includes cell systems which can be stimulated to enter the cell cycle from a nonproliferative state. *Selection synchrony* includes those systems where cells at one phase of the cell cycle are physically separated from an asynchronous culture. The selected cells are often subcultured as a separate synchronous culture. There are also *natural synchrony* systems.

In general, induction synchrony systems give higher yields of cells than the selection synchrony methods. On the other hand, the selection systems generally give less distortion of normal cell cycle events and for this reason selection synchrony is becoming more popular.

Yield, convenience, and degree of synchrony are all factors which must be considered in the evaluation or the selection of a synchronous system. Reviews, specific references, and discussions of cell cycle synchrony are found in Zeuthen (1964), Cameron and Padilla (1966), Helmstetter (1969), and Nias and Fox (1971).

Inhibitors of DNA synthesis stop cells in DNA synthesis and collect cells at the beginning of DNA synthesis (S phase). After the cells have all collected in or at the beginning of S phase the inhibitor is washed from the cells which should all be in S phase. This system works for excess thymidine, fluorodeoxyuridine, amethopterin, hydroxyurea, and excess deoxyadenosine. Problems which can be encountered include the possibility of an incomplete inhibition of DNA synthesis or that the inhibitor is causing side effects other than are expected. For example, because amethopterin is a folic acid antagonist it can interfere with a number of different biosynthetic pathways and may actually interfere with the cell's progress toward the S phase. Hydroxyurea is thought not only to interfere with cellular DNA synthesis but also to cause lesions to the cellular DNA (see Cameron and Jeter, 1973). And finally it should be realized that such DNA inhibitors usually allow cellular RNA, protein synthesis, and cell growth to continue producing an unbalanced growth of the inhibited cell.

Colchicine, Colcemid, and vinblastine are mitotic inhibitors. When placed in a culture of mitotically dividing mammalian cells, the cells accumulate in metaphase. If the inhibitor is not removed after 3 hours or so the first metaphase collected cells will become necrotic and eventually die. Because the generation time of mammalian cells is relatively long in relation to the time the inhibitor can be applied, the removal of the inhibitor produces only partial synchrony. In addition, the effects of colchicine are not entirely specific to cells in mitosis (see Kuzmich and Zimmerman, 1972).

Selective killing of cells in S phase can be accomplished by adding high specific activity tritiated ^{3}H-thymidine to a culture of cells. Cells in S phase incorporate the isotope into nuclear DNA in such high amounts as to cause radiation damage and death of the cell. The excess ^{3}H-thymidine is then washed from the cells and only unlabeled cells survive to grow as a synchronous culture. Hydroxyurea selectively kills some mammalian cells (Chinese hamster cells) in S phase and can be used in the same manner as ^{3}H-thymidine for production of synchrony in these cells, on the other hand some types of mammalian cells (HeLa cells) do not show such an S phase sensitivity to the action of hydroxyurea (see Nias and Fox, 1971; Cameron and Jeter, 1973).

Starvation and refeeding synchrony methods have found wide use with many workers using many different cell types. The starvation refeeding technique has been used to synchronize mammalian cells, fission and budding yeast, diatoms, bacteria, and the ciliated protozoan *Tetrahymena*. As indicated in Table I, the starvation may be general as occurs on removal of all nutrients or it may be caused by selective removal of any one or combination of several essential nutrients. For example, medium depleted of glutamic acid and isoleucine has proved to be very useful in bringing Chinese hamster cells into the pre-DNA synthetic or G_1 phase. Replacement of these specific amino acids then results in synchrony. In some bacterial and mammalian cell cultures, addition of fresh medium to a densely packed cell culture will lead to a degree of synchrony in the first cell division. This type of synchrony procedure usually creates both poor division synchrony and a sequence of cellular events quite different from the normal cell cycle.

Single or multiple changes in the environment were some of the classic approaches used to obtain synchronized cells. Light and dark cycles have been used to synchronize photosynthetic algae such as *Chlorella*, *Chlamydomonas*, and *Euglena*, whereas temperature cycles have been used to synchronize various protistans such as amoeba, the flagellate *Astasia longa* budding yeast, and the ciliate *Tetrahymena*.

Each cell type and synchrony procedure must be evaluated in order to make proper interpretations of the results. For example, a series of heat shocks followed by a rest period will produce excellent division synchrony in *Tetrahymena* but the treatment uncouples the nuclear DNA duplication cycle from the division cycle (see Jeffery et al., 1973). Those synchrony systems which use one change per cycle do not seem to cause uncoupling of these cell cycle events, however they still perturb the events away from that of a normal cell cycle. Actually the ability to use several different synchrony systems on the same cell can reveal important mechanisms which control gene expression and the cell cycle.

There are a number of metazoan cell systems which can be stimulated or induced by one means or another to enter into a sequence of events leading to cell division. Some examples of these systems are listed in Table I. Several factors should be kept in mind concerning these systems:

1. They usually consist of a heterogenous population of cell types; for example about half the cells in a regenerating liver are hepatocytes the remainder are divided between several other cell types.

2. Only a fraction of the cell population is stimulated to proliferate and therefore the system gives only a low degree of synchrony.

3. If the stimulus takes place *in vivo* it is usually subjected to a number of extraneous and different compensating control systems operating in the intact organism. Even though these stimulated cell systems are subject to criticism they have produced useful literature on the relationship between gene regulation and the cell cycle (see Cameron *et al.*, 1971b; Balls and Billett, 1973).

Turning to selection synchrony methods, the wash-off or shake-off method gives good synchrony for those mammalian cells which can be cultured on a solid surface. During mitosis mammalian cells round up and loosen their attachment to the surface. If an asynchronous culture of mammalian cells which is grown in a petri dish is swirled, the mitotic cells will collect in the vortex at the center of the dish. The cells can then be pipetted out of the culture and made to grow as a synchronous subculture. Low calcium in the culture medium helps loosen even more the attachment of the mitotic cells to the surface. Although the yield of cells is often low the synchrony is excellent and the cells have been subjected to a minimum of perturbation. Sometimes the yield is increased by collecting mitotic cells by use of a mitotic inhibitor or by prior synchronization by use of a DNA-inhibiting agent. Unfortunately these enhancers of the yield are each subject to perturbation criticisms as mentioned above.

Sedimentation separation of an asynchronous culture of different sized cells has been successfully used for fission and budding yeast, bacteria, mammalian cells, and *Chlorella*. Because small cells at the beginning of the cell cycle have lower sedimentation properties, they can be separated from the larger cells which are later in the cell cycle and the small cells can then be subcultured in synchrony. This system works especially well for *Chlorella* because the volume increases eightfold over the cell cycle. If a low molecular weight gradient material, such as sucrose, is used the cells may be osmotically perturbed. Gradients made from high molecular weight polymers (Ficoll or proteins) can be used to help overcome this osmotic problem. In addition this method assumes that there is complete correlation between cell size and the temporal position in the cell cycle.

Bacteria (such as *Escherichia coli*) can be synchronized by placing an asynchronous population of bacteria on filter paper. The filter paper is then turned over and nutrient medium is allowed to trickle through from the top. The run off through the filter contains only the very smallest newborn bacteria. Subculture of these newborn bacteria yield synchronous cultures.

In theory a number of mechanical or electronic devices can be devised to collect cells at one stage of the cell cycle based on some physical or chemical change which is known to occur during the cell cycle. Such cellular factors as net charge, size, light scattering ability, fluorescence, shape changes, etc., could be used to physically separate a synchronous population of cells.

Of the naturally synchronous systems the fertilized sea urchin egg is perhaps best known. The synchrony lasts up to the 8- or 16-cell stage. Because the eggs do not grow, the cell division synchrony occurs in the absence of cell growth. It is also becoming apparent that the embryonic cell cycle is markedly different from that in the adult organism (see Mitchison, 1971; Balls and Billett, 1973).

The anthers of some members of the lily family show a striking synchrony of meiosis which is directly related to the length of the anther. This system has proved of considerable value for some workers but it has not been given wide usage (Hotta *et al.*, 1966).

In another naturally synchronous system the microplasmodia of the acellular slime mould *Physarum* are grown in shaker flasks, when such microplasmodia are poured together onto a filter paper the microplasmodia fuse into one large macroplasmodium. If supplied with nutrition the multinucleated macroplasmodium continues to grow and the numerous nuclei divided periodically and in complete synchrony. Of course, there is no cell division (cytokinesis) in the system but the nuclear division synchrony is excellent and the system gives good yields of material for biochemical study.

Before continuing, it seems important to point out several of the ways that are used to determine the degree of synchrony. A sharp increase in cell number with time is perhaps the best index of synchrony, in other words, sharp doubling of cell number in a cell system which divides by binary fission demonstrates good synchrony.

The division or mitotic index is another useful measure of synchrony. However, because the duration of division may be influenced by the synchrony procedure it is subject to some criticism. In other words, it is possible to have a system where more cells are in division if division takes a longer time. This could give a misleading idea of the degree of synchrony. Another useful index is the number of cells in DNA synthesis as determined by ^{3}H-thymidine radioautography. Measuring the content of DNA per cell is a good way of checking for cell synchrony as it should double at each period of DNA replication. On the other hand, a word of caution and criticism is in order to those who put too much emphasis on using a peak of radioisotope (i.e., ^{14}C- or ^{3}H-thymidine) incorporation into DNA as an indication of synchrony. Such a procedure tells little about the number of cells involved in the response and may be influenced by pool size and permeability changes. A combination of morphological, radioautographic, chemical, and cell number changes as a measure of the degree of synchrony is not only helpful in evaluating changes that occur during the cell cycle but in some cases this information is essential.

III. ACIDIC PROTEINS OF THE NUCLEUS IN RELATION TO THE CELL CYCLE

Before we review the literature on the acidic nuclear (nonhistone) protein changes during the cell cycle, it seems important to hold in mind some problems that the various experimental methods and approaches encounter.

The method of nuclear or chromatin isolation must give clean nuclei or chromatin which is free of cytoplasmic contamination. The isolation procedure must retain those proteins which are to be studied and if one wants to do further characterization studies of the nuclear proteins they must be treated so as to preserve their functional and structural properties. The extraction sequence of the nuclei must give reproducible results and should not depend on the physical state of the chromatin, i.e., the extraction should work the same way for hetero- or euchromatin. The possibility of differential extractability of the same proteins must always be considered (Comings and Tuck, 1973).

Because regulatory proteins may be present in small numbers (perhaps only a very few per genome) it may be necessary to use sensitive and high resolution analytical techniques to follow the important role of the regulatory proteins in the cell cycle.

These problems are given more careful discussion in several other chapters in this volume.

A. General Synthesis of Acidic Nuclear Proteins

Evidence has been accumulating rapidly which suggests that, in contrast to the histones whose synthesis is closely coupled to that of DNA, the acidic nuclear proteins are synthesized throughout all phases of the cell cycle. Shapiro and Levina (1967) (IG, IA*) studied chromosomal protein synthesis in lymphocytes synchronized by phytohemagglutinin stimulation and either fluorodeoxyuridine or colchicine arrest. Radioautography demonstrated that these proteins incorporated ^{3}H-leucine throughout the entire cell cycle, including mitosis. In other studies Shapiro and Polikarpova (1969) (IB), using ^{3}H-tryptophan which is absent in histones (Crampton et al., 1955) found that the acidic chromosomal proteins of Colcemid arrest-synchronized hamster fibroblasts were synthesized during the entire cell cycle. Arrighi (1971) also demonstrated radioautographically that the chromosomal proteins were synthesized at all times during the cell cycle of nonsynchronous Chinese hamster cells.

Other laboratories, using biochemical techniques, have corroborated these studies. Stein and Baserga (1970a) (IA) have studied the synthesis of the acidic nuclear proteins during the S, G_2, and M phases of HeLa cells synchronized with a double thymidine block and found that the acidic nuclear proteins were synthesized in each of these three phases of the cell cycle. McClure and Hnilica (1970) (IB) have also reported that the acidic proteins obtained from the chromatin of Chinese hamster cells were synthesized in all phases of the cell cycle, including mitosis. The Chinese hamster cells were synchronized by combining Colcemid blockage of cell division with subsequent harvest of metaphase cells.

These papers indicate that in general the acidic nuclear acidic proteins are synthesized throughout the entire cell cycle. This is in contrast to the other major group of nuclear proteins, the histones, whose synthesis is closely coupled to DNA synthesis. The experimental work reviewed in this chapter will show that the acidic nuclear proteins are synthesized throughout the entire cell cycle, and that, in some cases, there are differential rates of synthesis of individual proteins and groups of proteins at various times during the cell cycle.

B. Acidic Nuclear Proteins in Cells Stimulated to Proliferate

Stimulation of cell populations to proliferate has been one method of synchronizing cells for studying variations in the activity of the acidic nuclear proteins during the G_1 and S phases.

Stellwagen and Cole (1969) have reported that an increase in the synthesis of the acidic nuclear proteins occurred at the time of DNA

---

* Indicates the technique(s) used to synchronize cells. Refers to the notations in Table I in Section II of this chapter.

synthesis in developing and lactating mammary gland. Additionally Malipoix (1971) (IG) found that when fetal mouse liver cells were stimulated with erythropoietin an increase in the synthesis of these proteins occurred which paralleled the increase in DNA synthetic activity. However, two different laboratories have demonstrated that following estrogen treatment an increase in the synthesis of specific rat or calf uterine acidic nuclear proteins occurred well before any increase in DNA synthesis (Teng and Hamilton, 1969, 1970; Smith et al., 1970) (IG). Teng and Hamilton (1969) (IG) demonstrated that not only was the increased synthesis of these proteins organ specific but the increased activity was restricted almost entirely to one protein band. In addition when the acidic protein fraction was added to histone inhibited DNA, RNA synthesis was restored. They suggested that the acidic nuclear proteins were involved in the control of gene activity because of their ability to counteract the inhibitory effects of the histones on gene transcription.

Chung and Coffey (1971) (IG) in somewhat similar experiments found that following castration and subsequent testosterone treatment the maximal rate of acidic nuclear protein synthesis occurred 24–48 hours prior to the peak of DNA synthesis in the ventral prostate of rats. Anderson et al. (1973) (IG) also using this system reported an early increase in the synthesis of the acidic proteins which was followed by a decline in synthetic activity at 12–24 hours after stimulation.

Levy et al. (1973) (IG) have noted that following phytohemagglutinin stimulation of lymphoid cells an immediate increase in the synthesis of the acidic nuclear proteins occurred. The synthetic activity of these proteins continued to increase throughout the first 10 hours following stimulation. In these cells an increase in DNA synthesis occurred 24 hours after phytohemagglutinin stimulation. In addition, electrophoresis showed that there were quantitative changes in the individual proteins during the process of cellular activation. Weisenthal and Ruddon (1972) (IG) have also noted correlations between changes in the synthetic activity of the acidic nuclear proteins and the initiation of DNA synthesis in lymphoid tissue. They reported a lack of higher molecular weight acidic nuclear proteins in nondividing chronic lymphocytic leukemia. However, when these cells were cultured and stimulated to divide with phytohemagglutinin the banding profile changed to one showing primarily middle and higher molecular weight proteins. In particular one middle molecular weight protein was increased in relative quantity two- to threefold. In addition they reported that the banding profile obtained from rapidly dividing Burkitt lymphoma cells showed a heterogeneous population consisting mainly of higher molecular weight proteins with a comparatively small quantity of lower molecular weight proteins. Myelocytic leukemias, which in comparison to lymphocytic leukemias are relatively more active synthetically, again showed more higher molecular weight proteins.

Changes in the acidic nuclear proteins similar to those described for lymphocytic leukemias (Weisenthal and Ruddon, 1972) (IG) have also been found in the ciliated protozoan *Tetrahymena pyriformis*. In experiments employing the starvation–refeeding method of synchronization Jeter (1973) (ID) noted differences in the electrophoretic profiles of the acidic nuclear proteins of stationary, log, and starved cells and at different times following refeeding. The banding patterns of the proteins synthesized by the starved cells showed most of the synthetic activity to be in the lower molecular weight proteins ($<32,000$). Following refeeding, as the cells progressed through G_1 and then into S, there was a shift in the electrophoretic profile from one showing mainly lower molecular weight proteins to one showing quantitatively more middle and higher molecular weight proteins. Additionally, 45 minutes after refeeding (early G_1) there was a threefold increase in the quantity of acidic protein per nucleus (Fig. 3). By late G_1, just prior to the initiation of DNA synthesis, the amount of protein per nucleus had returned to previous levels.

Other workers (Holbrook et al., 1962; Evans et al., 1962; Butler and Cohn, 1963; Pogo et al., 1968; Kostraba and Wang, 1970, 1973; Dastugue et al., 1971) (IG) have used regenerating liver as an experimental model for studying the temporal relationship of the synthesis of the nuclear proteins, primarily the histones, to DNA synthesis. Generally they have noted that an increase in the synthesis of the acidic proteins precedes by several hours any increase in the synthetic activity of either the histones or DNA. In the regenerating liver a synchronous wave of DNA synthesis begins about 18 hours after partial hepatectomy and reaches a peak by 22–28 hours (Holbrook et al., 1962) (IG). Kostraba and Wang (1970) (IG) found that in this system the acidic nuclear proteins showed an increase in isotope incorporation up to approximately 12 hours after lobectomy. This initial increase was followed by a slight decrease in synthetic activity ending at 18 hours. Following this, the synthesis of all nuclear proteins increased rapidly in conjunction with increased DNA synthesis, reaching a maximum about 25 hours after partial hepatectomy. The biosynthetic activity of the proteins then remained high throughout the S period, declining as the number of cells synthesizing DNA declined. Additionally, Kostraba and Wang (1973) (IG) demonstrated that when chromatin is reconstituted using the components of normal and/or 6 hour regenerating rat liver, the acidic nuclear proteins, not the histones or DNA, derepress the chromatin and are responsible for the specificity of transcription of that chromatin. It has previously been reported that at 6 hours after partial hepatectomy the regenerating liver exhibits a twofold increase in *in vitro* chromatin template activity over normal liver chromatin (Thaler and Villee, 1967) (IG). The RNA's transcribed by the chromatin obtained from the regenerating liver also contain new RNA's not found in the transcript from normal liver chromatin (Church and McCarthy, 1967) (IG).

Using induced *in vivo* or synchronized *in vitro* systems Baserga and co-workers (Stein and Baserga, 1970b; Rovera and Baserga, 1971; Stein et al., 1972; Tsuboi and Baserga, 1972) have shown that immediately following stimulation of cells to proliferate, synthesis of the acidic nuclear proteins takes place.

Stein and Baserga (1970b) (IG) found that within 30 minutes after isoproterenol treatment of mouse salivary glands an increase in the synthesis of the acidic nuclear proteins occurred. Following stimulation the synthetic activity of these proteins reached a peak at 12 hours, well in advance of DNA synthesis which occurs at 20 hours. SDS–gel electrophoresis showed that the increased synthetic activity of the acidic nuclear proteins occurs primarily in specific protein bands and that there were quantitative changes in three bands at different times following stimulation (Baserga and Stein, 1971) (IG). When RNA synthesis was inhibited by actinomycin D prior to isoproterenol stimulation the early increase in the rate of acidic nuclear protein synthesis was not affected. However, later nuclear acidic protein synthesis was inhibited by the actinomycin D as was the subsequent DNA synthesis. Cycloheximide administered 1 hour after isoproterenol stimulation also inhibited the later increase in acidic nuclear protein synthesis. Since cycloheximide inhibits protein synthesis in the salivary gland for only 2 hours, Stein and Baserga (1970b) (IG) suggested that an acidic nuclear protein(s) synthesized 1 hour after isoproterenol stimulation is (are) necessary for subsequent synthesis of both acidic proteins and DNA. In addition, the fact that the early acidic nuclear protein synthesis occurred in the presence of actinomycin D led them to conclude that at this time activation of previously existing templates, rather than synthesis of new templates, takes place in this system. On the other hand, Rovera and Baserga (1971) (ID) reported that in WI-38 fibroblasts stimulated to proliferate the synthesis of the acidic nuclear proteins occurring immediately after stimulation was sensitive to actinomycin D inhibition of RNA synthesis. However, as was the case in the isoproterenol-stimulated salivary gland, they found later DNA synthesis to be dependent on prior protein synthesis.

Rovera and Baserga (1971) (ID) noted that following activation of cultured WI-38 fibroblasts to proliferate a biphasic increase in the synthetic activity of the acidic nuclear protein occurred. The initial increase in the synthesis of the acidic proteins was evident within minutes after stimulation and peaked approximately 1 to 3 hours later. Subsequently there was a decline in synthesis which was followed by a second increase in the rate of synthesis of these proteins. This second peak of protein synthesis occurred in conjunction with the peak of DNA synthesis. In addition, pulse-chase experiments indicated that the acidic proteins obtained from stimulated cells had a faster turnover time than those from unstimulated cells.

Chromatin template activity in WI-38 fibroblasts following stimulation by refeeding was studied by Stein *et al.* (1972) and found to be due to differences in the acidic nuclear protein complement between the stimulated and unstimulated cells. They found that the template activity of chromatin reconstituted using acidic nuclear proteins from 1-hour stimulated WI-38 cells with HeLa DNA and pooled histones was higher than the template activity of chromatin reconstituted using acidic proteins from unstimulated cells with HeLa DNA and pooled histones.

Tsuboi and Baserga (1972) (ID) have also reported a biphasic increase in the synthetic activity of the acidic nuclear proteins of cultured fibroblasts stimulated to proliferate. They found that in activated 3T6 cells the changes in the synthesis of the acidic proteins were not evident for several hours after stimulation, at which time the incorporation showed a fourfold increase over control levels. Isotope incorporation then declined to approximately two-thirds of its initial peak activity, reaching a low point just prior to the initiation of DNA synthesis. Following initiation of DNA synthesis isotope incorporation again increased to previous levels and remained elevated throughout most of the S phase. Electrophoresis of the acidic nuclear protein fractions from both 3T6 and WI-38 cells showed differences in the synthetic activity of some of the individual proteins between stimulated and unstimulated cells. In addition, following stimulation, differences were noted in the banding patterns of the acidic proteins obtained from both cell lines. In each case electrophoreograms showed that after activation to proliferate, in both the 3T6 and WI-38 cell lines, there was a decrease in the synthesis of some higher molecular weight proteins and a concomitant increase in the synthetic activity of the middle and lower molecular weight proteins. Stein and Matthews (1973) (ID) have also reported changes between G_1 and S in the SDS–gel electrophoretic profile of the acidic nuclear proteins of WI-38 fibroblasts following stimulation of confluent monolayers to proliferate. Since Tsuboi and Baserga (1972) (ID) did not stain their gels, they were not able to ascertain whether or not there were changes in the individual proteins due to movement of preexisting proteins into the nucleus following differential gene activation as has been reported in other systems (Helmsing and Berendes, 1971; Helmsing, 1972; Nations et al., 1973).

Because the experiments presented thus far have used stimulation to proliferate as a means of achieving cell synchrony they have dealt primarily with the differential synthesis of the acidic nuclear proteins during the G_1 and S phases of the cell cycle. Since stimulatory methods restrict studies to G_1 and S only and because of the relatively low degree of cell cycle synchrony achieved by these procedures, it was impossible to obtain meaningful results for G_2 and M. Therefore the next section will deal with experiments in which techniques were used to achieve a higher degree of cell synchrony, thus allowing the experimenter to look at G_2 and M in addition to G_1 and S.

C. Acidic Nuclear Proteins in More Highly Synchronized Cell Populations

Synchronization techniques such as selective detachment of mitotic cells or double thymidine block produce a much higher degree of cell cycle phase specific synchrony than do the various stimulatory methods used in the above-mentioned experiments. In addition, since some of these methods synchronize the cell in the M or S phase of the cycle, even better synchrony may be obtained when combinations of these techniques are used. For example, usage of double thymidine block for obtaining large numbers of cells in S and G_2 combined with detachment of dividers to obtain cells in M and G_1 allows even more precise dissection of the various cell cycle steps and the events related to them. Several of the papers to be presented in this section use more than one synchronization method to study the relationship of the acidic nuclear proteins to the cell cycle.

Salas and Green (1971) (1A, 1D), using the DNA–cellulose column technique of Alberts et al. (1968) to detect proteins that have an affinity for homologous DNA, have reported the differential synthesis of ^{3}H-tryptophan-labeled proteins between resting (stationary) and exponentially growing (log) cells and during the cell cycle of mouse fibroblasts. Electrophoresis of the 0.15 M NaCl column eluate of whole cell protein extracts resolved the DNA-binding proteins into eight distinct fractions (p1–p8). Three of the eight peaks showed significant quantitative differences, depending on whether the initial cell extracts were obtained from resting or growing cells. The three protein bands of interest were p1, p2, and p6. In resting cells the two higher molecular weight proteins, p1 and p2, were actively synthesized whereas the intermediate molecular weight protein, p6, was practically absent. The converse relationship was found for the proteins obtained from growing cells, i.e., the rate of synthesis of p1 and p2 was greatly reduced while that of p6 was highly increased. In order to determine the relationship of the changes in these three proteins to the different phases of the cell cycle, log phase cells were arrested in S by exposure to excess thymidine. Electropherograms of the proteins obtained from these cells showed little evidence of synthesis in any of the three peaks, indicating that indeed the synthesis of these proteins was related to specific phases of the cell cycle. To further delineate this relationship cells were synchronized using the double thymidine block then released and allowed to pass through S phase. The electrophoretic profile obtained from these cells showed that the proteins synthesized during S resembled those obtained from growing cells with the exception that p6 was quantitatively larger in the synchronized cells than in the growing cells. This data suggested that the synthesis of p6 was coupled to DNA synthesis. Further evidence in support of this came from experiments in which contact-inhibited cells were stimulated to proliferate by addition of fresh serum and then the electrophoretic profiles of the DNA–binding proteins of resting, early G_1, and S phase were compared. The results indicated that during late G_1, p1 quantitatively decreased to approximately one-half its early G_1 amount and subsequently disappeared during S. Again, as shown in the earlier experiments the synthesis of p6 did not quantiatively increase until after the initiation of DNA synthesis. From these experiments Salas and Green suggested that not only was the synthesis of p6 coupled to DNA synthesis, but in addition, that p1, which is synthesized only in resting and serum starved cells and not in growing or S phase cells, might act as a repressor or inhibitor of DNA replication.

Fox and Pardee (1971) (ID, IIA) also used the DNA–cellulose column technique to study the synthetic activity of the DNA-binding proteins of CHO cells at specific times during the cell cycle. Their work supports some of the cell cycle findings of Salas and Green (1971) (IA, ID). When they synchronized cells by the addition of fresh serum, as had Salas and Green, electrophoresis of the eluates demonstrated that numerous proteins were differentially synthesized during G_1 and S, but not between early and late G_1 or early and late S phase. In one eluate (0.15 M NaCl) a significant quantitative decrease occurred during S in one protein of approximately 90,000–95,000 molecular weight. At the same time in two other eluates (0.6 M and 2.0 M NaCl) there was an increase in the synthesis of the intermediate and lower molecular weight proteins and a corresponding decrease in the synthesis of the higher molecular weight proteins. Quantitative differences were also noted in these last two elution fractions, one eluate (0.6 M NaCl) being considerably increased in relative magnitude during S while the other (2.0 M NaCl) was decreased. However, when cells were synchronized by harvest of mitotic cells and the eluates electrophoresed, no qualitative or relative quantitative differences were noted in the patterns of the acidic proteins above molecular weight 30,000 for the last elution fractions (0.6 M NaCl). They did find differential synthesis of some proteins of less than 25,000 molecular weight during S but not G_1. These proteins could possibly be histones since they incorporated relatively little ^{3}H-tryptophan. Although no variations were noted in the profiles of the proteins from these two eluates, quantitative differences were noted between G_1 and S in the percent of total protein eluted from the DNA–cellulose columns. Electrophoretic profiles obtained from the other eluate (0.15 M NaCl) showed some quantitative decrease in two individual proteins of approximately 90,000 molecular weight during S as compared to G_1.

Becker and Stanners (1972) (ID, IIA) have also done a comparative study of acidic nuclear protein synthesis between log and stationary phase cells and during the cell cycle. Their work supports some of the findings of both Salas and Green (1971) (IA, ID) and Fox and Pardee (1971) (ID, IIA). In agreement with the work of Salas and Green, they noted differences in the synthesis of the acidic proteins of log and stationary phase cells. However, when the cells were synchronized by selection of dividers Becker and Stanners (1972) (IIA) found, as had Fox and Pardee, that there were no differences in the synthetic activity of the individual protein bands of a molecular weight of 30,000 or greater throughout the cell cycle. Additionally, they reported that the electrophoretic profiles of the acidic proteins of log cells resembled those obtained from cells in either the G_1, S, or G_2 phases of the cell cycle.

Comparison of the differences in experimental findings from the three laboratories are difficult due to differences in the cell systems used and to variations in techniques employed for synchronization and isolation of proteins. For example, Salas and Green (1971) (IA, ID) synchronized cells by excess thymidine arrest, double thymidine block, and stimulation of cells to proliferate, while Becker and Stanners (1972) (ID, IIA) and Fox and Pardee (1971) (IIA) both used selection of dividers. In addition, Fox and Pardee also employed stimulation of cells to proliferate. Furthermore, Becker and Stanners and Fox and Pardee generally excluded proteins of less than 30,000 molecular weight from their studies even though Fox and Pardee noted some cell cycle-related changes in these lower molecular weight proteins. Salas and Green, on the other hand, made no attempt to determine the molecular weight of the proteins they found to be associated with specific phases of the cell cycle.

Mueller (1969) (IA), Karn et al. (1973) (IA), and Bhorjee and Pederson (1972) (IA) have examined the qualitative and quantitative changes of the individual acidic nuclear proteins in relation to the cell cycle of double thymidine block-synchronized HeLa cells. Mueller (1969) (IA), in a comparison of the acidic nuclear proteins of G_1 and S phases of HeLa cells, found no striking differences in their electrophoretic pattern, but he did find that there was a quantitative increase in all of the proteins as the cells proceeded through S.

Karn et al. (1973) (IA) reported that the electrophoretic profiles of the phenol-soluble nuclear acidic proteins from HeLa were relatively uniform throughout the cell cycle with only some minor differences being detectable. However, they did find significant variations in the nuclear protein to DNA ratio during the cell cycle. The protein to DNA ratios ranged from 2.19:1 to 3.27:1, the ratio being lowest at M and gradually increasing through G_1 to late S where it begins to decline. Since the ratio of the histones to DNA remains constant throughout the cell cycle these variations must be due to the changes in the nuclear acidic proteins. The ratio of the phenol soluble nuclear acidic proteins to DNA also varied during the cell cycle, paralleling the changes in the total nuclear protein to DNA ratios.

Bhorjee and Pederson (1972) (IA) detected several relative quantitative changes in the densitometric scans of the nuclear acidic proteins obtained from various stages of the HeLa cell cycle. They noted a relative reduction of approximately 50% in one band during mid-S and G_2 while another was significantly reduced or absent in late G_1-early S but had reappeared by mid-S. Actual changes in the synthetic activity or amount of the proteins per nucleus was not determined. The differences in the results obtained between these three groups could in part be due to variations in the degree of synchrony from one laboratory to another even though the same line of cells and the same synchronization techniques were used. Thus, if the degree of synchrony were greater in one laboratory than in another, quantitative changes seen in the culture having a lower degree of synchrony would be relatively small or even nonexistent. The fact that HeLa is an aneuploid cell line must also be taken into account when comparing results from different laboratories.

In still more sophisticated attempts to determine the relationship of the nuclear acidic proteins to the cell cycle Gerner and Humphrey (1973) (IA, IB) have combined different methods of synchronizing CHO cells in order to obtain as high a degree of synchrony as possible for each phase of the cell cycle. Early G_1 or S phase cells were obtained by treating asynchronous populations with a single thymidine block followed by detachment of Colcemid-blocked metaphase cells. Early S phase cells were also obtained by blocking cells at the G_1/S interface with hydroxyurea. The double thymidine block was used for synchronizing cells in middle and late S and in G_2. Electrophoretic profiles of the nuclear acidic proteins from cells in the various stages of the cell cycle show the existence of two distinct groups of proteins. One group has a molecular weight range of 35,000 to 150,000 and the other 9000 to 30,000. A few quantitative but no qualitative changes in individual proteins were noted throughout the cell cycle. The synthesis of these proteins in early G_1, S, and G_2 phases was approximately equal; however, in late G_1 a threefold increase in synthetic activity of the acidic nuclear proteins was seen. The high synthetic rate occurring in late G_1 was found to be characteristic of proteins of molecular weight of 45,000 or higher. Lower molecular weight proteins were synthesized at approximately equal rates in all phases of the cell cycle. When G_1 cells were subjected to hydroxyurea treatment in order to prevent them from entering S the higher molecular weight group of proteins continued to demonstrate high rates of synthetic activity. However, when the cells were allowed to enter S the synthetic activity of the proteins returned to the early G_1 level, indicating that though the nuclear acidic proteins were synthesized in all phases of the cell cycle, variations in the rate of synthesis may be temporally linked to the different phases of the cell cycle. Analysis of the individual bands in SDS–gel electrophoretic profiles shows that two higher molecular weight proteins (78,000) exhibited little if any synthesis in early G_1 while two proteins of intermediate (40,000–50,000) molecular weight were actively synthesized in G_1 but not in S or G_2. Another protein of approximately 65,000 molecular weight which is the predominant protein in the electrophoretic profile in all phases of the cell cycle also reaches maximal synthetic activity in late G_1 and then declines through S and G_2.

Stein and Borun (1972) (IA, IIA) have also reported significant increases in the specific activity of the acidic proteins during G_1 of synchronized HeLa cells. Using cells synchronized by either the double thymidine block or by selective detachment of dividers they noted that the rate of synthesis and accumulation of these proteins began increasing immediately after mitosis and reached a maximum in late G_1 just before the initiation of DNA synthesis. The specific activity of the nuclear acidic proteins was higher than that of either the total cellular or crude histone fractions throughout the cell cycle. SDS–polyacrylamide gel electropherograms of the nuclear acidic proteins demonstrated both qualitative and quantitative changes in the synthetic activity of the individual proteins during the cell cycle. In addition, pulse-chase experiments (Borun and Stein, 1972) showed that about two times as much protein was synthesized, transported, and retained in the nuclear acidic proteins in early and late G_1 than in either S or G_2. During G_1 and G_2 about 40% of the proteins synthesized and transported to the nuclear acidic protein fraction left during the chase period while about 20–25% of the proteins entering during S left during the subsequent chase period. Electrophoretic profiles of these proteins demonstrated that at various times after pulse labeling the individual polypeptides exhibited stage-specific differences in the kinetics of turnover and synthesis that were very different from those of the bulk fractions of which they were a part. In G_1 the high and middle molecular weight proteins tended to be retained while the lower molecular weight proteins were chased out by 120 minutes. During S polypeptides of intermediate molecular weight were chased from the nucleus. Finally, in G_2 the higher and middle molecular weight protein fractions exhibited the highest rate of turnover.

In the same system Stein and Matthews (1973) (IA, IIA) have shown that actinomycin D applied prior to mitosis inhibited the synthesis of only some of the acidic nuclear proteins. In addition to the inhibition of synthesis of various individual polypeptides the lower molecular weight group of acidic proteins were completely inhibited in the presence of the antimetabolite. Altogether this indicates that for some acidic nuclear proteins there are stable messenger RNA templates transcribed before mitosis which may be reactivated in G_1, while some other acidic proteins require the transcription of new messenger RNA for their synthesis. The individual acidic nuclear proteins are themselves the product of differential gene activity.

Experiments with inhibitors of DNA synthesis in HeLa (Stein and Borun, 1972), (IA, IIA) and in WI-38 fibroblasts stimulated to proliferate (Stein and Thrall, 1973) (ID) have established that, unlike the histones, the synthesis of the acidic nuclear proteins continues independently of DNA synthesis. Stein and Borun (1972) (IA, IIA) reported that when DNA synthesis was inhibited by cytosine arabinoside, neither the rate of synthesis or the synthetic activity of the individual polypeptides was affected in any phase of the cell cycle. In the WI-38 fibroblasts similar results were obtained when the cells were exposed to cytosine arabinoside in G_1 or in S (Stein and Thrall, 1973) (ID).

Gerner and Humphrey (1973) (IA, IB) in the same type of experiment have interpreted their results differently. When they blocked DNA synthesis in HeLa with hydroxyurea so that the cells remained in G_1 they noted that the synthetic activity of the acidic nuclear proteins remained at G_1 levels until the hydroxyurea was removed and the cells allowed to initiate DNA replication, at which time the specific activity of the proteins dropped to its normal S phase levels. They construed this to mean that acidic nuclear protein synthesis was temporally coupled to DNA synthesis. However, they did not look at the individual polypeptides and thus did not determine the effects of inhibition of DNA synthesis on the individual polypeptides.

LeStourgeon and Rusch (1971, 1973) (IIIC) using the acellular slime mould *Physarum polycephalum,* which undergoes synchronous karyokinesis, have reported that the maximal amount of nuclear acidic protein synthesis occurred during the S phase. In addition, they found no stage-specific difference in the individual proteins as the plasmodium traverses the cell cycle. In spite of the advantages of using a naturally synchronous system such as *Physarum,* investigations into the events occurring in the prereplicative phase of the cell cycle are hampered because the organism lacks a separate defined G_1 period.

Many of the above experiments have shown that in addition to changes individual acidic nuclear proteins there are also changes in the total quantity of these proteins per nucleus throughout the cell cycle. M. E. McClure and L. S. Hnilica (personal communication) (IB) have attempted to correlate these changes in the amounts of the acidic nuclear proteins with variations in the *in vitro* chromatin-directed template activity throughout the cell cycle. Using synchronized Chinese hamster cells, they have demonstrated that the quantity of acidic proteins per unit chromatin initially increased during late G_1. This was followed by a second increase coincident with DNA synthesis which reached a maximum in early to middle-S phase. The proteins then decreased to their initial level through G_2 and M. The *in vitro* chromatin template efficiency, however, showed an inverse relationship to the amount of acidic protein per unit chromatin, i.e., template efficiency was maximal during G_1 and G_2–M and minimal during all parts of the S phase. At present the relationship between the two phenomena is not understood. However, it may lead one to further speculation about the role of the acidic nuclear proteins in the nucleus. One possible explanation is that many of the proteins associated with the chromatin during S are in fact involved in the repression, rather than activation, of specific genes and thus are responsible for the decreased *in vitro* chromatin template activity. Stein and Farber (1972) have indicated that the acidic nuclear proteins are responsible for the repression of chromatin template activity during M phase and Hnilica (1972) has postulated that some of these proteins may act as specific repressors of variable gene expression.

D. Phosphorylation of Acidic Nuclear Proteins

The acidic nuclear proteins have been found to be associated with high concentrations of protein-bound phosphate. Since phosphorylation has been proposed as a mechanism of regulation of gene transcription, (see chapters by Kleinsmith and by Magun in this volume) changes in the phosphate content of proteins relative to one another during various times in the cell cycle might indicate differential activation or suppression of gene activity. On the other hand changes in phosphorylation in which the total quantitites of phosphate change but in which the ratios are invariant might indicate a less specific function such as control over general RNA synthesis throughout the cell cycle.

Cell cycle-related phosphorylation of the acidic nuclear proteins has been studied in synchronized HeLa cells by Platz et al. (1973) and by Karn et al. (1973). Platz et al. (1973) (IA, IIA) used both the selective detachment of mitotic cells and the double thymidine block methods of synchronization to study all five phases of the cell cycle. They reported changes in rate and in the quantitative and qualitative pattern of phosphorylation of the acidic proteins during the cell cycle. The overall rate of phosphorylation was maximal during G_1 and G_2, somewhat decreased during the S phase and depressed by almost 90% during M. Examination of SDS-gel electropherograms showed that not only were there significant differences in the incorporation of ^{32}P into the individual proteins during the cell cycle but there were bands that appeared to be selectively phosphorylated during specific phases of the cycle. One intermediate molecular weight polypeptide was selectively phosphorylated during G_1 and to a lesser extent in G_2 while that of another protein of lower molecular weight was associated with M and to a small extent G_1.

Karn et al. (1973) (IA) also using synchronized HeLa cells, in agreement with Platz et al. (1973) (IA, IIA,) found two peaks in the rate of phosphorylation of the acidic nuclear proteins during the cell cycle. However, they noted that the maximal rates of ^{32}P incorporation occurred in G_1 and S instead of G_2 and S as had been reported by Platz et al. (1973) (IA, IIA). By late S a marked decrease in phosphorylation had occurred and the rate remained low during G_2 and M. Electrophoretic profiles of these proteins showed significant quantitative, but no qualitative differences in the uptake of ^{32}P into the individual polypeptides throughout the cell cycle. In some cases peaks of ^{32}P activity were found not to coincide with stained bands. This suggests that some of the major acidic nuclear phosphoproteins may exist in relatively low concentrations in the nucleus. Other experiments involving long-term

8. Acidic Nuclear Proteins and the Cell Cycle

(23 hours) labeling with ^{32}P- and ^{14}C-leucine followed by a cold chase indicate that the phosphoryl groups can be removed without a corresponding degradation of the proteins. The average half-life of the phosphoryl groups was 6.7 hours while that of the proteins was 25 hours. Additionally, the rate of turnover of ^{32}P was found to vary between the individual polypeptides. The half-life of ^{32}P in the proteins ranged from 5 to 12 hours (Karn et al., 1973).

In part, the differences in results between these two laboratories may be due to the different methods used both for synchrony and for isolation of the proteins. The combinations of synchronization technquies employed by Platz et al. (1973) (IA, IIA) yield a higher degree of stage-specific synchrony than did the single method used by Karn et al. (1973) (IA) and thus maximizes the cell cycle-related differences in the nuclear acidic phosphoproteins.

Changes in phosphorylation may be one mechanism by which the acidic nuclear proteins may control differential gene activity during the cell cycle. For a more in-depth discussion of their possible role as regulators of transcription see the chapters by Kleinsmith and by Magun in this volume.

E. Electron Microscopic Evidence for Gene Activation

Paul (1972) has recently proposed a model for gene activation in the eukaryotic nucleus in which the acidic nuclear proteins confer specificity to the control of transcription. He has speculated that these macromolecules act to reduce the supercoiling of the chromatin and in this way make new areas of the genome available for transcription.

Recent morphological and biochemical evidence from this laboratory is presented in support of Paul's hypothesis (Jeter et al., 1974) (ID) (also see chapter by Gilmour in this volume). The evidence has been obtained from cultures of Tetrahymena pyriformis which were synchronized by the starvation–refeeding method as outlined by Cameron and Jeter (1970) (ID). This technique involves starving the cells in sterile inorganic phosphate buffer for 24 hours, which arrests the cells in G_1, followed by refeeding in sterile growth medium. The cells enter nuclear DNA synthesis beginning at 150 minutes after refeeding. Subsequently, cytokinesis begins at 240 minutes after refeeding (Fig. 1). Previous studies (Cameron et al., 1971a; Rudick and Cameron, 1972) (ID) have shown an increase in the rate of protein synthesis at 30 minutes and in the rate of RNA synthesis at 90 minutes after refeeding. Activation of the codon necessary for synthesis of DNA polymerase was found to occur just prior to the time the cells enter DNA synthesis. After

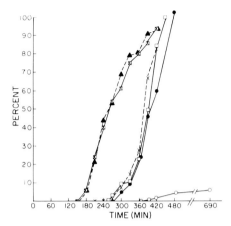

Fig. 1. Abscissa: time after refeeding (minutes); ordinate: percent increase in number of cells which have incorporated ^{3}H-thymidine into their macronucleus as determined in radioautographs (results from two separate experiments are plotted ▲---▲, ⊠——⊠), or percent increase in cell number in the absence (results from three separate experiments are plotted ×---×, ●——●, □——□) or in the presence (○——○) of hydroxyurea. The hydroxyurea (50 mM) was added at time of refeeding. Cells were continuously exposed to ^{3}H-thymidine (10 μCi/ml).

refeeding, concomitant changes in the nuclear and cellular volume and in the amount of phenol-soluble acidic nuclear proteins per nucleus were noted which also corresponded to the previously described increase in transcriptional activity seen in G_1 (Cameron et al., 1971a; Rudick and Cameron, 1971) (ID). Figure 2 illustrates the volume changes occurring in both the nucleus and the whole cell following refeeding. It was found that when the cells were placed in the refeeding medium, there was approximately a 50% decrease in both the cell and nuclear volume. However, by 90 minutes the volumes of both were similar to that of the starved cell. The cell volume then increased at a constant rate to 360 minutes while the nuclear volume increased at a somewhat slower rate from 90 to 240 minutes. After 240 minutes a higher rate of increase in nuclear volume was observed. Concomitant with these volume changes significant quantitative differences were noted in the phenol-soluble acidic nuclear protein to DNA ratios (Fig. 3). There is no DNA synthesis prior to 180 minutes, so the time from 0 to 3 hours in Fig. 3 can be interpreted as what is happening in a single cell with a given amount of DNA. The results indicate that there was a threefold increase in the phenol-soluble acidic nuclear protein to DNA ratio within 45 minutes after refeeding; at a time when the cell is becoming more active synthetically. By 135 and 240 minutes the ratios had decreased to approximately the same levels seen in the starved cells.

8. Acidic Nuclear Proteins and the Cell Cycle

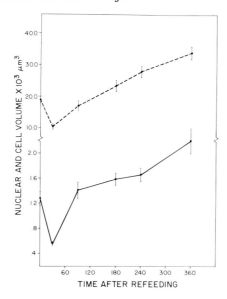

Fig. 2. Abscissa: time after refeeding (minutes); ordinate: volume of cells (●---●) and nucleus (●——●) in cubic micrometers. Graph shows volume of both the cell and the nucleus at various times after refeeding as cells progress temporally through the cell cycle (0–360 minutes). Graph indicates that when cells were refed there was a 50% decrease in cell and nuclear volume. By 90 minutes the volumes of both were similar to that of starved cells (0 minute). Cell volume then increased at a constant rate to 360 minutes while nuclear volume increased at a slower rate from 90 to 240 minutes.

Electron microscopy revealed that in addition to the events which have just been described changes in the ultrastructure of the nucleus occurred after refeeding. Cells were taken at various times following refeeding and fixed for electron microscopy. Representative electron micrographs of cells sampled at 0, 30, 90, and 180 minutes are shown in Fig. 4. Notice that in the starved cells there are a number of chromatin bodies with rather distinct and somewhat smooth boundaries, however, at 30 minutes and at later times the shape of these chromatin bodies becomes somewhat irregular and they appear to radiate much more fibrous material. One must keep in mind that cells fixed at 30 minutes (Fig. 4B) show chromatin bodies much closer together than is shown in starved cells or at later time periods after refeeding. This concentration of chromatin bodies is due to the 50% decrease in nuclear volume that is known to occur at this time (Fig. 2). In the same frame of reference notice that the ribosomes in the cytoplasm of this cell are much more densely packed than in the cytoplasm of the starved cells. Once again this indicates that the total cell volume has shrunk, concentrating the

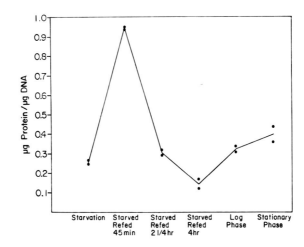

Fig. 3. Abscissa: experimental situation; ordinate: μg phenol-soluble acidic nuclear protein/μg DNA in isolated *Tetrahymena* nuclei. Graph shows quantitation of amount of phenol-soluble acidic nuclear protein per unit DNA for each experimental situation as cells progress temporally through the cell cycle (45 minutes–4 hours) and in different physiological states (starved, stationary, and log phase). Graph indicates that 45 minutes after refeeding a threefold increase in the amount of protein per nucleus occurs. By 2¼ hours following refeeding the protein/DNA ratio returned to the same level as in starved cells. Each point represents one experiment.

intracellular materials. At 30 minutes and at later time periods after refeeding there is a marked increase in the amount of nuclear material between the chromatin bodies. Although some of the increase in the amount of material between the chromatin bodies can be accounted for by the shrinking of the nucleus there is even more material between the chromatin bodies than might be expected by a volume shift, indicating that there is an accumulation of material in the nucleus at 30 minutes after refeeding. At 90 minutes after refeeding the nuclear volume has increased to slightly above that seen in the starved cell and the chromatin bodies are further apart than at 30 minutes. However, the chromatin bodies still exhibit irregular boundaries and have a considerable amount of fibrous and granular material between them. Higher power electron micrographs indicate that at 90 minutes after refeeding there appear to be numerous granules approximately 200–250 Å in diameter closely associated with the boundaries of the chromatin bodies. By 180 minutes after refeeding the nuclear and cell volume is slightly increased over that of the starved cells and the nucleus shows chromatin bodies to have the same spatial relationship between themselves as was seen in the 90-minute refed cells. However, at 180 minutes the 200–250 Å unit

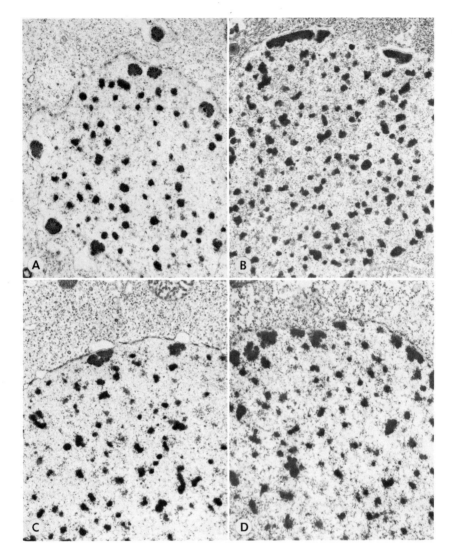

Fig. 4. Electron micrographs of sections through the macronucleus of *Tetrahymena pyriformis*. Figure 4A is a section through a starved cell. 4B, 4C, and 4D are sections through cells which have been refed for 30, 90, and 180 minutes, respectively. A portion of the cytoplasm the nuclear envelope and the nucleus is shown in each micrograph. The dense bodies in the center of each nucleus are referred to as chromatin bodies.

granules, which were found in close proximity to the chromatin bodies at 90 minutes, are scattered throughout the nucleoplasm. It is at this latter time that the transcription of mRNA for DNA polymerase, histone accumulation, and nuclear DNA synthesis are all occurring (Cameron et al., 1971a; Rudick and Cameron, 1972) (ID). Also note that though the cell volumes of the starved and 180-minute refed cells are approximately equal there are considerably more ribosomes per unit area in the cytoplasm of the 180-minute cells than in cytoplasm of the starved cell.

Drawings based on the micrographs in Fig. 4, as shown in Fig. 5, indicate our interpretation of the micrographs in starved and refed cells. We feel that the chromatin bodies represent heterochromatin while the radiating fibrous material is euchromatin. These studies suggest that the correlation between the increase in electron dense material in the nucleus and the apparent unraveling of the chromatin bodies is correlated with the threefold increase in phenol-soluble acidic nuclear proteins which occurs within 45 minutes of refeeding. This increase in uncoiling of the chromatin bodies is accompanied by an increase in transcriptional activity or RNA synthesis which is known to take place between 60 and 90 minutes following refeeding (Cameron et al., 1971a). We feel that the fibrous material between the chromatin bodies as seen in the starved cells represents the amount of active euchromatin which is necessary to maintain the synthetic activity of the cells during the period of starvation. On the other hand, refeeding causes the cell to shrink which concentrates the material in the cell, perhaps initiating a migration of acidic nuclear proteins from the cytoplasm into the nucleus where they are concentrated in the areas immediately surrounding the chromatin bodies. These proteins then, according to Paul's (1972) hypothesis, act to destabilize the chromatin bodies and cause the unraveling and subsequent increased transcriptional activity seen in the refed cells. Our interpretation, then, is that the chromatin bodies represent transcriptionally inactive heterochromatin which can be made to unwind progressively perhaps due to the destabilizing influence of specific acidic nuclear proteins, thus becoming transcriptionally active euchromatin.

One further point of interest is that the accumulation of acidic nuclear proteins occurs at a time when the nucleus is decreased in size by approximately 50%. This is at variance with other systems in which nuclear swelling has been proposed to be necessary for gene activation (Harris, 1970; Kraemer and Coffey, 1970).

IV. SUMMARY AND CONCLUSIONS

In theory study of acidic nuclear protein changes during the cell cycle should give us valuable information on how the eukaryotic cell regulates

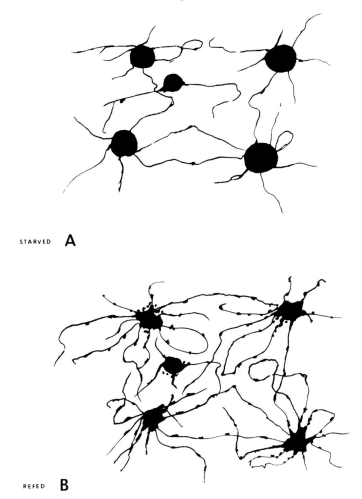

STARVED **A**

REFED **B**

Fig. 5. Interpretation of the micrographs of the starved and refed cells (Fig. 4). Figure 5A represents chromatin in starved cells. Figure 5B represents chromatin in a refed cell that is actively synthesizing RNA. The chromatin bodies represent heterochromatin and the radiating fibrous material euchromatin. After refeeding the chromatin bodies progressively unwind (perhaps due to the destabilizing influence of the acidic nuclear proteins) to become transcriptionally active euchromatin.

gene expression. Since masses of well-synchronized cells are needed to obtain enough material for analysis we have devoted part of this chapter to a review of the various methods for obtaining synchronized cells and the advantages and disadvantages of the different methods.

Review of the literature on the acidic nuclear proteins during the cell cycle allow us to make a few generalizations: Synthesis of the acidic nuclear proteins occurs throughout all phases of the cell cycle, even during mitosis. Acidic nuclear proteins are synthesized at varying rates and accumulated into the nucleus in different amounts in the different stages of the cell cycle. Stimulation of quiescent cells to proliferate is accompanied by an early accumulation of acidic proteins in the nucleus. There is some electron microscopic evidence for the accumulation of such material in the nucleus. Although there are quantitative and even qualitative changes in a few specific individual proteins during various phases of the cell cycle, it is at first somewhat surprising that these changes are not more numerous, especially if some of the acidic nuclear proteins are really acting as regulators of gene expression. This paucity of reports on specific changes is perhaps even more perplexing in view of the evidence that hormonally stimulated systems do show changes in specific individual acidic nuclear proteins (see the chapter by Spelsberg in this volume).

It would appear that several factors may mitigate against the finding of changes in individual acidic nuclear proteins during the cell cycle. Such factors may include the fact that the homogeneity of the cell population may not be good enough (contamination by cells with various degrees of ploidy or contamination with other cell types). The isolation, extraction, separation, and detection procedures may be too variable or insensitive. For example, it is quite possible that the regulatory proteins may be in too small a number per genome to be detected or that the protein changes may involve slight molecular modifications such as phosphorylation which were not measured and therefore go undetected. As yet we can not even rule out the possibility that the acidic nuclear proteins do not play a role in the fine control of gene expression.

Unfortunately at the present time, we are forced to conclude that, because of inadequate cell cycle synchrony, because the regulatory proteins may be present in the nucleus in minute quantities, and because of the lack of resolution of the present analytical techniques, substantial judgments cannot yet be made about whether these proteins are involved in the fine control of gene expression during the cell cycle.

REFERENCES

Alberts, B. M., Amodio, F. J., Jenkins, M., Gutmann, E. D., and Ferris, F. L. (1968). *Cold Spring Harbor Symp. Quant. Biol.* 33, 289.

Allfrey, V. G., Littau, V. C., and Mirsky, A. E. (1963). In "Histones and Nucleohistones" (D. M. P. Phillips, ed.), pp. 241–294. Plenum, New York.

Anderson, K. M., Slavik, M., Evans, A. K., and Couch, R. M. (1973). *Exp. Cell Res.* 77, 143–158.

Arrighi, F. E. (1971). *Proc. Int. Cancer Congr., 10th, 1970* Abstracts, p. 365.
Balls, M., and Billett, F. S. (1973). "The Cell Cycle in Development and Differentiation." Cambridge Univ. Press, London and New York.
Baserga, R., and Stein, G. (1971). *Fed. Proc., Fed. Amer. Soc. Exp. Biol.* **30**, 1752–1759.
Becker, H., and Stanners, C. P. (1972). *J. Cell. Physiol.* **80**, 51–62.
Benjamin, W., and Gellhorn, A. (1968). *Proc. Nat. Acad. Sci. U.S.* **59**, 262–268.
Bhorjee, J. S., and Pederson, T. (1972). *Proc. Nat. Acad. Sci. U.S.* **69**, 3345–3349.
Borun, T. W., and Stein, G. S. (1972). *J. Cell Biol.* **52**, 308–315.
Butler, J. A. V., and Cohn, P. (1963). *Biochem. J.* **87**, 330–334.
Cameron, I. L., and Jeter, J. R., Jr. (1970). *J. Protozool.* **17**, 175–181.
Cameron, I. L., and Jeter, J. R., Jr. (1973). *Cell Tissue Kinet.* **6**, 289–301.
Cameron, I. L., and Padilla, G. M., eds. (1966). "Cell Synchrony." Academic Press, New York.
Cameron, I. L., Griffin, E. E., and Rudick, M. J. (1971a). *Exp. Cell Res.* **65**, 262–272.
Cameron, I. L., Padilla, G. M., and Zimmerman, A. M., eds. (1971b). "Developmental Aspects of the Cell Cycle." Academic Press, New York.
Chung, L. W. K., and Coffey, D. S. (1971). *Biochim. Biophys. Acta* **247**, 584–596.
Church, R. B., and McCarthy, B. J. (1967). *J. Mol. Biol.* **23**, 459–475.
Chytyl, F., and Spelsberg, T. C. (1971). *Nature (London), New Biol.* **233**, 215–218.
Comings, D. E., and Tack, L. O. (1973). *Exp. Cell Res.* **82**, 175–191.
Crampton, C. F., Moore, S., and Stein, W. H. (1955). *J. Biol. Chem.* **215**, 787–801.
Dastugue, B., Hanoune, J., and Kruh, J. (1971). *FEBS Lett.* **19**, 65–68.
Dingman, C. W., and Sporn, M. B. (1964). *J. Biol. Chem.* **239**, 3483–3492.
Elgin, S. C. R., and Bonner, J. (1970). *Biochemistry* **9**, 4440–4447.
Elgin, S. C. R., and Bonner, J. (1972). *Biochemistry* **11**, 772–781.
Enea, V., and Allfrey, V. G. (1973). *Nature (London)* **242**, 265–267.
Evans, J. N., Holbrook, D. J., and Irvin, J. L. (1962). *Exp. Cell Res.* **28**, 126–132.
Fox, T. O., and Pardee, A. B. (1971). *J. Biol. Chem.* **246**, 6159–6165.
Frenster, J. H. (1965). *Nature (London)* **206**, 680–683.
Gerner, E. W., and Humphrey, R. M. (1973). *Biochim. Biophys. Acta* (in press).
Gilmour, R. S., and Paul, J. (1969). *J. Mol. Biol.* **40**, 137–139.
Harris, H. (1970). "Nucleus and Cytoplasm" Oxford Univ. Press (Clarendon), London and New York.
Helmsing, P. J. (1972). *Cell Differentiation* **1**, 19–24.
Helmsing, P. J., and Berendes, H. D. (1971). *J. Cell Biol.* **50**, 893–896.
Helmstetter, C. E. (1969). *In* "Methods in Microbiology" (J. R. Norris and D. W. Ribbons eds.), Vol. 1, pp. 327–369. Academic Press, New York.
Hnilica, L. S. (1967). *Mol. Biol.* **7**, 25–106.
Hnilica, L. S. (1972). "The Structure and Function of Histones," Chapter 9. Chem. Rubber Publ. Co., Cleveland, Ohio.
Holbrook, D. J., Evans, J. H., and Irvin, J. L. (1962). *Exp. Cell Res.* **28**, 120–125.
Hotta, Y., Ito, M., and Stern, H. (1966). *Proc. Nat. Acad Sci. U.S.* **56**, 1184–1191.
Huang, R. C., and Bonner, J. (1962). *Proc. Nat. Acad. Sci. U.S.* **48**, 1216–1222.
James, T. W. (1966). *In* "Cell Synchrony" (I. L. Cameron and G. M. Padilla, eds.), pp. 1–13. Academic Press, New York.
Jeffery, W. R., Frankel, J., Debault, L. E., and Jenkins, L. M. (1973). *J. Cell Biol.* **59**, 1–11.

Jeter, J. R., Jr. (1973). Ph.D. Thesis, University of Texas Health Sciences Center at San Antonio.
Jeter, J. R., Jr., Pavlat, W. A., and Cameron, I. L. (1974). Submitted for publication.
Kamiyama, K., and Wang, T. (1971). *Biochim. Biophys. Acta* **228**, 563–576.
Karn, J., Johnson, E. M., Vidali, G., and Allfrey, V. G. (1974). *J. Biol. Chem.* **249**, 667–677.
Kleinsmith, L. J., Allfrey, V. G., and Mirsky, A. E. (1966). *Science* **154**, 780–781.
Kostraba, N. C., and Wang, T. Y. (1970). *Int. J. Biochem.* **1**, 327–334.
Kostraba, N. C., and Wang, T. Y. (1972). *Biochim. Biophys. Acta* **262**, 169–180.
Kostraba, N. C., and Wang, T. Y. (1973). *Exp. Cell Res.* **80**, 291–296.
Kraemer, R. J., and Coffey, D. S. (1970). *Biochim. Biophys. Acta* **224**, 568–578.
Kuzmich, M. J., and Zimmerman, A. M. (1972). *Exp. Cell Res.* **72**, 441–452.
LeStourgeon, W. M., and Rusch, H. P. (1971). *Science* **174**, 1233–1236.
LeStourgeon, W. M., and Rusch, H. P. (1973). *Arch. Biochem. Biophys.* **155**, 144–158.
Levy, R., Levy S., Rosenberg, S. A., and Simpson, R. T. (1973). *Biochemistry* **12**, 224–228.
Loeb, J. E., and Creuzet, C. (1969). *FEBS Lett.* **5**, 37–40.
MacGillivray, A. J., Paul, J., and Threlfall, G. (1972). *Advan. Cancer Res.* **15**, 93–162.
Malpoix, P. J. (1971). *Exp. Cell Res.* **65**, 393–400.
Marushige, K., and Bonner, J. (1971). *Proc. Nat. Acad. Sci. U.S.* **68**, 2941–2944.
Marushige, K., and Dixon, F. H. (1969). *Develop. Biol.* **19**, 397–414.
Marushige, K., and Ozaki, H. (1967). *Develop. Biol.* **16**, 474–488.
McClure, M. E., and Hnilica, L. S. (1970). *Proc. Int. Cancer Cong., 10th*, 494–509.
McClure, M. E., and Hnilica, L. S. (1972). *Sub. Cell. Biochem.* **1**, 311–332.
Mitchison, J. M. (1971). "The Biology of the Cell Cycle." Cambridge Univ. Press, London and New York.
Mueller, G. C. (1969). *Fed. Proc., Fed. Amer. Soc. Exp. Biol.* **28**, 1780–1789.
Nations, C., LeStourgeon, W. M., Magun, B. E., and Rusch, H. P. (1974). *Exp. Cell Res.* (in press).
Nias, A. H. W., and Fox, M. (1971). *Cell Tissue Kinet.* **4**, 351–375.
Paul, J. (1972). *Nature (London)*. **238**, 444–446.
Phillips, D. M. P., ed. (1971). "Histones and Nucleohistones." Plenum, New York.
Platz, R. D., Stein, G. S., and Kleinsmith, L. J. (1973). *Biochem. Biophys. Res. Commun.* **51**, 735–740.
Pogo, B. G. T., Pogo, A. O., Allfrey, V. G., and Mirsky, A. E. (1968). *Proc. Nat. Acad. Sci. U.S.* **59**, 1337–1344.
Rovera, G., and Baserga, R. (1971). *J. Cell. Physiol.* **77**, 201–212.
Rudick, M. J., and Cameron, I. L. (1972). *Exp. Cell Res.* **70**, 411–516.
Salas, J., and Green, H. (1971). *Nature (London), New Biol.* **229**, 165–169.
Shapiro, I. M., and Levina, L. (1967). *Exp. Cell Res.* **47**, 75–85.
Shapiro, I. M., and Polikapova, S. T. (1969). *Chromosoma* **28**, 188–198.
Shelton, K. R., and Allfrey, V. G. (1970). *Nature (London)* **228**, 132–134.
Smith, J. A., Martin, L., King, R. J. B., and Vertes, M. (1970). *Biochem. J.* **119**, 773–784.
Stein, G., and Baserga, R. (1970a). *Biochim. Biophys. Res. Commun.* **41**, 715–722.
Stein, G., and Baserga, R. (1970b). *J. Biol. Chem.* **245**, 6097–6105.
Stein, G., and Baserga, R. (1972). *Advan. Cancer Res.* **15**, 287–330.
Stein, G. S., and Borun, T. W. (1972). *J. Cell Biol.* **52**, 292–307.

Stein, G., and Farber, J. (1972). *Proc. Nat. Acad. Sci. U.S.* **69**, 2918–2921.
Stein, G., and Matthews, D. E. (1973). *Science* **181**, 71–83.
Stein, G. S., and Thrall, C. L. (1973). *FEBS Lett.* **34**, 35–39.
Stein, G., Chaudhuri, S., and Baserga, R. (1972). *J. Biol. Chem.* **247**, 3918–3922.
Stellwagen, R. H., and Cole, R. D. (1969). *J. Biol. Chem.* **244**, 4878–4887.
Teng, C. S., and Hamilton, T. H. (1969). *Proc. Nat. Acad. Sci. U.S.* **63**, 645–672.
Teng, C. S., and Hamilton, T. H. (1970). *Biochem. Biophys. Res. Commun.* **40**, 1231–1238.
Teng, C. T., Teng, C. S., and Allfrey, V. G. (1970). *Biochem. Biophys. Res. Commun.* **41**, 690–696.
Teng, C. S., Teng, C. T., and Allfrey, V. G. (1971). *J. Biol. Chem.* **246**, 3597–3609.
Thayler, M. M., and Villee, C. A. (1967). *Proc. Nat. Acad. Sci. U.S.* **58**, 2055–2062.
Tsuboi, A., and Baserga, R. (1972). *J. Cell. Physiol.* **80**, 107–118.
Wang, T. Y. (1967). *J. Biol. Chem.* **242**, 1220–1226.
Wang, T. Y. (1968). *Exp. Cell Res.* **53**, 288–291.
Weisenthal, L. M., and Ruddon, R. W. (1972). *Cancer Res.* **32**, 1009–1017.
Zeuthen, E., ed. (1964). "Synchrony in Cell Division and Growth." Wiley, New York.

man, 1966b; Wagner, 1970; Swaneck et al., 1970; King and Gordon, 1967). Both histone and acidic (nonhistone) chromatin proteins have been suggested as the site of hormone binding. However, the cytosol receptor for hormones should not be ruled out in this case since the hormone enters the nucleus bound to this receptor and can be extracted from the nucleus in a complex with a similar type protein. As mentioned earlier, all steroid receptors thus studied, with the possible exception of that for cortisol (Morey and Litwack, 1969; Litwack and Morey, 1970), are acidic in nature. If the radioactive steroids are bound to chromatin through this receptor, then the steroids would behave as if they were complexed to acidic chromatin proteins. In this case, the hormone bound to chromatin would be labile to proteases even if the hormone-receptor complex was bound directly to the DNA.

VI. CHROMATIN-BINDING SITES FOR HORMONE–RECEPTOR COMPLEXES: "THE ACCEPTOR MOLECULES"

A. Hormone Binding to Histones

Since corticosteroid hormones were reported to cause an increase in the templating efficiency of rat liver chromatin and biosynthesis of new RNA species in liver (Kidson and Kirby, 1964; Kidson, 1967), it was suggested (Tata, 1966; Sekeris and Lang, 1965; Sluyser, 1966a,b) that corticosteroid hormones may associate with and affect histones thereby producing a stimulation of RNA synthesis.

It was first reported by Sluyser (1966a,b) that histones can form stable complexes with hydrocortisone when incubated together in NaCl solutions of low ionic strength. Arginine-rich histones were found to bind more hormone than the lysine-rich fractions. The *in vitro* binding of arginine-rich and other histones by cortisol, tetrahydrocortisol, cortisone, tetrahydrocortisone, progesterone, estradiol-17β, and testosterone was confirmed by Sunaga and Koide (1967a,b,c). However, it is improbable that the cell nucleus is biologically affected by free hormones since the cytosol receptors *in vivo* play such a dominant role in binding, retaining, and transporting the steroid hormones into the nucleus. Much more meaningful experiments are those performed *in vivo* as well as those *in vitro* with the hormone complexed with its cytoplasmic receptor. After injecting labeled cortisone into rats, it was still found that the radioactivity was associated with histones (Sekeris and Lang, 1965). The *in vivo* binding of hydrocortisone to histones was shown by Sluyser (1966a) who found the arginine-rich histone fraction F_3 associated with the hormone more extensively than all the histones. Similarly, injected testosterone was found associated with histones in prostate and in levator ani muscle of rats and to a much lesser extent with liver or spleen histones of the same animals (Sluyser, 1966a). In contrast to hydrocortisone, labeled testosterone was bound preferentially to the very lysine-rich histone fraction F_1. In spleen, the arginine-rich histone accepted more and retained labeled testosterone for a longer time than the other histone fractions. However, the fact that the cortisol receptor has been reported to be a basic protein (Litwack and Morey, 1970) may explain its extraction in acid solutions from the chromatin along with the histones as reported by King and Gordon (1967). Another problem possibly overlooked by these investigations is the fact that the histones (especially the arginine-rich) are notorious for being contaminated with acidic proteins (Hnilica and Bess, 1965; Stellwagen and Cole, 1968a,b).

In this respect, Tsai and Hnilica (1971) investigated the *in vivo* binding of corticosteroid hormones to histones and examined the role that contaminating acidic proteins played in the hormone binding. They isolated nuclei from rats injected with labeled hydrocortisone or cortisone. Four different techniques for the isolation of nuclei were employed. Only nuclei contaminated by cytoplasmic particles yielded histones associated with significant amounts of hormone. Hypertonic sucrose procedure (Blobel and Potter, 1966; Chaveau *et al.*, 1956) produced hormone-labeled histones only if the isolated nuclei were extracted directly with diluted acid (without partial fractionation into the arginine-rich and lysine-rich histones). This technique is known to result in some contamination of histones by acid-soluble nonhistone proteins. The major part of radioactive hormone in the nucleus was associated with the acid-insoluble residue. It was concluded from these experiments that the *in vivo* labeling of histones with radioactive corticosteroid hormones in liver was due to their contamination by other proteins, most likely of cytoplasmic origin.

In addition to these studies, Tsai and Hnilica (1971), after incubating labeled hydrocortisone with calf thymus histone *in vitro*, found that 65% of the hormone to be associated with the arginine-rich F_3 histones. The remaining radioactivity was associated with the very lysine-rich histone (F_1). The F_3 histone fraction bound with tritiated hydrocortisone was subjected to polyacrylamide gel electrophoresis, the gels sliced, and each slice measured for radioactivity. It was found that less than 1 μmole of hydrocortisone was bound per milligram of protein. Assuming a molecular weight of 15,000 for the F_3 histone, the binding represented 15 μmoles of hydrocortisone bound per mole of F_3 histone or 1 molecule of hormone for every 6.6×10^4 molecules of F_3 histone. This ratio of hormone bound to F_3 histone is similar to the data published by Sluyser (1969). It has been suggested that this extremely low binding may be due to the actual association of the hormone with acidic proteins contaminating the F_3 histone (Tsai and Hnilica, 1971). Fingerprint analysis of the tryptic digests of F_3 histone which contained bound hormone revealed that most of the hormone remained at the origin. This suggests that the hormone was bound to a trypsin-resistant protein, resembling the trypsin-resistant acidic proteins. In addition, the arginine-rich histones are known to contain significant amounts of contaminating acidic proteins which are difficult to remove (Hnilica and Bess, 1965; Stellwagen and Cole, 1968a,b). Consequently, the *in vitro* bound hydrocortisone is probably complexed to the acidic proteins of the F_3 fraction. Although this conclusion cannot be generalized without further experimental evidence, it appears that histones are not very likely candidates for the tissue-specific interactions of steroid hormones with chromatin. Other *in vitro* studies (described below) show that oviduct chromatin devoid of histones binds much greater amounts of progesterone–receptor complex than intact chromatin (containing histones). These results, based on receptor-bound hormone and DNA-bound proteins, also reduce the probability that histones are the "acceptor" sites for hormone–receptor complexes.

B. Hormone Binding to Acidic (Nonhistone) Proteins

Since histones are unlikely to be involved in the binding of the steroid hormone–receptor complexes with chromatin, the DNA and acidic proteins remain the potential "acceptor" sites. As in the studies with histones, early *in vitro* experiments dealing with the binding of hormones with the acidic proteins of chromatin utilized only free steroid molecules. Several reports of *in vivo* experiments present evidence that the steroid hormones are complexed to the chromatin of target tissues through the nonhistone (acidic) proteins (King et al., 1966; Maurer and Chalkley, 1967; King and Gordon, 1969; Bruchovsky and Wilson, 1968b; Wagner, 1970; Tsai and Hnilica, 1971; Talwar et al., 1964; King et al., 1965a,b, 1969; Beato et al., 1969; Swaneck et al., 1970). Many of these studies involve proteases which could use the hormone receptor (an acidic protein) as a substrate as opposed to using a chromatin acidic protein as a substrate. Therefore, the release of steroids from nuclei or chromatin by proteases at present cannot be interpreted as a binding of the steroid receptor complex to chromatin acidic protein. Furthermore, the hormone bound to its receptor protein (an acidic protein) would readily associate with and behave like the acidic proteins. Therefore, the extraction of the steroid from nuclei or chromatin by various solvents which differentiate between histones and acidic proteins cannot be used as a sole criteria for binding sites on chromatin. It is evident that such a problem as this one (identifying the nature of the site on chromatin which binds the steroid hormone–receptor complex) is not readily resolvable with present techniques using *in vivo* systems. More direct evidence is obtainable using *in vitro* systems. The results of some of these studies will be described in more detail below.

C. Hormone Binding to DNA

From the previous discussion on steroid hormone binding to chromatin, it is apparent that the binding of a steroid, complexed with its nuclear receptor, to chromatin may well be mediated through the DNA. Since the steroid hormones are complexed to receptor proteins, which are acidic in nature, studies using proteases and procedures for isolation of histones and acidic proteins do not eliminate the DNA as the "acceptor" site. As discussed previously, the hormone receptor in this case presents an interference in many techniques for detecting the chromatin-binding site. In addition, since the steroid hormone–receptor complexes are dissociated from the chromatin at relatively low ionic strengths (0.2–0.4 M KCl), the isolation of the DNA from chromatin without the simultaneous loss of the labeled hormone–receptor complexes is impossible. Harris (1971) was able to release labeled estradiol from the uterine nuclei of immature mice (previously injected with the hormone) by mild treatment with deoxyribonuclease. The released hormone was complexed with a protein which sedimented similar to the cytoplasmic receptor in uteri for estradiol. These results contradict other reports which claim a resistance of releasing of steroid hormones from their nuclear-binding sites with nucleases. Harris (1971) did not add divalent cations to the incubations so the DNase, using only the nuclear retained ions added during the isolation of nuclei, probably degraded only part of the DNA. Her experiments did not involve as extensive a DNA removal as the other studies (Fanestil and Edelman, 1966b; King and Gordon, 1967). It is probable that a more thorough removal of the DNA would cause a greater release of histones and acidic proteins and the hormone–receptor complex could be lost through ionic complexing with the basic histones and hydrophobic binding with the acidic proteins. In any case, the release of the estradiol–protein complex by DNase could be interpreted as: (1) the hormone–receptor complex was bound to the DNA and was released due to the nucleolytic degradation of the DNA, or

(2) the hormone–receptor complex was bound to a protein or other entity on chromatin which requires the presence of DNA to maintain the ability to bind the hormone–receptor complex.

Toft (1972, 1973a,b) has studied the binding of the estradiol–receptor complex of both calf and rat uteri to purified DNA of various sources and under various conditions. Both the 4 S and 8 S receptors for estradiol from calf and rat uteri could bind DNA. The binding was determined both by sedimentation through sucrose gradients and by column chromatography using DNA bound to cellulose. As much as 50% of the total labeled (hormone-bound) receptor could be bound to the DNA. The binding to the DNA was stable in 0.1 M KCl but dissociated at 0.2 M KCl and higher salt concentrations. No specificity of binding to the DNA of various sources or to synthetic DNA was observed. However, denatured (single strand) DNA did bind more effectively than native (double strand) DNA. Contrary to these results, Clemens and Kleinsmith (1972) reported that the binding of the 4 S estrogen receptor from rat uteri to DNA was specific for the source (type) of DNA. Rat DNA was able to bind more (4%) of the total labeled receptor than fish or bacterial DNA (0.5–1.5% of the total labeled receptor). However, in contrast to the binding of 50% of the total receptor by Toft (1973a,b), the report of Clemens and Kleinsmith (1972) is only dealing with a few percent of the total labeled receptor. The discrepancy may reside in the possibility that a few percent of the hormone-bound protein sedimenting at 4 S may have a specificity for homologous DNA. Other work by King et al. (1971a,b) and King and Gordon (1972) gives evidence that DNA does bind the estrogen–receptor complex with high affinity ($Kd = 2 \times 10^{-10}$ M) but without any specificity as to the source of DNA. Baxter et al. (1972) suggest from similarities in the binding of dexamethasone to HTC cells, DNA, and nuclei that the DNA may represent the binding site on chromatin for the hormone–receptor complex.

The controversy may be partially resolved on considering the work of Schrader et al. (1972). The chick oviduct cytoplasmic and nuclear receptor for progesterone can be fractionated into two types of receptors: "A" and "B." The "A" receptor shows significant binding to DNA but not to whole chromatin, while the "B" receptor shows significant binding to target tissue chromatin but not pure DNA. Further discussion on the ramifications of these findings are described later. In any case, whether the DNA represents none, part, or all of the actual binding site for steroid hormone–receptor remains to be determined. Confusion that exists in this area is partly due to the various bits of information available only on a variety of different hormone–target tissue systems.

VII. IDENTIFICATION AND CHARACTERIZATION OF THE "ACCEPTOR" MOLECULE IN CHICK OVIDUCT CHROMATIN WHICH BINDS THE PROGRESTERONE–RECEPTOR COMPLEX

A. *In Vitro* Binding of the Progesterone–Receptor Complex with Isolated Chromatin

As demonstrated in Fig. 4, once labeled progesterone enters a cell of the chick oviduct it binds to the cytoplasmic receptor and is transported into the nucleus. As with most steroid hormone–target cell systems studied, the whole process has several requirements. First, the hormone must be bound to its receptor. Second, it requires an elevated temperature. Modification of the uterine estrogen receptor as well as cell entrance appear to require elevated temperatures (Williams and Gorski, 1971). In the case of the progesterone receptor of the chick oviduct, there appears to be an additional reason for increase temperature and that is transport into the cell nucleus (Buller *et al.*, 1974). A third requirement for nuclear binding discovered from similar *in vitro* experiments is the presence of oviduct nuclei (as opposed to other nontarget tissue nuclei) as discussed earlier and shown in Fig. 5. The complex of the progesterone and the oviduct receptor (prepared by incubating the 100,000 g super-

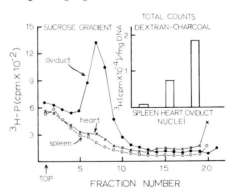

Fig. 5. Uptake of bound radioactivity into the nuclei of several organs of immature chicks after incubation of isolated nuclei in the oviduct cytosol containing labeled progesterone (^{3}H-P) (50.3 Ci/mmole). Incubation was performed at 25°C for one hour. The nuclei were then sedimented by centrifugation and washed in 10 volumes of a buffered 0.5 M sucrose solution containing 0.2% Triton-X100. The radioactivity in the nuclei was extracted with 0.4 M KCl, and bound radioactivity measured by dextran-charcoal assay and ultracentrifugation in a sucrose gradient. Reproduced with permission of Dr. D. Toft (Buller *et al.*, 1974).

nate of oviduct homogenates with labeled progesterone) binds to oviduct nuclei but fails to bind to the nuclei of other tissues (heart and spleen) (Buller et al., 1974; O'Malley et al., 1971a). Contradictory results have been found with the estrogen–target tissue systems as discussed earlier. The progesterone–chick oviduct experiments were extended one step further wherein chromatin was isolated from the nuclei of various tissues of the chick and incubated with the oviduct cytosol containing the labeled progesterone (Spelsberg et al., 1971a; Steggles et al., 1971a). Figure 6 shows that the chromatin from oviduct (a target tissue for progesterone) binds the progesterone–receptor complex (obtained from oviduct) more extensively than does the chromatin from spleen of immature chicks or from the erythrocytes of mature hens. The oviduct chromatin also bound more of the complex than heart, kidney, lung, and to a lesser extent liver chromatin. These results, supporting the experiments involving isolated nuclei, suggest that the oviduct requires not only a cytoplasmic receptor to complex with the hormone but also a programing of the genetic material in order to bind more extensively

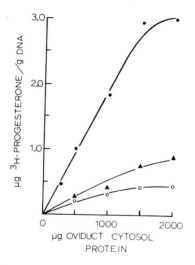

Fig. 6. Binding of ³H-progesterone, bound to its oviduct cytosol receptor, to the chromatins of various organs of the chicken. The hormone–receptor complex was obtained by incubating the labeled progesterone (44 Ci/mmole) at 1.5×10^{-8} M for 1 or more hours at 4°C in oviduct cytosol (100,000 g supernatants of oviduct homogenates which contains about 10 mg protein per ml). Incubation of the chromatin and hormone–receptor complex were carried out at 4°C for 1 hour in 0.15 M NaCl buffered to pH 7.5. Forty micrograms of (●) oviduct chromatin from 15-day DES-treated immature chicks, (○) spleen chromatin from untreated immature chicks, and (▲) erythrocyte chromatin from 2-year-old hens were incubated with various levels of labeled cytosol protein (Spelsberg et al., 1971a).

the progesterone–receptor complex. Figure 7 shows the results of the substitution of the oviduct cytosol (containing the progresterone receptor) with spleen and liver cytosols (containing no receptor) in similar *in vitro* incubations with various chromatins. Clearly, labeled progesterone not complexed to its receptor (spleen or liver cytosols) displays much less binding to oviduct chromatin compared to the binding by progesterone complexed to the receptor (oviduct cytosol). Thus as in the experiments with isolated oviduct nuclei, the extensive chromatin binding by labeled progesterone requires that the hormone be bound to its receptor.

This *in vitro* binding of the progesterone–receptor complex to oviduct chromatin resembles the nuclear binding of the hormone *in vivo* in two other respects. First, after incubation of oviduct chromatin with labeled progesterone in the presence of oviduct cytosol, the radioactivity can be extracted with 0.3 M NaCl (Fig. 9B). Second, this radioactivity extracted from chromatin with salt is associated with a protein which

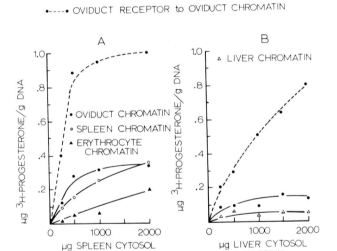

Fig. 7. Binding of ^{3}H-progesterone to the chromatin of various chick tissues in the presence of cytosols from various tissues of the chick. The experiment is the same as that described in the legend of Fig. 6 except that in (A) the labeled hormone was first incubated in the spleen cytosol and in (B) the labeled hormone was first incubated in liver cytosol before incubation with chromatin. The chromatins used in these experiments were from (●) oviduct from 15-day DES-treated chicks, (○) spleen and (△) liver from untreated immature chicks, and (▲) erythrocytes from 2-year-old hens. The broken line represents binding of ^{3}H-progesterone to oviduct chromatin in the presence of oviduct cytosol for comparison (Spelsberg et al., 1971a).

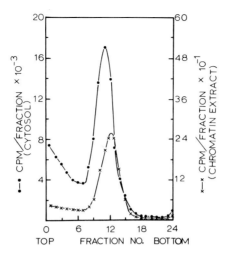

Fig. 8. Sucrose gradient analysis of a 0.3 M KCl extraction of oviduct chromatin which had been incubated with the oviduct cytosol–^{3}H-progesterone mixture. Oviduct chromatin (3 mg) was incubated with 100 mg of oviduct cytosol protein containing ^{3}H-progesterone. The chromatin was sedimented by centrifugation, washed with 0.15 M NaCl, and then extracted with 2 ml of 0.3 M KCl at 4°C for 1 hour. This solution was centrifuged, and 0.2 ml of the supernatant was applied to 5 to 20% sucrose gradients containing 0.3 M KCl in the Tris–EDTA buffer. The cytosol supernatant was labeled with 1.5×10^{-8} ^{3}H-progesterone. After incubation at 4°C for 1 hour, 0.2-ml aliquots were layered over 5 to 20% sucrose gradients containing 0.3 M KCl in Tris–EDTA buffer, pH 7.4. Centrifugation was at 45,000 rpm for 16 hours at 5°C in a Spinco SW-65 rotor, after which 0.2-ml fractions were collected and the radioactivity was measured (Spelsberg et al., 1971a).

has a sedimentation constant of 4 S (Fig. 8). This is similar to the nuclear receptor extractable from tissues after injection of progesterone in chicks. One difference between the *in vitro* binding using isolated chromatin and the *in vitro* nuclear binding using whole tissue or isolated nuclei is the lack of a temperature dependency (20° to 25°C incubations) for chromatin binding. The binding of the progesterone–receptor complex to isolated oviduct chromatin is maximal at 0° to 4°C, slightly higher but more variable at 20°C, and greatly diminished at 37°C (T. C. Spelsberg, unpublished). Explanations for this lack of higher temperature requirement for chromatin binding include an instability of the progesterone receptor at higher temperatures and the fact that hormone receptors may be modified by incubations in solutions of higher ionic strength (0.1 M NaCl, 4°C) resulting in similar modifications induced by the increased temperature treatment. This salt activation has been observed with dexamethasone by Baxter et al. (1972) and with estrogen by De-

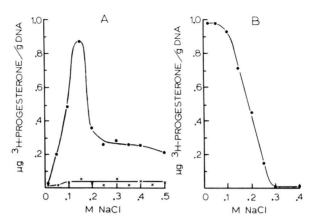

Fig. 9. A, the association of ³H-progesterone, first incubated with (●) oviduct or (×) spleen cytosol, with oviduct chromatin under conditions of increasing ionic strength. The incubations and amounts of cytosol and chromatin are the same as those described in the legend of Fig. 6 except that the incubations were carried out under varying ionic strength and the chromatin was washed and collected on Millipore filters. The radioactivity and DNA per filter were measured as described in the reference cited below and the micrograms of ³H-progesterone bound per gram of DNA were calculated. B, the dissociation of ³H-progesterone from oviduct chromatin by salt. Aliquots of oviduct chromatin (40 μg) were incubated for 1 hour at 4°C with 2 mg of oviduct cytosol, incubated first with 1.5×10^{-8} M ³H-progesterone, in 0.15 M NaCl in the Tris–EDTA buffer. The chromatin was sedimented by centrifugation for 10 minutes at 1200 g, and resuspended in the respective cold NaCl solutions. The chromatin was then collected on Millipore filters and treated as described to obtain the micrograms of ³H-progesterone bound per gram of DNA (Spelsberg et al., 1971a).

Sombre et al. (1972). The labeled progesterone–receptor complex shows maximum binding to isolated chromatin at about 0.15 M NaCl (Fig. 9A). This ionic strength is known to prevent the major portion of soluble proteins in the oviduct cytosol from binding to and covering the chromatin during the incubations (Spelsberg et al., 1971a). However, it may also modify the progesterone receptor to enhance its interaction with chromatin. This salt activation which mimics the temperature activation (for nuclear uptake) with the progesterone receptor has recently been shown to occur (Toft and Spelsberg, in preparation). Further studies indicate that other hormones complexed with their receptors display a more extensive binding to the chromatin of their target tissues than to chromatin of nontarget tissues (Steggles et al., 1971b). In Fig. 10C, labeled estradiol bound to its rat uterine cytosol receptor binds to rat uterine chromatin more extensively than to the chromatin of other rat tissues. Similarly, dihydrotestosterone (Fig. 10B) bound to its rat prostate receptor binds the rat prostate chromatin more extensively than the chromatins of other tissues. Maximum binding to target tissue chromatin by these hormones requires that the hormones be complexed with their receptors (i.e., require the presence of target tissue cytosol) (Steggles et al., 1971b) (Fig. 10).

Mainwaring and Peterkin (1971) using essentially the same conditions as described elsewhere (Spelsberg et al., 1971a; Steggles et al., 1971b) have shown binding of the labeled dihydrotestosterone–receptor complex to prostate chromatin *in vitro*. The prostate chromatin bound more of the hormone than did other (nontarget) chromatins. Hormone complexed to its receptor was required for the extensive and specific binding to the prostate chromatin. McGuire et al. (1972) using rat uteri and labeled estradiol reported that maximal chromatin binding by the estrogen–receptor complexes required elevated temperatures (20°C) during

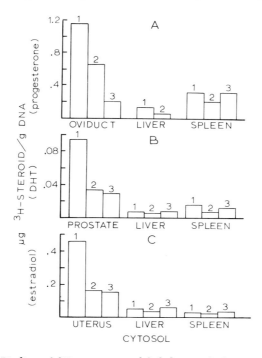

Fig. 10. A, Binding of ³H-progesterone-labeled cytosols from oviduct, liver, and spleen to oviduct (1), liver (2), and spleen (3) chromatin. B, Binding of ³H-DHT-labeled cytosols from prostate, liver, and spleen to prostate (1), liver (2), and spleen (3) chromatin. C, Binding of ³H-estradiol-labeled cytosols from uterus, liver, and spleen to uterine (1), liver (2), and spleen (3) chromatin. In all instances, 2 mg of cytosol protein was incubated with 50–60 μg of chromatin. See the legend of Fig. 9 for details on the chromatin-binding experiments (Steggles et al., 1971b).

the incubations of the hormone, uterine cytosol, and uterine chromatin. Heat pretreatment of the chromatin alone or of the estrogen–uterine cytosol mixture (hormone–receptor) alone before the combined incubations at 4°C resulted in lower binding than when elevated temperatures were used in the combined incubations (hormone, receptor, and chromatin). However, a possible explanation for these results is the fact that the crude cytosol used as the source of estradiol receptor contains high levels of proteases. Incubations of this cytosol at elevated temperatures with uterine chromatin would result in destruction and removal from the chromatin of the very proteolytic-sensitive histones. As discussed later, removal of histones from a target or nontarget tissue chromatin increases the binding of the hormone receptor complex 10- to 20-fold (Spelsberg et al., 1971a; Mainwaring and Peterkin, 1971). In any case, it is not known whether the increased binding of hormone–receptor complexes by the chromatin of target as opposed to nontarget tissues plays a major role in the selective response of target tissues to steroid hormones.

B. Component of Oviduct Chromatin Responsible for the Binding of the Progesterone–Receptor Complex

The identification of the components of chromatin which bind the progesterone–receptor complex was investigated using techniques to dissociate and reconstitute chromatin based on original work reported in Bonner's laboratory (Bekhor et al., 1969) and Paul's laboratory (Gilmour and Paul, 1969). These techniques were modified by Spelsberg and Hnilica (1970) and Spelsberg et al. (1971b) to reduce proteolytic activity and to allow selective dissociation of the histones from the majority of the acidic proteins.

Native (untreated) oviduct chromatin and reconstituted oviduct chromatin with and without the acidic proteins were incubated with the progesterone–receptor complex (the oviduct cytosol containing labeled progesterone). As shown in Fig. 11, the hormone bound to the reconstituted oviduct chromatin (containing all the acidic proteins) as well as to the native oviduct chromatin. However, the chromatin devoid of the acidic proteins showed a marked reduction in the binding of the hormone–receptor complex. This suggested that the acidic proteins were in some manner responsible for the extensive hormone binding by the oviduct (target tissue) chromatin. In support of these results, "hybrid chromatins" (Spelsberg et al., 1971b) were prepared by switching the histones from the chromatin of spleen with those of the oviduct chro-

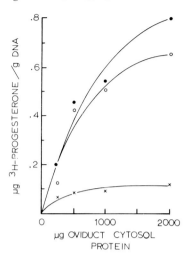

Fig. 11. Binding of ³H-progesterone-labeled oviduct cytosol to 30 to 40 µg of the following: (○) oviduct chromatin; (●) hybrid chromatin reconstituted with de-histonized oviduct chromatin + purified calf thymus histone; and (×) reconstituted chromatin composed of histone of oviduct + purified chick DNA (nonhistone protein removed with phenol treatment) (see legend of Fig. 12 for details on reconstitution) (Spelsberg et al., 1971a).

matin. The progesterone–receptor complex still showed more extensive binding to the reconstituted chromatin containing the spleen histone and oviduct DNA and acidic proteins (Spelsberg et al., 1971a). Since the DNA among the various tissues of an organism is the same, the acidic proteins again appeared to be responsible for the hormone binding. As a final test for which component in chromatin was responsible for the hormone binding *in vitro*, the acidic proteins were exchanged between the oviduct chromatin and other (nontarget) tissue chromatins of the chick (Spelsberg et al., 1972b). The reconstituted erythrocyte or oviduct chromatins bound the progesterone–receptor complex to the same extent as the native (untreated) chromatins. However, as Fig. 12 demonstrates, the substitution of the acidic proteins (but not histones) of oviduct chromatin for those in the erythrocyte chromatin caused a marked increase in the capacity of the former erythrocyte chromatin to bind the hormone. The substitution of the acidic proteins of erythrocyte chromatin for those in the oviduct chromatin caused a marked decrease in the hormone-binding capacity in the former oviduct chromatin. These results again point to the acidic proteins of oviduct chromatin or some entity contained in them as responsible for the extensive hormone binding by that chromatin.

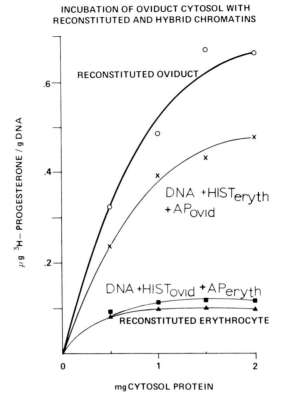

Fig. 12. Binding of ^{3}H-progesterone, previously incubated with oviduct cytosol, to intact and reconstituted chromatins of chick. Reconstituted oviduct chromatin (○); reconstituted erythrocyte chromatin (▲); reconstituted hybrid chromatin containing the DNA and the histones from erythrocyte chromatin and acidic protein from oviduct chromatin (×): and reconstituted hybrid chromatin containing the DNA and the histone from oviduct chromatin and acidic protein from erythrocyte chromatin (■) were used.

The reconstituted homologous and hybrid chromatins were prepared as follows. Chromatin was suspended in dilute buffer (2 mM Tris-HCl + 0.1 mM EDTA, pH 7.5) with a Teflon pestle-glass homogenizer at a concentration of about 0.6 mg of chromatin DNA per ml of buffer. The solution is stirred for 30 minutes at 4°C and 2 volumes of NaCl–urea solution (−20°C) (3.0 M NaCl, 7.5 M urea, 15 mM phosphate buffer, pH 6.0) were added. The solution was mixed vigorously and allowed to stand at 15°C for 2 to 4 hours. The solution was then centrifuged at 110,000 max g for 36 hours at 0°C in a 50.1 angle rotor (Beckman) to sediment the DNA. The supernatant fraction containing the histones and 9% of the acidic proteins was removed and stored for a brief period at −20°C. The DNA pellets and tube walls were rinsed with cold deionized water and the pellet resuspended in a small volume (1.0 mg of DNA per ml) of cold 2 mM Tris-HCl, pH 7.5, with a Teflon pestle-glass homogenizer. The DNA solution was then stirred for 1 hour at 4°C, and then 2 volumes of

C. Purification and Properties of the "Acceptor" Molecule

A fractionation procedure of the acidic chromatin proteins is depicted in Fig. 13 (for further details, see Spelsberg et al., 1972b). Figure 14 represents patterns of these protein fractions separated by SDS–polyacrylamide gel electrophoresis and shows a different profile for each of the fractions. In one series of experiments, the various fractions of partially deproteinized oviduct chromatin, deficient in one or more of the acidic protein fractions (e.g., DNA–$AP_{2,3,3}$; DNA–$AP_{3,4}$; DNA–AP_4), were reconstituted with histone + AP_1 to form reconstituted chromatins lacking

TABLE I
Chemical Analysis and Template Capacity of Untreated and Reconstituted Oviduct Chromatins Devoid of Selected Acidic Protein Fractions[a]

Chromatin[b]	Histone (mg protein/mg DNA)	Acidic protein (mg protein/mg DNA)	Template capacity (nmoles ^{14}C-UMP incorporated/mg DNA)
Control (all AP present) (AP_1, AP_2, AP_3, AP_4)	0.90 ± 0.10	0.86 ± 0.04	48.7 ± 2.3
Minus AP_2	1.00 ± 0.06	0.63 ± 0.02	22.4 ± 3.5
Minus AP_2 plus AP_3	1.16 ± 0.04	0.31 ± 0.02	17.7 ± 1.2
Minus AP_2, AP_3, and AP_4	1.20 ± 0.09	0.20 ± 0.06	18.1 ± 3.0

[a] Spelsberg et al. (1972b).
[b] These reconstituted chromatins were prepared by reconstituting the histone and AP_1 back to DNA preparations containing either (a) AP_2, AP_3, and AP_4, (b) AP_3 and AP_4, or (c) no protein.

the concentrated NaCl–urea solution described above were added to produce a DNA solution containing 2.0 M NaCl, 5.0 M urea, 1.0 mM $NaHSO_3$, 1.0 mM ETDA, 10.0 mM Tris-HCl, pH 8.5. This DNA solution was then mixed with the dissociated histone and acid protein from the same chromatin (to produce reconstituted "homologous" chromatin) or with the proteins from the chromatin of another tissue (to produce reconstituted "hybrid" chromatin). The mixture containing DNA, histone, and acidic protein in a 50-ml solution containing 2.0 M NaCl, 5.0 M urea, 1.0 mM EDTA, 1.0 mM $NaHSO_3$, and 10.0 mM Tris-HCl (pH 8.5) was then placed in washed dialysis tubing. The salt was first dialyzed to 0.01 M NaCl. The urea and the remaining NaCl were then removed by dialysis against 0.01 M phosphate buffer, pH 6.0, and the aggregated chromatin was subjected to the chromatin purification procedure as described (Spelsberg et al., 1971b) The reconstituted homologous or hybrid chromatins were then analyzed for chemical composition and hormone–receptor binding capacity (Spelsberg et al., 1972b).

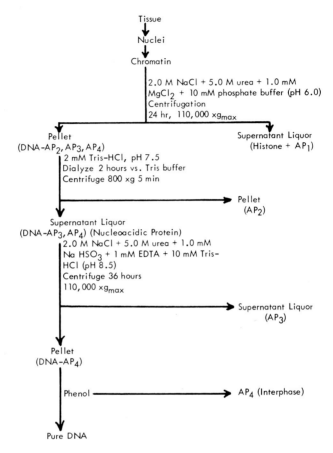

Fig. 13. Flow sheet for the isolation of acidic protein subfractions from the chick oviduct chromatin (Spelsberg et al., 1972b).

one or more of these acidic protein fractions. The reconstituted chromatins were then subjected to the hormone-binding assay. Figure 15 demonstrates that the removal of AP_3 results in major loss of hormone-binding capacity to the reconstituted oviduct chromatins. That this loss of hormone-binding capacity is not due to decreased availability of the DNA in chromatin is shown in Table I. The removal of AP_2 from the chromatin causes about 60% decrease in DNA available for transcription while removal of AP_3 causes only a 10% further decrease.

The AP_3 fraction of oviduct chromatin exhibits from 15 to 20 bands on SDS–polyacrylamide gel patterns (Fig. 14). Antisera was prepared

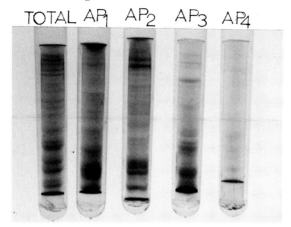

Fig. 14. Polyacrylamide gel electrophoresis of the acidic proteins of the oviduct chromatin of chicks injected for 12 days with DES. From left to right: total acidic proteins, AP_1, AP_2, AP_3, and AP_4. Each gel received about 150 μg of protein. The proteins, were run on 5% acrylamide gels (5 cm long) with a 3% spacer gel (½ cm long) at 3 mA per column until the bromophenol blue marker band reached the bottom of the gels. Migration was toward the anode. After electrophoresis, the gels were washed in 50% (v/v) methanol containing 10% (v/v) acetic acid overnight, stained 1 hour in Coomassie blue and destained and stored in 7% (v/v) acetic acid. The total protein was isolated by an acid–phenol treatment described elsewhere (Spelsberg et al., 1973).

against AP_3 by injecting purified nucleoacidic protein ($DNA–AP_3–AP_4$) into rabbits (Chytil and Spelsberg, 1971). Figure 16 demonstrates that using the technique of microcomplement fixation, the $DNA–AP_3–AP_4$ complex from oviduct, used as an immunogen, is readily recognized by the antibodies (increasing percent complement fixed). The treatment of the $DNA–AP_3–AP_4$ with trypsin decreases the antigenicity somewhat while complete removal of the protein from the DNA eliminates all antigenicity (Fig. 16) (F. Chytil and T. C. Spelsberg, unpublished). Since the AP_3 fraction represents 90% of the total protein in the $DNA–AP_3–AP_4$ complex (Spelsberg et al., 1973), the major portion of the antigenicity should represent the AP_3 fraction. This has recently been shown to be the case in that AP_3 anneals to DNA and shows a high complement fixation (F. Chytil and T. C. Spelsberg, unpublished). Testing nucleoacidic proteins ($DNA–AP_3–AP_4$) from other organs of the chick, it was found that the AP_3 fraction is, antigenically at least, very specific (Fig. 17). These proteins are different even between the undeveloped and fully developed oviduct. Most interesting is the fact that the antiserum against the oviduct AP_3 fraction reacts well with intact chromatin (Fig. 18). Comparison of the pattern of

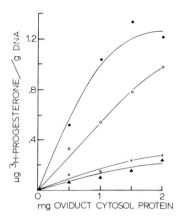

Fig. 15. Binding of ^{3}H-progesterone, preincubated with oviduct cytosol, to the reconstituted oviduct chromatins devoid of selected acidic protein fractions. The histones and acidic protein fraction AP_1 were removed by high salt and urea concentrations at low pH as depicted in Fig. 13. After centrifugation, the DNA–acidic protein pellet was dissolved in dilute buffer, and the insoluble acidic protein fraction AP_2 removed by low speed centrifugation. The insoluble DNA with the remaining acidic proteins was then treated with high salt–urea concentrations at pH 8.5 to dissociate the acidic protein fraction AP_3. Ultracentrifugation sediments the DNA bound to fraction AP_4. This last fraction (AP_4) can only be removed from the DNA with phenol or protease treatment. These reconstituted chromatins were prepared by reconstituting the histone and AP_1 back to DNA preparations containing either (1) AP_2, AP_3, and AP_4; (2) AP_3 and AP_4; or (3) no protein. In this procedure, the relative distributions of total acidic protein into the various subfractions are as follows: AP_1, 15%; AP_2, 40%; AP_3, 35%; and AP_4, 10%. The binding of ^{3}H-progesterone preincubated with oviduct cytosol to the reconstituted oviduct chromatins devoid of selected acidic protein fractions was then assessed. Approximately 50 µg (expressed as DNA) of reconstituted oviduct chromatin containing all acidic proteins (●), minus AP_2 (○), AP_2 and AP_3 (×), or minus AP_2, AP_3, and AP_4 (▲) were each incubated with increasing amounts of labeled cytosol (O'Malley et al., 1972).

fixation obtained when using the nucleoacidic protein and the whole chromatin indicates that about five times as much whole chromatin is needed to obtain the same degree of fixation as the nucleoacidic protein. This suggests that about 80% of the total antigenic sites (represented by the AP_3 fraction) are unavailable (possibly covered by histones) for reaction with the antibodies.

Table II summarizes many properties of the AP_3 fraction which have been obtained in past experiments (T. C. Spelsberg, unpublished). These properties suggest that the "acceptor" in AP_3 which binds the progesterone-receptor complex is a protein and must be bound to DNA to exhibit hormone binding.

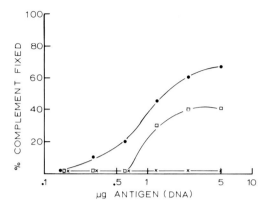

Fig. 16. Complement fixation by varying quantities of treated and untreated nucleoacidic protein (DNA–AP_3–AP_4, see Fig. 13) from the oviduct of chicks injected for 15 days with DES.

The immunogen (chick oviduct nucleoacidic protein) was prepared from immature chicks (Rhode Island Reds) which also received daily injections of diethylstilbestrol (5 mg/day) for 15 days (Chytil and Spelsberg, 1971). After cervical dislocation of the animals, the organs were immediately frozen ($-20°C$) and used for the preparation of nuclei and chromatin. The chromatins were isolated from the nuclei and analyzed for content of DNA and histone and nonhistone protein. The nucleoacidic protein was prepared from the chromatin as described in Fig. 13. (●) represents the untreated oviduct nucleoacidic protein, (□) the nucleoacidic protein which was treated with trypsin for a few minutes, and (×) pure DNA (nucleoacidic protein in which all protein was removed by phenol and pronase treatment) (F. Chytil and T. C. Spelsberg, unpublished).

D. Affinity of the "Acceptor" Molecule for DNA

Isolated AP_3 from oviduct chromatin can be reconstituted to DNA using a gradient dialysis of NaCl in the presence of 5.0 M urea followed by the removal of the urea. The DNA–AP_3 complex binds the progesterone-receptor complex to a greater extent than does DNA. To test the affinity of the "acceptor" entity in the AP_3 fraction for DNA, a series of dialysis systems was designed. Into each dialysis bag was put oviduct AP_3 and chick DNA in 2.0 M NaCl and 5.0 M urea buffered at pH 8.5. The salt in each dialysis bag was then gradually lowered by dialysis against solutions containing 5.0 M urea with succeedingly lower concentrations of salt. The dialysis was stopped at various levels of NaCl, the retentate (containing the DNA and AP_3) removed, layered over 1.75 M sucrose, and centrifuged at 100,000 g for 45 hours to separate unbound protein from the DNA and attached protein. Unbound (free) protein will float on top of the heavy sucrose since 1.75 M sucrose (at 4°C) is more dense (sp. gr. 1.28 gm/cm^3) than protein. The pellet of DNA

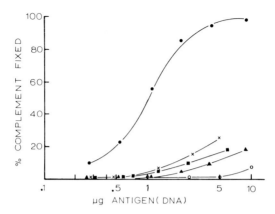

Fig. 17. Complement fixation by varying quantities of nucleoacidic protein from chick oviduct and other organs in the presence of rabbit antiserum. Nucleoacidic protein from chick oviduct prepared as described in the legend of Fig. 15 was used for immunizing rabbits (Chytil and Spelsberg, 1971). The preparation was stored at −20°C, thawed, and homogenized in a Teflon pestle–glass homogenizer before use. Equal volume of complete Freund's adjuvant was added and the mixture was homogenized again. One hundred ten micrograms of nonhistone protein (185 μg of DNA) was injected into male New Zealand rabbit toe pads (hind feet) once a week for two weeks and for six more weeks intramuscularly. The rabbits were bled by cardiac puncture 7 days following the last injection and resulting serum (diluted 1/400) was used for determination of microcomplement fixation by varying amounts of different antigens. The anticomplementarity (binding of the complement in the absence of antiserum) was tested in the whole range of concentration and subtracted from the total complement fixation. (●) Oviduct; (■), liver; (▲), spleen; (×), heart; all from immature chicks injected for 15 days with DES. The oviduct is fully developed in this case. The (○) represents oviducts from untreated chicks in which the oviduct is undifferentiated (Chytil and Spelsberg, 1971).

and attached acidic protein was resuspended in dilute aqueous buffer, analyzed for proteins and DNA, and used in the binding experiment with the progesterone–receptor complex. Figure 19 shows that while the majority of protein begins to anneal to the DNA at 0.05 M and lower NaCl levels, the binding capability of the DNA–protein complex is greatest after dialysis to 0.25 M NaCl (S. H. Socher, T. C. Spelsberg, and B. W. O'Malley, unpublished). Consequently, it appears that the "acceptor" for the progesterone–receptor complex binds to the DNA under condition in which the major portion of the proteins in the AP_3 fraction remain unbound. The proteins of the AP_3 fraction which reconstituted to the DNA at 0.25 M NaCl and at 0.01 M NaCl were isolated and assayed on SDS gel electrophoresis by the method of Wilson and Spelsberg (1973). In Fig. 20, it can be seen that some of the higher

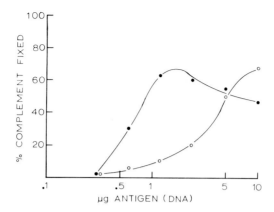

Fig. 18. Complement fixation by varying quantities of chick oviduct chromatin (○) and nucleoacidic protein (●) prepared as described in Fig. 13 in the presence of antisera against the nucleoacidic protein. For details, see legend of Figs. 16 and 17 (Chytil and Spelsberg, 1971).

TABLE II
Properties of the AP_3 Fraction of Acidic Proteins from Chromatin of Fully Mature Oviduct[a]

	Percent	Yes	No	Renaturable by gradient dialysis vs. urea
Percent DNA in AP_3 fraction	0	—	—	—
Percent RNA in AP_3 fraction	9.4	—	—	—
Percent AP_3 in total acidic protein	40–50	—	—	—
Denaturates of hormone binding				
Heat labile	—	X		Yes
Acid labile	—	X		Yes
Organic solvent labile	—	X		Yes
Detergent labile	—	X		Yes
Protease labile	—	X		No
DNase labile	—	X		Yes if DNA added
RNase labile	—		X	—
Antigenic	—	X		—
Tissue specific	—	X		—
Heterogeneous	—	X		—
High affinity for DNA	—	X		—
Requires DNA for binding hormone–receptor complex	—	X		—
Requires homologous DNA for hormonal binding	—		X	—
Soluble in aqueous solutions	—	X		—

[a] Spelsberg (unpublished, 1973).

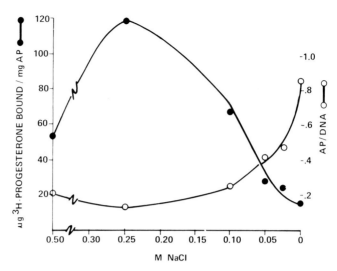

Fig. 19. Reconstitution of chick oviduct AP_3 to chick DNA at various ionic strengths: Effects on progesterone binding. AP_3 was isolated from chick oviduct chromatin by the method shown in Fig. 13. It was reannealed to the DNA by methods described in the legend of Fig. 12 except that certain groups were removed when the ionic strength reached specific salt levels. After removal of salt, the DNA-acidic protein complex still in 5.0 M urea was separated from the unbound AP_3 protein by layering the whole mixture over and sedimenting through 1.8 M sucrose. The pellet of DNA and bound AP_3 proteins was resuspended in buffered 0.15 M NaCl, analyzed for protein and DNA levels, and assayed for hormone binding (S. H. Socher, T. C. Spelsberg, and B. W. O'Malley unpublished).

and a few of the lower molecular weight proteins are absent from the group of proteins which bind the DNA at 0.25 M NaCl compared to the 0.01 M NaCl. It is of interest that the phosphoproteins, isolated according to Shelton and Allfrey (1970), contain many species of protein of the same molecular weight range as the proteins of the AP_3 fraction which bind the DNA at 0.25 M NaCl in 5.0 M urea and which contain the "acceptor" for the progesterone–receptor complex (Fig. 20).

E. Requirements of DNA for "Acceptor" Action of the AP_3 Proteins

Preliminary studies have been largely unsuccessful in attempts to bind the progesterone–receptor complex to free AP_3 (not bound to DNA) (S. H. Socher, W. T. Schrader, T. C. Spelsberg, and B. W. O'Malley, unpublished). Figure 21 shows that when the AP_3 of chick oviduct

9. Steroid Hormone Binding to Acidic Proteins

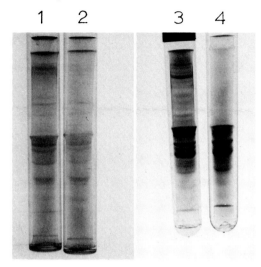

Fig. 20. Polyacrylamide gel electrophoretic patterns of column (1), AP_3 proteins bound to DNA at 0.005 M NaCl in the presence of 5.0 M urea; column (2), AP_3 proteins bound to DNA at 0.25 M NaCl in the presence of 5.0 M urea; column (3), total acidic proteins isolated by the method of Wilson and Spelsberg (1973); and column (4), phenol-soluble proteins (the phosphoproteins) isolated by the method of Shelton and Allfrey (1970). The proteins were run on triphasic SDS–polyacrylamide gels using (from top to bottom): 0.5 cm of 3% acrylamide, 3 cm of 5% acrylamide, and 4 cm of 8.5% acrylamide gels. See legend of Fig. 14 for more details (Wilson and Spelsberg, 1973).

chromatin is annealed to the DNA of various sources under a gradient dialysis of salt to 0.005 M NaCl, the quantity of AP_3 proteins bound shows no preference to the type of DNA. Similarly, the studies on the abilities of these complexes to bind the progesterone–receptor complex based on unit mass of the protein demonstrate a greater binding capacity by the DNA–AP_3 complexes composed of nonhomologous DNA. However, when the DNA–AP_3 complex are formed under the conditions of 0.25 M NaCl minimum salt level, the chick DNA displays the least quantitative binding of protein of the three DNA preparations but shows the greatest capacity to bind the progesterone–receptor complex (again on a protein basis) (S. H. Socher, T. C. Spelsberg, and B. W. O'Malley, unpublished). These results are readily explained by the fact that at the 0.25 M NaCl reconstitution conditions, the "acceptor" molecules reanneal to the DNA of all sources but the majority of other AP_3 proteins bind more to the nonhomologous DNA than to homologous (chick) DNA. Therefore, calculations of hormone-binding capacity per milligram

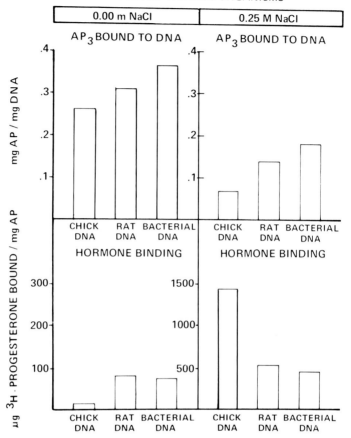

Fig. 21. Level of AP$_3$ protein bound to DNA and progesterone-binding capacity of the DNA–acidic protein complex, reconstituted to 0.005 M NaCl or 0.25 M NaCl in the presence of urea. See legend of Fig. 19 for details of making the reconstituted DNA–acidic protein complex (S. H. Socher, T. C. Spelsberg, and B. W. O'Malley, unpublished).

acidic protein of the reconstituted DNA–AP$_3$ complexes would show a greater binding to the chick DNA (since it has less protein reannealed to it). Unfortunately, no DNA–AP$_3$ complexes could be formed which mimic the high binding capacity per unit DNA for the progesterone–receptor complex as the dehistonized oviduct chromatin or nucleo-acidic protein from oviduct chromatin. Either many "acceptors" are lost or denatured during the handling or the reconstitution methods do not yield sufficiently nativelike complexes.

F. DNA or Protein as the "Acceptor"

The elucidation of whether the "acceptor" for the progesterone–receptor complex is a protein or DNA or both is still unsolved. In all of the studies with AP_3 in the chick oviduct the protein fraction had to be bound to DNA for the binding of the hormone–receptor complex to occur. The AP_3 fraction could serve to maintain certain DNA sites open for binding of the hormone–receptor complex since the free DNA binds much less of the hormone than does dehistonized chromatin or the DNA–AP_3 complex (Table III). The author believes that both are involved. Either the DNA gives rise to a specific stereostructure of an "acceptor" protein which is to be complexed by the progesterone–receptor complex or the AP_3 protein instead establishes specific configurations of certain sequences of DNA to which the hormone–receptor complexes bind.

Interesting results were obtained by Schrader and O'Malley (1972) (see also O'Malley et al., 1972). The cytoplasmic receptor of oviduct for progesterone was separated by ion-exchange chromatography into

TABLE III
Chemical Composition, Template Efficiency, and Binding Capacity of Various Treated Chromatin Preparations[a,b]

Chromatin type	Ratio of histone to DNA	Ratio of nonhistone protein to DNA	% of open template	Amount (μg) ^{3}H-progesterone bound per gm DNA
Oviduct	0.95 ± 0.10	0.50 ± 0.10	3.8 ± 0.3	3.3 ± 0.4
Spleen	0.95 ± 0.06	0.28 ± 0.07	1.5 ± 0.2	1.3 ± 0.2
Heart	0.96 ± 0.12	0.70 + 0.09	8.4 ± 0.6	1.6 ± 0.3
Erythrocyte	0.95 ± 0.05	0.14 ± 0.03	1.6 ± 0.2	1.0 ± 0.1
Dehistonized oviduct	0.06 ± 0.06	0.50 ± 0.05	64 ± 4	22.4 ± 2.0
Dehistonized spleen	0.08 ± 0.05	0.30 ± 0.05	73 ± 7	9.5 ± 0.8
Deproteinized liver chromatin	<0.005	<0.005	100 ± 6	3.8 ± 0.3
Deproteinized oviduct chromatin	<0.005	<0.005	100 ± 8	5.1 ± 1.2
Deproteinized spleen chromatin	<0.005	<0.005	102 ± 4	3.9 ± 0.4

[a] Spelsberg et al. (1971a).
[b] Oviduct cytosol protein (10 mg) containing ^{3}H-progesterone was incubated for 1 hour with 100 μg of each chromatin, the solution centrifuged 18 hours at 120,000 g, and the sedimented chromatin gently washed and collected on Millipore filters. Each value represents the average of two replications.

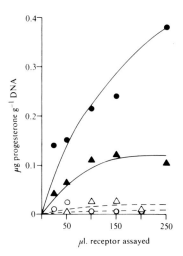

Fig. 22. Binding of purified progesterone receptor components (A) and (B) to chromatin. Increasing amounts of labeled receptor components A and B prepared by DEAE–cellulose chromatography (Schrader and O'Malley, 1972) were incubated for 1 hour at 4°C with 50 μg (expressed as DNA) of chromatin. The figure shows binding of component B to oviduct (●) and spleen (▲) chromatin and of component A to oviduct (○) and spleen (△) chromatin in the hormone-binding assay described in the legend of Fig. 6 (O'Malley et al., 1972).

two receptors "A" and "B." The two receptors display identical hormone-binding kinetics, steroid specificities, and sedimentation values in sucrose gradients. The primary difference is that the "A" receptor is more unstable than the "B" receptor. It also has been shown that the nuclear progesterone receptor is composed of "A" and "B" receptors (Schrader et al., 1972). The "A" and "B" cytoplasmic receptors can be bound in vitro to oviduct nuclei together or individually. Figure 22 shows results of in vitro binding of progesterone receptor "A" and "B" to chromatin. The progesterone receptor "A" of oviduct binds pure DNA but not chromatin while receptor "B" displays a preference for chromatin binding. The receptor "B" also displays the specific binding to the oviduct chromatin compared to other (nontarget) tissue chromatins (Schrader et al., 1972; Schrader and O'Malley, 1972; O'Malley et al., 1972). Consequently, the cytoplasmic and subsequent nuclear receptor of progesterone in the chick oviduct is composed of two components: an "A" component which recognizes DNA and a "B" component which recognizes the chromatin "acceptor." Whether these two components are attached and act coordinately or act at different nuclear sites remains to be determined.

G. Sequence of Events of Progesterone Action on Oviduct Cells

Figure 23 depicts the incomplete series of events which occur for progesterone or any steroid hormone (H) action on target cell. Within minutes after a hormone enters the vascular system, it enters the cytoplasm of a target cell via a temperature-dependent process and binds to the cytoplasmic receptor. The intracellular location of an unbound cytoplasmic receptor is not known. It could be: (1) bound to the cell membrane; (2) bound to the endoplasmic reticulum; (3) free in the cytoplasm; or (4) bound to the nuclear envelope. Once complexed to the hormone, the receptor is released from its restriction of only cytoplasmic existence. The receptor undergoes a temperature-dependent (and possibly salt-dependent) modification before migration to the nucleus. The complex then migrates either to the inner nuclear envelope or inside the nucleus where it binds to chromatin. The progesterone studies in the chick suggest several possible oviduct intranuclear fates for the progesterone receptor. It may: (1) bind to acidic protein of the chromatin in the nucleoplasm or at the nuclear envelope; (2) divide into two

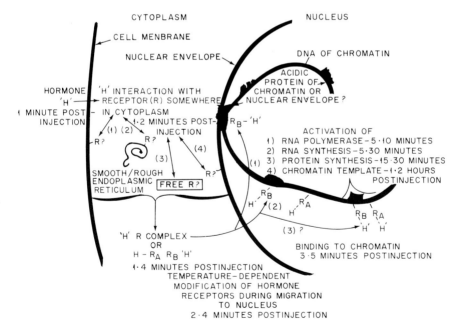

Fig. 23. Model showing the sequence of events which occurs after a steroid hormone enters a target cell. The R_A and R_B refers to receptors "A" and "B" found to exist in the oviduct receptor for progesterone (T. C. Spelsberg, unpublished).

different receptors with one binding the acidic proteins (receptor "B") and the other binding the DNA (receptor "A"); or (3) bind the chromatin through a combination of acidic protein- and DNA-binding sites with both receptor species complexed and working coordinately. In any case, the results of this interaction appears to be an alteration of RNA polymerase activity and RNA synthesis in general followed by increases in protein synthesis in the cytoplasm and finally followed by changes (usually a derepression) of the chromatin restriction of the DNA (see Fig. 1).

In conclusion, I am of the opinion that the steroid hormone receptors are intracellular gene regulators and that there are many other cytonuclear proteins of a similar nature in cells. These other receptors may interact with metabolites and other compounds and relay the information of the presence of these compounds to the control center of the cell (the genetic material) by traveling to the nucleus and interacting with the chromatin. It appears that the steroid receptors serve as positive regulators (usually activating gene expression) and are held in the cytoplasm possibly bound to some cytoplasmic structure. When a steroid enters the cell and binds the receptor, it causes the release of cytoplasmic restricted hormone receptor and allows the receptor to undergo some form of modification and to enter the nucleus. As long as the steroid and receptor are complexed, they are protected but probably not indefinitely. Once dissociated, they are both destroyed or at least the receptor returned immediately to the cytoplasm. The acidic chromatin proteins may serve as "acceptors" through which the steroid hormone-receptor complexes are bound to chromatin. These acidic protein "acceptors" may also be involved in the hormone-induced alterations of gene transcription, but evidence to this possible role remains to be shown.

ACKNOWLEDGMENTS

The author would like to thank Drs. D. Toft, B. W. O'Malley, F. Chytil and S. H. Socher for the use of certain figures. Thanks is also give to Drs. David Toft, John Knowler, and Robert Krueger for their review and criticisms of this article. The author is a fellow of the National Genetics Foundation. The work on this chapter was also sponsored by the Mayo Foundation and Public Health Service grant CA-14920-01.

REFERENCES

Alberga, A., Jung, I., Massol, N., Raynaud, J. P., Raynaud-Jammet, C., Rochefort, H., Truong, H., and Baulieu, E. E. (1971). *Advan. Biosci.* **7**, 45–74.
Anderson, K. M., and Liao, S. (1968). *Nature (London)* **219**, 277–279.

Ausiello, D. A., and Sharp, G. W. G. (1968). *Endocrinology* **82**, 1163–1169.
Barry, J., and Gorski, J. (1971). *Biochemistry* **10**, 2384–2390.
Barton, R. W., and Liao, S. (1967). *Endocrinology* **81**, 409–412.
Baxter, J. D., and Tomkins, G. M. (1970a). *Proc. Nat. Acad. Sci. U.S.* **65**, 709–715.
Baxter, J. D., and Tomkins, G. M. (1970b). *Advan. Biosci.* **7**, 331–344.
Baxter, J. D., Rousseau, G. G., Bensen, C. M., Garcea, R. L., Ito, J., and Tomkins, G. M. (1972). *Proc. Nat. Acad. Sci. U.S.* **69**, 1892–1896.
Beato, M., Biesewig, D., Braendle, W., and Sekeris, C. E. (1969). *Biochim. Biophys. Acta* **192**, 494–507.
Bekhor, I., Kung, G. M., and Bonner, J. (1969). *J. Mol. Biol.* **39**, 351–364.
Billing, R. J., Barbiroli, B., and Smellie, R. M. S. (1968). *Biochem. J.* **109**, 705–706.
Blobel, G., and Potter, V. R. (1966). *Science* **154**, 1662–1665.
Brecher, P. I., Vigersky, R., Wotiz, H. S., and Wotiz, H. S. (1967). *Steroids* **10**, 635–651.
Brecher, P. I., Numata, M., DeSombre, E. R., and Jensen, E. V. (1970). *Fed. Proc., Fed. Amer. Soc. Exp. Biol.* **29**, A249.
Bruchovsky, N., and Wilson, J. D. (1968a). *J. Biol. Chem.* **243**, 2012–2021.
Bruchovsky, N., and Wilson, J. D. (1968b). *J. Biol. Chem.* **243**, 5953–5960.
Buller, R. F., Toft, D. O., and O'Malley, B. W. (1974). In preparation.
Busch, H. (1965). "Histones and Other Nuclear Proteins," pp. 227–243. Academic Press, New York.
Byers, T. J., Platt, D. B., and Goldstein, L. (1963a). *J. Cell Biol.* **19**, 453–466.
Byers, T. J., Platt, D. B., and Goldstein, L. (1963b). *J. Cell Biol.* **19**, 467–476.
Castles, T. R., and Williamson, H. E. (1965). *Proc. Soc. Exp. Biol. Med.* **119**, 308–311.
Chamness, G. C., and McGuire, W. L. (1972). *Biochemistry* **11**, 2466–2572.
Chamness, G. C., Jennings, A. W., and McGuire, W. L. (1973). *Nature (London), New Biol.* **241**, 458–460.
Chauveau, J., Moulé, Y., and Rouiller, C. (1956). *Exp. Cell Res.* **11**, 317–321.
Church, R. B., and McCarthy, B. J. (1970). *Biochim. Biophys. Acta* **199**, 103–114.
Chytil, F., and Spelsberg, T. C. (1971). *Nature (London), New Biol.* **233**, 215–218.
Clark, J. H., and Gorski, J. (1969). *Biochim. Biophys. Acta* **192**, 508–515.
Clemens, L. E., and Kleinsmith, L. J. (1972). *Nature (London), New Biol.* **237**, 204–206.
Crabbé, J., and DeWeer, P. (1964). *Nature (London)* **202**, 298–299.
DeAngelo, A. B., and Gorski, J. (1970). *Proc. Nat. Acad. Sci. U.S.* **66**, 693–700.
DeSombre, E. R., Puca, G. A., and Jensen, E. V. (1969). *Proc. Nat. Acad. Sci. U.S.* **64**, 148–154.
DeSombre, E. R., Mohla, S., and Jensen, E. V. (1972). *Biochem. Biophys. Res. Commun.* **48**, 1601–1608.
Drews, J., and Brawerman, G. (1967). *J. Biol. Chem.* **242**, 801–808.
Edelman, I. S. (1971). *Advan. Biosci.* **7**, 267–275.
Edelman, I. S., and Fimognari, G. M. (1968). *Recent Progr. Horm. Res.* **24**, 1–44.
Edelman, I. S., Bogoroch, R., and Porter, G. A. (1963). *Proc. Nat. Scad. Sci.* **50**, 1169–1177.
Falk, R. J., and Bardin, C. W. (1970). *Endocrinology* **86**, 1059–1063.
Fanestil, D. D., and Edelman, I. S. (1966a). *Fed. Proc., Fed. Amer. Soc. Exp. Biol.* **25**, 912–916.
Fanestil, D. D., and Edelman, I. S. (1966b). *Proc. Nat. Acad. Sci. U.S.* **56**, 872–879.

Fang, S., and Liao, S. (1969). *Mol. Pharmacol.* **5**, 428–431.
Fang, S., Anderson, K. M., and Liao, S. (1969). *J. Biol. Chem.* **244**, 6584–6595.
Feigelson, P., and Feigelson, M. (1964). *In* "Actions of Hormones on Molecular Processes" (G. Litwack and D. Kritchevshy, eds.), pp. 218–233. Wiley, New York.
Forte, L., and Landon, E. J. (1968). *Biochim. Biophys. Acta* **157**, 303–309.
Garren, L. D., Howell, R. R., and Tomkins, G. (1964). *J. Mol. Biol.* **9**, 100–108.
Gelehrter, T. D., and Tomkins, G. (1967). *J. Mol. Biol.* **29**, 59–76.
Giannopoulos, G., and Gorski, J. (1971). *J. Biol. Chem.* **246**, 2524–2529.
Gilmour, R. S., and Paul, J. (1969). *J. Mol. Biol.* **40**, 137–139.
Glascock, R. F., and Hoekstra, W. G. (1959). *Biochem. J.* **72**, 673–682.
Glasser, S. R., Chytil. F., and Spelsberg, T. C. (1972). *Biochem. J.* **130**, 947–957.
Gorski, J. (1964). *J. Biol. Chem.* **239**, 889–892.
Gorski, J., and Nicolette, J. (1963). *Arch. Biochem. Biophys.* **103**, 418–423.
Gorski, J., Toft, D., Shyamala, G., Smith, D., and Notides, A. (1968). *Recent Progr. Horm. Res.* **24**, 45–80.
Greenman, D. L., Wicks, W. D., and Kenney, F. T. (1965). *J. Biol. Chem.* **240**, 4420–4426.
Hamilton, T. H. (1964). *Proc. Nat. Acad. Sci. U.S.* **51**, 83–89.
Hamilton, T. H. (1968). *Science* **161**, 649–661.
Hamilton, T. H., Widnell, C. C., and Tata, J. R. (1965). *Biochim. Biophys. Acta* **108**, 168–172.
Hamilton, T. H., Widnell, C. C., and Tata, J. R. (1968). *J. Biol. Chem.* **243**, 408–417.
Harris, G. S. (1971). *Nature (London), New Biol.* **231**, 246–248.
Herman, T. S., Fimognari, G. M., and Edelman, I. S. (1968). *J. Biol. Chem.* **243**, 3849–3856.
Hnilica, L. S., and Bess, L. G. (1965). *Anal. Biochem.* **12**, 421–436.
Jensen, E. V. *Proc. Int. Congr. Biochem., 4th, Vienna, 1958.* Published as abstract in 1960, Vol. 15, p. 119.
Jensen, E. V. (1966). *Proc. Can. Cancer Res. Conf.* **6**, 143–165.
Jensen, E. V., and DeSombre, E. R. (1972). *Annu. Rev. Biochem.* **41**, 203–230.
Jensen, E. V., and Jacobson, H. I. (1960). *In* "Biological Activities of Steroids in Relation to Cancer" (G. Pincus and E. P. Vollmer, eds.), pp. 161–178. Academic Press, New York.
Jensen, E. V., and Jacobson, H. I. (1962). *Recent Progr. Horm. Res.* **18**, 387–414.
Jensen, E. V., Hurst, J. D., DeSombre, E. R., and Jungblut, P. W. (1967). *Science* **158**, 385–387.
Jensen, E. V., Suzuki, T., Kawashima, T., Stumpf, W. E., Jungblut, P. W., and DeSombre, E. R. (1968). *Proc. Nat. Acad. Sci. U.S.* **59**, 632–638.
Jensen, E. V., Numata, M., Smith, S., Suzuki, T., Brecher, P. I., and DeSombre, E. R. (1969a). *Develop. Biol., Suppl.* **3**, 151–171.
Jensen, E. V., Suzuki, T., Numata, M., Smith, S., and DeSombre, E. R. (1969b). *Steroids* **13**, 417–427.
Jensen, E. V., Numata, M., Brecher, P. I., and DeSombre, E. R. (1971). *Biochem. Soc. Symp.* **32**, 133–159.
Jungblut, P. W., Hatzel, I., DeSombre, E. R., and Jensen, E. V. (1967). *Colloq. Ges. Physiol. Chem.* **18**, 58–86.
Kenney, F. T., Wicks, W., and Greenman, D. (1965). *J. Cell. Comp. Physiol.* **66**, Suppl. 1, 125–136.

Kidson, C. (1967). *Nature (London)* 213, 770–782.
Kidson, C., and Kirby, K. S. (1964). *Nature (London)* 203, 599–603.
King, R. J. B., and Gordon, J. (1967). *J. Endocrinol.* 39, 533–542.
King, R. J. B., and Gordon, J. (1969). *Biochem. J.* 112, 32.
King, R. J. B., and Gordon, J. (1972). *Nature (London), New Biol.* 240, 185–187.
King, R. J. B., Gordon, J., and Martin, L. (1965a). *Biochem. J.* 97, 28.
King, R. J. B., Gordon, J., and Inman, D. R. (1965b). *J. Endocrinol.* 32, 9–15.
King, R. J. B., Gordon, J., Cowan, D. M., and Inman, D. R. (1966). *J. Endocrinol.* 36, 139–150.
King, R. J. B., Gordon, J., and Steggles, A. W. (1969). *Biochem. J.* 114, 649–657.
King, R. J. B., Gordon, J., Marx, J., and Steggles, A. W. (1971a). *In* "Basic Actions of Sex Steroids on Target Organs" (P. O. Hubinont, F. Le Roy, and P. Galand, eds.), pp. 21–43. Karger, Basel.
King, R. J. B., Beard, V., Gordon, J., Pooley, A. S., Smith, J. A., Steggles, A. W., and Vertes, M. (1971b). *Advan. Biosci.* 7, 21–44.
Knowler, J. T., and Smellie, R. M. S. (1971). *Biochem. J.* 125, 605–614.
Korenman, S. G. (1969). *Steroids* 13, 163–177.
Kroeger, H., Jacob, J., and Sirlin, J. L. (1963). *Exp. Cell Res.* 31, 416–423.
Kyser, K. A. (1970). Ph.D. Dissertation, University of Chicago, Chicago, Illinois.
Liang, T., and Liao, S. (1972). *Biochim. Biophys. Acta* 277, 590–594.
Liao, S., and Fang, S. (1969). *Vitam. Horm. (New York)* 27, 17–90.
Liao, S., and Lin, A. H. (1967). *Proc. Nat. Acad. Sci. U.S.* 57, 379–386.
Liao, S., Barton, R. W., and Lin, A. H. (1966). *Proc. Nat. Acad. Sci. U.S.* 55, 1593–1600.
Liao, S., Tymoczko, J. L., Liang, T., Anderson, K. M., and Fang, S. (1971). *Advan. Biosci.* 7, 155–160.
Litwack, G., and Morey, K. S. (1970). *Biochem. Biophys. Res. Commun.* 38, 1141–1148.
McGuire, J. L., and DeSella, C. (1971). *Endocrinology* 88, 1099–1103.
McGuire, W. L., Huff, K., and Chamness, G. C. (1972). *Biochemistry* 11, 4562–4565.
Mainwaring, W. I. P. (1969a). *J. Endocrinol.* 43, XXXVII–XXXVIII.
Mainwaring, W. I. P. (1969b). *J. Endocrinol.* 44, 323–333.
Mainwaring, W. I. P. (1969c). *J. Endocrinol.* 45, 531–541.
Mainwaring, W. I. P., and Mangan, F. R. (1971). *Advan. Biosci.* 7, 165–172.
Mainwaring, W. I. P., and Peterkin, B. M. (1971). *Biochem. J.* 125, 285–295.
Maurer, H. R., and Chalkley, G. R. (1967). *J. Mol. Biol.* 27, 431–441.
Means, A. R., and Hamilton, T. H. (1966). *Proc. Nat. Acad. Sci. U.S.* 56, 686–693.
Milgrom, E., Atger, M., and Baulieu, E. E. (1970). *Steroids* 16, 741–754.
Morey, K. S., and Litwack, G. (1969). *Biochemistry* 8, 4813–4821.
Munck, A., and Brinck-Johnson, T. (1968). *J. Biol. Chem.* 243, 5556–5565.
Munck, A., and Wira, C. (1971). *Advan. Biosci.* 7, 301–327.
Munns, T. W., and Katzman, P. A. (1971). *Biochemistry* 10, 4941–4948.
Musliner, T. A., Chader, G. J., and Villee, C. A. (1970). *Biochemistry* 9, 4448–4453.
Nicolette, J. A., and Mueller, G. C. (1966). *Biochem. Biophys. Res. Commun.* 24, 851–857.
Noteboom, W. D., and Gorski, J. (1963). *Proc. Nat. Acad. Sci. U.S.* 50, 250–255.
Noteboom, W. D., and Gorski, J. (1965). *Arch. Biochem. Biophys.* 11, 559–568.
O'Malley, B. W. (1969). *Trans. N.Y. Acad. Sci.* [2] 31, 478–503.
O'Malley, B. W. (1971). *Metab., Clin. Exp.* 20, 981–988.

O'Malley, B. W., McGuire, W. L., Kohler, P. O., and Korenman, S. G. (1969). *Recent Progr. Horm. Res.* **25**, 105–160.
O'Malley, B. W., Sherman, M. R., and Toft, D. O. (1970). *Proc. Nat. Acad. Sci. U.S.* **67**, 501–508.
O'Malley, B. W., Toft, D. O., and Sherman, M. R. (1971a). *J. Biol. Chem.* **246**, 1117–1122.
O'Malley, B. W., Sherman, M. R., Toft, D. O., Spelsberg, T. C., Schrader, W. T., and Steggles, A. W. (1971b). *Advan. Biosci.* **7**, 213–231.
O'Malley, B. W., Spelsberg, T. C., Schrader, W. T., Chytil, F., and Steggles, A. W. (1972). *Nature (London)* **235**, 141–144.
Porter, G. A., Bogoroch, R., and Edelman, I. S. (1964). *Proc. Nat. Acad. Sci. U.S.* **52**, 1326–1333.
Prescott, D. M., and Bender, M. A. (1963). *J. Cell. Comp. Physiol.* **62**, 175–194.
Puca, G. A., and Bresciani, F. (1968). *Nature (London)* **218**, 967–969.
Puca, G. A., and Bresciani, F. (1970). *Nature (London)* **225**, 1251–1252.
Puca, G. A., Nola, E., Sica, V., and Bresciani, F. (1971a). *Biochemistry* **10**, 3769–3780.
Puca, G. A., Nola, E., Sica, V., and Bresciani, F. (1971b). *Advan. Biosci.* **7**, 97–118.
Raina, P. N., and Rosen, F. (1968). *Biochim. Biophys. Acta* **165**, 470–475.
Rao, B. R., and Wiest, W. G. (1971). *Fed. Proc., Fed. Amer. Soc. Exp. Biol.* **30**, 1213.
Raspé, G. ed. (1971). Advances in Biosciences, Vol. 7. Pergamon, Oxford.
Reif-Lehrer, L., and Amos, H. (1967). *Biochem. J.* **106**, 425–430.
Reti, I., and Erdos, T. (1971). *Biochimie* **53**, 435–437.
Rochefort, H., and Baulieu, E. E. (1969). *Endocrinology* **84**, 108–116.
Rogers, A. W., Thomas, G. H., and Yates, K. M. (1966). *Exp. Cell Res.* **40**, 668–670.
Sar, M., Liao, S., and Stumpf, W. E. (1969). *Fed. Proc., Fed. Amer. Soc. Exp. Biol.* **28**, 707.
Sar, M., Liao, S., and Stumpf, W. E. (1970). *Endocrinology* **86**, 1008–1010.
Sarff, M., and Gorski, J. (1971). *Biochemistry* **10**, 2557–2563.
Schrader, W. T., and O'Malley, B. W. (1972). *J. Biol. Chem.* **247**, 51–59.
Schrader, W. T., Toft, D. O., and O'Malley, B. W. (1972). *J. Biol. Chem.* **247**, 2401–2407.
Sekeris, C. E., and Lang, N. (1965). *Hoppe Seyler's Z. Physiol. Chem.* **340**, 92–94.
Sharp, G. W. G., and Alberti, K. G. M. (1971). *Advan. Biosci.* **7**, 281–295.
Shelton, K. R., and Allfrey, V. G. (1970). *Nature (London)* **228**, 132–134.
Sherman, M. R., Corvol, P. L., and O'Malley, B. W. (1970). *J. Biol. Chem.* **245**, 6085–6086.
Shyamala, G., and Gorski, J. (1969). *J. Biol. Chem.* **244**, 1097–1103.
Sluyser, M. (1966a). *J. Mol. Biol.* **19**, 591–595.
Sluyser, M. (1966b). *J. Mol. Biol.* **22**, 411–414.
Sluyser, M. (1969). *Biochim. Biophys. Acta* **182**, 235–244.
Spelsberg, T. C., and Hnilica, L. S. (1970). *Biochem. J.* **120**, 435–437.
Spelsberg, T. C., Steggles, A. W., and O'Malley, B. W. (1971a). *J. Biol. Chem.* **246**, 4188–4197.
Spelsberg, T. C., Hnilica, L. S., and Ansevin, A. T. (1971b). *Biochim. Biophys. Acta* **228**, 550–562.
Spelsberg, T. C., Wilhelm, J. A., and Hnilica, L. S. (1972a). *Sub-Cell. Biochem.* **1**, 107–145.

Spelsberg, T. C., Steggles, A. W., Chytil, F., and O'Malley, B. W. (1972b). *J. Biol. Chem.* **247**, 1368–1374.
Spelsberg, T. C., Mitchell, W. M., Chytil, F., Wilson, E. M., and O'Malley, B. W. (1973). *Biochim. Biophys. Acta* **312**, 765–778.
Stancel, G. M., Leung, K. M. T., and Gorski, J. (1973a). *Biochemistry* **12**, 2130–2136.
Stancel, G. M., Leung, K. M. T., and Gorski, J. (1973b). *Biochemistry* **12**, 2137–2141.
Steggles, A. W., Spelsberg, T. C., and O'Malley, B. W. (1971a). *Biochem. Biophys. Res. Commun.* **43**, 20–27.
Steggles, A. W., Spelsberg, T. C., Glasser, S. R., and O'Malley, B. W. (1971b). *Proc. Nat. Acad. Sci. U.S.* **68**, 1479–1482.
Stellwagen, R. H., and Cole, R. D. (1968a). *J. Biol. Chem.* **243**, 4452–4455.
Stellwagen, R. H., and Cole, R. D. (1968b). *J. Biol. Chem.* **243**, 4456–4462.
Stone, G. M., and Baggett, B. (1965). *Steroids* **6**, 277–299.
Stumpf, W. E. (1968). *Endocrinology* **83**, 777–782.
Stumpf, W. E. (1969). *Endocrinology* **85**, 31–37.
Sunaga, K., and Koide, S. S. (1967a). *Arch. Biochem. Biophys.* **122**, 670–673.
Sunaga, K., and Koide, S. S. (1967b). *Biochem. Biophys. Res. Commun.* **26**, 342–348.
Sunaga, K., and Koide, S. S. (1967c). *Steroids* **9**, 451–456.
Swaneck, G. E., Highland, E., and Edelman, I. S. (1969). *Nephron* **6**, 297–316.
Swaneck, G. E., Chu, L. L. H., and Edelman, I. S. (1970). *J. Biol. Chem.* **245**, 5382–5389.
Szego, C. M. (1965). *Fed. Proc., Fed. Amer. Soc. Exp. Biol.* **24**, 1343–1352.
Talwar, G. P., Segal, S. J., Evans, A., and Davidson, O. W. (1964). *Proc. Nat. Acad. Sci. U.S.* **52**, 1059–1066.
Tata, J. R. (1966). *Progr. Nucl. Acid Res. Mol. Biol.* **5**, 191–250.
Teng, C. S., and Hamilton, T. H. (1968). *Proc. Nat. Acad. Sci. U.S.* **60**, 1410–1417.
Terenius, L. (1966). *Acta Endocrinol.* *(Copenhagen)* **53**, 611–618.
Terenius, L. (1971). *Acta Endocrinol.* *(Copenhagen)* **66**, 431–447.
Toft, D. (1972). *J. Steroid Biochem.* **3**, 515–522.
Toft, D. O. (1973a). *In* "Annual of Obstetrics and Gynecology, 1973" (R. M. Wynn, ed.), pp. 405–430. Appleton, New York.
Toft, D. O. (1973b). *Advan. Exp. Med. Biol.* **36**, 85–96.
Toft, D., and Gorski, J. (1966). *Proc. Nat. Acad. Sci. U.S.* **55**, 1574–1581.
Toft, D., Shyamala, G., and Gorski, J. (1967). *Proc. Nat. Acad. Sci. U.S.A.* **57**, 1740.
Trachewsky, D., and Segal, S. J. (1967). *Biochem. Biophys. Res. Commun.* **27**, 588–594.
Trachewsky, D., and Segal, S. J. (1968). *Eur. J. Biochem.* **4**, 279–285.
Tsai, Y. H., and Hnilica, L. S. (1971). *Biochim. Biophys. Acta* **238**, 277–287.
Tveter, K. J., and Attramadal, A. (1968). *Acta Endocrinol.* *(Copenhagen)* **59**, 218–226.
Tveter, K. J., and Attramadal, A. (1969). *Endocrinology* **85**, 350–354.
Wagner, T. E. (1970). *Biochem. Biophys. Res. Commun.* **38**, 890–893.
Wiest, W. G., and Rao, B. R. (1971). *Advan. Biosci.* **7**, 251–264.
Williams, D., and Gorski, J. (1971). *Biochem. Biophys. Res. Commun.* **45**, 258–264.
Williamson, H. E. (1963). *Biochem. Pharmacol.* **12**, 1449–1450.
Wilson, E. M., and Spelsberg, T. C. (1973). *Biochim. Biophys. Acta* (in press).
Wilson, J. D. (1963). *Proc. Nat. Acad. Sci. U.S.* **50**, 93–100.
Wilson, J. D., and Loeb, P. M. (1965). *In* "Developmental and Metabolic Control

Mechanisms and Neoplasia," pp. 375–391. Williams & Wilkins, Baltimore, Maryland.

Wira, C., and Munck, A. (1969). *Fed. Proc., Fed. Amer. Soc. Exp. Biol.* **28**, 702.

Wira, C., and Munck, A. (1970a). *Fed. Proc., Fed. Amer. Soc. Exp. Biol.* **29**, A832.

Wira, C., and Munck, A. (1970b). *J. Biol. Chem.* **245**, 3436–3438.

Young, D. A. (1970). *Fed. Proc., Fed. Amer. Soc. Exp. Biol.* **29**, 3006.

10

The Role of Acidic Proteins in Gene Regulation

R. STEWART GILMOUR

I. Introduction 297
II. Isolation and Characterization of Acidic Proteins 298
III. Distribution and Specificity of the Acidic Proteins 300
IV. Some Metabolic Aspects of Acidic Proteins 301
V. The Biological Assessment of the Acidic Proteins 303
VI. Gene Regulation in Eukaryotes 307
References 313

I. INTRODUCTION

The idea that chromosomal proteins might be involved in the control of gene expression is almost 30 years old. The original proposal by Stedman and Stedman (1950) that histones might act as gene repressors has been supported by subsequent experimental data (Allfrey et al., 1963; Huang and Bonner, 1962); however, their suggestion that histones might be responsible for the repression of specific genes is not generally accepted. This is due mainly to the demonstration that, with a few exceptions, there are no qualitative differences in the types of histones found in different tissues. Also the more recent investigations of Paul and Gilmour (1968) in which chromatin was reconstituted from its component parts suggest that the ability of histones alone to repress DNA transcription occurs in a nonspecific manner.

It is against this background that the possible role of acidic proteins as specific gene regulators is set. The scope of the present review is to consider whether the extent of our present knowledge of these proteins is compatible with a regulatory function and then to consider models which speculate on their possible mode of action.

II. ISOLATION AND CHARACTERIZATION OF ACIDIC PROTEINS

It is important to appreciate that our present knowledge of chromosomal acidic proteins has come from studies using a variety of methods of isolation and analysis. The reader is referred to the recent review by MacGillivray and Rickwood (1973a) for a more detailed discussion of the various procedures. These can be summarized briefly as follows.

After removal of histones from chromatin with acid, acidic proteins have been separated from DNA by phenol (Teng et al., 1971 Shelton and Neelin, 1971), detergent (Elgin and Bonner, 1970), or salt (Benjamin and Gellhorn, 1968). Alternatively chromatin has been dissociated in SDS (sodium dodecyl sulfate) (Shirey and Huang, 1969) or in a variety of salt solutions (Graziano and Huang, 1971; Hill et al., 1971; Levy et al., 1972; Spelsberg et al., 1972, 1973; Umansky et al., 1971; Wang, 1967; Richter and Sekeris, 1972; Arnold and Young, 1972; Yoshida and Shimura, 1972; Shaw and Huang, 1970) and the DNA removed by centrifugation (Levy et al., 1972; Umansky et al., 1971; Richter and Sekeris, 1972; Arnold and Young, 1972; Shaw and Huang, 1970), gel filtration (Graziano and Huang, 1971; Hill et al., 1971; Levy et al., 1972; Shaw and Huang, 1970), or precipitation with histones (Wang, 1967) or heavy metal ions (Yoshida and Shimura, 1972). The remaining acidic proteins are isolated pure (Wang, 1967) or subsequently separated from contaminating histones by ion-exchange chromatography (Graziano and Huang, 1971; Hill et al., 1971; Levy et al., 1972; Umansky et al., 1971; Richter and Sekeris, 1972; Arnold and Young, 1972; Yoshida and Shimura, 1972) or by electrophoresis (Shirey and Huang, 1969; Shaw and Huang, 1970). In this laboratory a high recovery of acidic proteins is achieved by a single hydroxylapatite column fraction of salt-urea dissociated chromatin (MacGillivray et al., 1972).

The important point to emphasize is that a number of these methods achieve only a selective extraction of acidic proteins with no guarantee that the extracted material is representative of the whole or that the selection is constant when applied to different tissues. Such criticism can be leveled at the salt dissociation procedures of Wang (1967) and Benjamin and Gellhorn (1968), and phenol procedure as used by a number of workers (Teng et al., 1971; Shelton and Allfrey, 1970; Le-

Stourgeon and Rusch, 1971; Tsuboi and Baserga, 1972), and the methods for the isolation of the phosphoprotein component of acidic proteins (Langan, 1967; Kleinsmith and Allfrey, 1969; Gershey and Kleinsmith, 1969). On the other hand, those procedures which dissociate chromatin in SDS (Elgin and Bonner, 1970; Shirey and Huang, 1969) or high concentrations of salt–urea or guanidine hydrochloride (MacGillivray et al., 1972; Hill et al., 1971; Levy et al., 1972; Shaw and Huang, 1970) followed by removal of DNA and histones usually give a good recovery of acidic proteins representative of the total species present in the original chromatin.

The characterization of acidic proteins is complicated by a tendency to aggregate with themselves and with histones and nucleic acid. Hence conventional electrophoresis of acidic proteins proved difficult and even those investigators who have reported reasonable electrophoretic separations still find considerable amounts of material fail to enter the electrophoresis matrix (Graziano and Huang, 1971; Levy et al., 1972; Arnold and Young, 1972; Shaw and Huang, 1970; Kostraba and Wang, 1970; Elgin and Bonner, 1972; Wang, 1971). Hence many workers completely dissociate acidic protein preparations in SDS (often in the presence of urea) and electrophorese in SDS containing polyacrylamide gels (MacGillivray et al., 1972; Shelton and Neelin, 1971; Elgin and Bonner, 1970). This achieves a separation based on differences in the molecular weights of the denatured proteins; however, any differences in primary structure between proteins of the same molecular weight would not be detected. In order to obtain better resolution various workers have combined ion-exchange chromatography and electrophoresis in SDS and other systems (Levy et al., 1972; Richter and Sekeris, 1972; Elgin and Bonner, 1972; MacGillivray and Richwood, 1973b). For example, in this laboratory (MacGillivray and Richwood, 1973b) fully reduced acidic proteins have been separated by a two-dimensional method involving isoelectric focusing in polyacrylamide gels in the first dimension and then SDS electrophoresis in the second dimension. The increased resolution achieved emphasizes the limitations of the one-dimensional SDS–gel system by revealing a wide range of isoelectric points and molecular weights within acidic protein preparation.

Despite the criticisms concerning the isolation procedures and the limitations of the characterization methods, the evidence to date shows clearly that the acidic proteins of chromatin are extremely complex and largely of high molecular weight. The possibility that cytoplasmic contamination contributes to the heterogeneity has been raised (Hill et al., 1971; Wang, 1967; Johns and Forrester, 1969; Goodwin and Johns, 1972; Harlow et al., 1972; Bhorjee and Pederson, 1972) and in some cases eliminated by careful preparative procedures (MacGillivray *et al.*, 1972; Bhorjee and Pederson, 1972; Wilhelm *et al.*, 1972b). It should also be noted from hormone localization (Means *et al.*, 1973) and cell activation studies *in vivo* (Carlsson *et al.*, 1973) that some proteins can migrate from the cytoplasm to the nucleus during the course of normal metabolism.

III. DISTRIBUTION AND SPECIFICITY OF THE ACIDIC PROTEINS

It might be thought on *a priori* grounds that if acidic proteins control the specificity of DNA transcription then marked differences should be observed between different tissues and species. As mentioned in the preceding chapter the methods for isolating acidic proteins can be classified according to whether they yield a partial or representative selection of species.

Isolation precedures which select only a fraction of the acidic proteins usually show considerable qualitative differences between tissues (Kostraba and Wang, 1970; Teng *et al.*, 1971; Kruh *et al.*, 1969, 1970). In contrast, methods which yield a representative extraction of acidic proteins from chromatin show little if any tissue specificity. Thus, while Elgin and Bonner (1970) found considerable heterogeneity in the acidic proteins from several tissues and species as judged by SDS–gel electrophoresis, the patterns obtained were remarkably similar in all cases. A similar lack of specificity has been reported in the acidic proteins of pig (Shaw and Huang, 1970), chicken (Shelton and Neelin, 1971), bovine, and mouse tissues (MacGillivray *et al.*, 1972).

This absence of cell specificity has led to a number of comparisons of the acidic proteins in fractionated chromatin, in the hope that active euchromatin would show distinct differences from inactive heterochromatin. The original experiments of Frenster (1965) indicated that there was twice the amount of acidic protein and four times the amount of phosphoprotein in the euchromatin fraction. However recent investigations have failed to show significant differences in the SDS–gel patterns of nonhistone proteins of eu- and heterochromatin (Wilhelm *et al.*, 1972a; Gronow, 1972; D. Rickwood and A. J. MacGillivray, unpublished results).

Because of the variation in preparative procedures in general use it is obviously difficult to make direct comparisons between results from different laboratories. In addition, objective conclusions can only be drawn from experiments employing total extraction methods. With these provisos it can be tentatively concluded that little tissue specificity can be found among these proteins within the limits of currently available analytical procedures. This could be explained on the basis that the bulk of the acidic proteins comprise enzymes and structural proteins which may be common to all chromatins, while specific regulatory proteins, if they exist, might be present in such small quantities so as to remain undetected by the usual analyses.

IV. SOME METABOLIC ASPECTS OF ACIDIC PROTEINS

An alternative approach to the identification of specific acidic proteins is provided by biological systems undergoing developmental change or gene activation. For example, during development in the sea urchin, varying degrees of quantitative and qualitative changes in the acidic proteins of embryo chromatin have been reported between blastomere and hatching blastula stages (Cognetti et al., 1972; Connor and Patel, 1972), while the transition from blastula to pluteus is accompanied only by an increase in the amount of all acidic protein species (Hill et al., 1971).

There have been a number of reports of changes in the acidic protein complement of target organs after treatment with hormones. Teng and Hamilton (1970) found that estrogen caused the appearance of new acidic proteins in uterine tissue, while Shelton and Allfrey (1970) showed that cortisol caused the appearance of a newly synthesized acid protein in liver chromatin. A similar situation was found in the polytene chromosomes of *Drosophila* during specific puffing in response to ecdysone or temperature shock, except that here the additional protein did not arise from *de novo* synthesis (Helmsing and Berendes, 1971; Helmsing, 1972).

The appearance of a specific group of phenol-soluble acidic proteins during differentiation of a slime mold has also been reported by LeStourgeon and Rusch (1971).

There has been considerable interest in the phosphoprotein fraction of acidic proteins since it has been shown from ^{32}P-labeling experiments that acidic proteins are phosphorylated to a greater extent than the histones (Rickwood et al., 1973; Schiltz and Sekeris, 1969, 1971; Ahmed and Ishida, 1971; Lurguin et al., 1972). In a number of cases increase in phosphorylation can be shown to follow specific stimuli. For example, testosterone has been found to cause increased phosphorylation of ventral prostate acidic proteins when isolated nuclei are incubated with ^{32}P-ATP (Ahmed and Ishida, 1971). Similar effects have been observed during phytohemagglutinin stimulation of lymphocytes (Kleinsmith et al., 1966), prolactin activation of mammary glands (Turkington and Riddle, 1969), gonadotropin activation of ovaries (Jungmann and Schweppe, 1972), and during the cell cycle of synchronously dividing HeLa cells (Platz et al., 1973). Conversely during maturation of avian red blood cells there is a decrease in the levels of phosphorylation (Gershey and Kleinsmith, 1969).

The possibility that the phosphorylation of acidic proteins shows a tissue-specific pattern was investigated by Rickwood et al. (1973) by incubating isolated nuclei from a number of tissues with ^{32}P-ATP. SDS–gel analysis showed that many of the phosphoproteins were common to several mouse tissues; however, a few tissue specific proteins were observed. Interestingly, the labeled phosphoproteins of Landschutz ascites chromatin differed markedly from those of other mouse tissues. Similar results have been reported by Platz et al. (1970).

Another aspect of acidic protein metabolism which has received considerable attention is their appearance and turnover during the cell cycle. While histones are mainly synthesized along with DNA during S phase, acidic proteins appear to be synthesized throughout the cell cycle at a high rate (Stein and Borun, 1972; Cross, 1972; McClure and Hnilica, 1970; Stein and Baserga, 1970).

In HeLa cells it has been shown that a number of individual proteins are specifically synthesized at different stages of the cell cycle (Bhorjee and Pederson, 1972; Stein and Borun, 1972; Borun and Stein, 1972); however, this contrasts with the failure to find any such variations in the cell cycles of a slime mold (LeStourgeon and Rusch, 1971) and hamster fibroblasts (Becker and Stanners, 1972).

It is clear from the foregoing discussion that while a general tissue comparison of the acidic proteins may not reveal marked differences, some qualitative differences can be found in special instances of gene activation or growth response. Some of these characteristics would be compatible with acidic proteins performing a regulatory role. However, the finding that different acidic proteins appear in response to changes in the differentiated state or growth pattern of a cell is, on reflection, not surprising. Considering the heterogeneous nature of the acidic proteins and the fact that they probably contain numerous important nuclear enzymes, the findings could be explained equally well by the induction or increased synthesis of one or more enzyme proteins. While this approach may prove useful in highlighting possible regulatory molecules, any argument concerning a regulatory function will be indecisive unless there also exists a specific test for the biological importance of the proteins in question.

V. THE BIOLOGICAL ASSESSMENT OF THE ACIDIC PROTEINS

Numerous attempts have been made to assess the biological relevance of the acidic proteins by studying the *in vitro* transcription of RNA from chromatin. Paul and Gilmour (1968) isolated RNA transcribed by a bacterial polymerase from different chromatin templates and compared them by hybridization to homologous DNA. These studies showed distinct differences between RNA transcribed from DNA and RNA derived from chromatin in that only certain specific sequences in DNA were transcribed in chromatin. Moreover, it could be shown by competitive hybridization that the RNA transcribed from chromatins of different tissues showed significant differences in the specific DNA sequences transcribed. Based on observations of this kind attempts have been made to identify by reconstitution experiments which components of chromatin are responsible for conferring specificity of transcription. Individual components of chromatin were recombined by a reconstitution procedure which involved dissociating the components in high salt–urea solutions followed by a progressive lowering of the salt concentration. These experiments showed that whereas hardly any hybridizable RNA was obtained when DNA was combined with histones alone, species of RNA similar to those transcribed from native chromatin were obtained when DNA and histones were reconstituted in the presence of acidic proteins (Gilmour and Paul, 1969). In later experiments (Gilmour and Paul, 1970) evidence was obtained for organ-specific reconstitution by combining acidic proteins of different tissues with DNA and histones. Similar reconstitution experiments carried out by Bekhor et al. (1969), and Huang and Huang (1969) also showed that organ-specific reconstitution was a property of the nonhistone fraction of chromatin; however, these workers claimed that a particular species of RNA (chromosomal RNA) rather than acidic protein was responsible.

In the experiments of Spelsberg and Hnilica (1970) and Spelsberg et al. (1973) histones were removed from chromatin to give a DNA–acidic protein complex. They found that when histones were added back to the complex, the organ specificity was restored and that this was entirely dependent on the source of the DNA–acidic protein complex and independent of the histone source. This is also supported by the findings that the DNA–acidic protein complex has antigenic specificity which is characteristic of its tissue of origin (Chytil and Spelsberg, 1971).

Additional evidence that acidic proteins can antagonize histones in a specific way has come from the work of Wang (1970) in which a fraction of rat liver acidic protein when added to condensed chromatin could cause the transcription of new sequences of DNA. Further studies of Kamiyama and Wang (1971) identified a phosphoprotein fraction of the acidic protein as being responsible for this effect.

Reconstitution experiments have also been used to investigate the qualitative changes seen in the acidic proteins after estrogen action. Andress et al. (1972) have been able to correlate the estrogen-induced stimulation of endometrium chromatin template activity with phosphorlation of acidic protein. Similarly, Chaudhuri et al. (1972) employed chromatin reconstitution to show that the increased chromatin template activity found after fibroblasts are stimulated to proliferate is associated with the changes in the acidic protein fraction. In further experiments Stein and Farber (1972) found that chromatin reconstituted with acidic proteins from tissue culture cells in S phase gave a higher rate of synthesis than chromatin reconstituted with the acidic proteins from the mitotic phase of the same cells.

In some of these studies the rate of RNA synthesis from chromatin has been used as a measure of the amount of DNA template it contains. The validity of this assumption has been challenged by Paul and Gilmour (1966) on the basis that rates can be affected by many other factors and that DNA-RNA hybridization offers a more direct measurement of template capacity. However the use of low C_0t DNA-RNA hybridization as a test for template capacity has also been criticized because it only yields information about the repetitive fraction of the DNA. It is often inferred from hybridization experiments that tissue specificity in chromatin can be equated with the transcription of specific messenger RNA molecules; however, it is not certain whether these RNA molecules would show up in such a test. Indeed the recent findings of Harrison et al. (1972) and Bishop et al. (1972; Bishop and Rosbach, 1973) show that globin messenger RNA, a tissue-specific messenger RNA, is transcribed from a unique DNA sequence. The question of whether or not all DNA sequences which code for proteins are repetitive or unique is still a moot point; however, if most messenger RNAs are similar to that for globin, then differences in messenger RNA populations transcribed from chromatin templates would not be detected by the usual hybridization procedures.

This problem has been considered recently by Gilmour and Paul (1973) by determining directly whether globin messenger RNA sequences can be detected in the RNA transcribed by E. coli RNA polymerase from the chromatin of hemopoietic and nonhemopoietic tissues. As a probe for globin messenger RNA sequences, complementary DNA (cDNA) was synthesized *in vitro* from globin messenger RNA using the reverse transcriptase of avian myeloblastosis virus (Verma *et al.*, 1972; Ross *et al.*, 1972; Kacian *et al.*, 1972). From a theoretical consideration of the kinetics of DNA–RNA hybridization, conditions were defined which permit the titration of ^{3}H-labeled globin cDNA against RNA in which globin messenger RNA sequences might be present in extremely small amounts. The problem of being unable to detect hybridization to unique sequences when total nuclear DNA is used is obviated by this approach. Figure 1 shows a comparison of RNA transcribed from mouse brain and mouse fetal liver (hemopoietic) chromatins by hybridization to cDNA. It was found that the RNA transcribed from hemopoietic chromatin hybridized to 40% of the cDNA at input ratio of 25,000, while the RNA from brain chromatin did not show significant hybridization above background levels. In a control incubation containing mouse fetal liver chromatin and all the other components of the system except GTP, no hybridization to cDNA was detected. This would suggest that the hybridizing RNA from liver chromatin was derived from polymerase action rather than by contamination with endogenous RNA. In a more critical test of this point RNA was prepared from liver chromatin as before but incorporating highly labeled ^{32}P-ATP. After hybridizing the labeled RNA to cDNA the hybrid was analyzed by centrifugation to equilibrium in a CsCl gradient as described in Fig. 2. Most of the ^{3}H-label appeared in the lower half of the gradient at a density of 1.78, at which DNA/RNA hybrid would be expected, while unhybridized

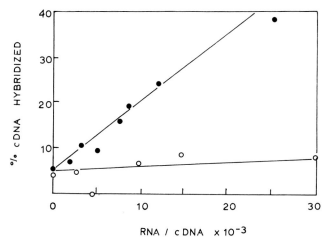

Fig. 1. Titration of RNA transcribed from chromatin with *E. coli* RNA polymerase against globin cDNA. Mouse fetal liver chromatin, ●——●; mouse brain chromatin ○——○.

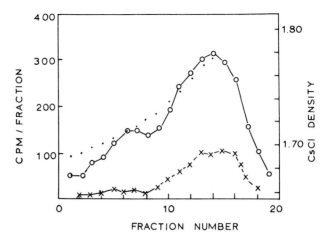

Fig. 2. Isopycnic banding in CsCl of the hybrid formed between ^{3}H-cDNA and ^{32}P-RNA transcribed from mouse fetal liver chromatin. Counts in ^{3}H, O———O; counts in ^{32}P, ×———×; density of CsCl, · · · · .

cDNA formed a smaller band at a density of about 1.71. ^{32}P-label was found to be associated with the hybridized cDNA in amounts which indicated that at least 80% of the RNA contained label. This analysis strongly suggests that the cDNA hybrid contains RNA transcribed *de novo* by the polymerase from liver chromatin. An identical conclusion has been reached by Axel *et al.* (1973) from the hybridization of globin cDNA to RNA transcribed *in vitro* by *E. coli* polymerase from duck reticulocyte and liver chromatins.

These results provide a rigorous proof that the proteins of chromatin restrict DNA transcription in a very specific manner and that the structures determining the restriction are not lost during chromatin isolation. The nature of the molecular mechanism which endows chromatin with tissue specificity has been investigated by reconstitution experiments of the type mentioned previously.

It is possible to show that when fetal liver is dissociated by high salt/urea and reconstituted by gradient dialysis the RNA transcribed from the resulting chromatin still possesses the ability to hybridize to cDNA to about the same extent as the RNA transcribed *in vitro* from native liver chromatin. This suggests that the structure which confers transcriptional specificity in chromatin is capable of self-assembly and as such must involve a recognition process. The acidic protein fraction of chromatin has been implicated in this process by the following experiment. The acidic protein fraction from fetal mouse liver chromatin was prepared by the hydroxyapatite method of MacGillivray *et al.* (1972).

A sample of mouse brain chromatin was reconstituted in the presence of a fixed amount of liver acidic proteins. The RNA transcribed from this template could be shown to contain globin messenger RNA sequences when hybridized to cDNA while the RNA transcribed from brain chromatin reconstituted in the absence of liver acidic proteins was incapable of hybridization to cDNA (Gilmour et al., 1973). This experiment confirms in a more stringent manner the previous findings of Gilmour and Paul (1970) that the acidic protein fraction of chromatin promotes transcription from organ-specific sites in the chromatin.

The nature of the active components in the acidic protein fraction is a matter of considerable current interest. Total acidic protein preparations can contain quite high proportions of RNA, and as has been mentioned previously two groups of workers have proposed that chromosomal RNA is responsible for directing organ-specific reconstitution. A number of investigators have found, however, that this RNA may be degradation products (Heyden and Zachau, 1971) or contamination by other species of nuclear RNA (Hill et al., 1971; Artman and Roth, 1971; Szeszak and Phil, 1971). Nevertheless on the basis of available evidence a number of models have been constructed to explain gene regulation in eukaryotes, and in many cases protein rather than RNA is favored as the controlling element.

VI. GENE REGULATION IN EUKARYOTES

While much of the preceding evidence points to the involvement of the acidic proteins in the genetic regulation of eukaryotic cells, their mechanism of action is virtually unknown. Most of the current theories on gene regulation supplement the scant experimental evidence with analogies to bacterial and dipteran systems. In addition, there are a number of unique characteristics peculiar to the genomes of eukaryotic cells which are thought to be of direct or indirect significance.

Most of the known biosynthetic pathways in eukaryotes are already represented in unicellular organisms, yet the mammalian genome contains 750 times more DNA than that of E. coli. In particular, much of the DNA of eukaryotes consists of repetitive sequences (Britten and Kohne, 1968; Flamm et al., 1966), many of which are not transcribed (Southern, 1970). The repetitive sequences appear to be intimately dispersed with the unique sequences of DNA (Grouse et al., 1972) which probably comprise the bulk of the protein-specifying DNA. Genetic evidence also suggests that some structural genes which are coordinately transcribed are not linked (Epstein and Motulsky, 1966; Nabholz et *al.,* 1969). Indeed, Britten and Davidson (1969) have argued that integration of gene activity in eukaryotes could not be based on the operation of polycistronic tissue-specific operons. It is currently held that most species of eukaryotic RNA are synthesized as larger precursor molecules (Darnell, 1968). These precursor molecules undergo specific processing, before being transported to the cytoplasm, while other species of RNA appear to be confined entirely to the nucleus (Shearer and McCarthy, 1967; Drews *et al.*, 1968).

Before considering models it is also useful to review some of the structural studies on chromatin. Several methods have been devised to determine the amount of free DNA in chromatin. Miura and Ohba (1967) developed a method for the titration of "free phosphate groups" in DNA using basic dyes and found that almost 50% of the phosphate groups in chromatin could bind toluidine blue. Itzhaki (1971a,b) provided similar figures and also found that polylysine of fairly high molecular weight could also bind to almost half the DNA in chromatin. This led to the conclusion that some parts of the DNA, although protected by histones, were not directly bound to them and hence were accessible to small molecules. Clark and Felsenfeld (1971) found that approximately half of the DNA in chromatin was susceptible to nuclease attack. They also postulated that about half the DNA in chromatin was not covered by proteins. Mirsky (1971) has pointed out that there are alternative explanations of these findings. Most of the space-filling models demonstrating DNA–histone association do not have all the backbone phosphate groups involved in salt linkages with basic amino acids in the histones. Indeed, only about half the phosphates need to be satisfied in this way. In addition, it is currently thought that histones bind in the larger DNA groove, leaving the smaller groove free to bind polylysine (Olins, 1969), and indeed there is experimental evidence to show that it preferentially binds there (Carroll and Botchan, 1972). These arguments are also supported by extensive electron microscopic evidence which fails to show long stretches of free DNA in chromatin and also from the titration of chromatin with specific anti-DNA sera which suggest that only 1–5% of the total DNA is exposed (Stollar, 1970). Thus, it seems that most of the DNA in chromatin is complexed with protein such that it is not available for transcription. Nevertheless, the idea that some DNA might be free in the sense of being biochemically reactive is worth consideration, because it is possible to postulate that control mechanisms similar to those in bacteria exist in eukaryotic cells.

One of the simplest models proposed was that of Georgiev (1969). He has suggested that the repetitive sequences found among the unique DNA could have a regulatory function and that the repetitious nature

10. Role of Acidic Proteins in Gene Regulation

of the regulatory DNA is due to the presence of a linked series of related regulatory protein-binding sites.

Each structural gene would be preceded by a large number of such sites (operator loci) which could be transcribed. Binding of a repressor to any one of these sites would prevent transcription of the gene. In this model the transcribable DNA would exist as "free DNA" and the control exercised by the repressive action of acidic proteins. Histones would presumably be involved in the permanent repression of other regions of DNA.

A similar model has been proposed by Britten and Davidson (1969) in an effort to explain how multiple changes in differentiation often are mediated by a simple signal (e.g., a hormone) and how in a given state of differentiation a large number of noncontiguous genes can achieve integrated activity. Here it is implied that the repressed state of the higher cell genome is histone mediated and that regulation is accomplished by the activation of otherwise repressed sites. It is postulated that initially an inducing agent binds to a specific sensor gene. This in turn causes an integrator gene to produce activator RNA which can in turn activate receptor genes, thus permitting linked structural genes to be transcribed. By postulating redundancy in either receptor or integrator genes it is possible for a single regulatory molecule to influence more than one structural gene. It can also be seen how a complex series of integrated functions can arise from the interplay of overlapping batteries of control elements. In this model it is proposed that the initial activation of the sensor gene might involve a protein and that the regulatory product of the integrator and receptor genes is RNA. However, the activation could also be carried out by proteins coded by these RNA's without changing the essence of the model.

More recent models of Paul (1972) and Crick (1971) are argued from the basis of chromosome morphology, as deduced from studies of the giant interphase chromosomes of Diptera.

These giant chromosomes consist of densely coiled band regions and less dense interband regions (Dupraw and Rae, 1965), each band plus interband apparently corresponding to a complementation group (Judd et al., 1972). Interband regions actively involved in DNA transcription appear as "puffs." Moreover, the distribution of these puffs depends on the cell type (Berendes and Beermann, 1969). Mammalian mitotic chromosomes can be shown by staining methods to exhibit a banding structure which although less defined than that of giant chromosomes does show species specificity (Caspersson et al., 1970).

Paul proposes that band and interband regions have different chemical compositions. A general model to explain the action of acidic proteins might propose that complexes of DNA and histones form tightly supercoiled structures into which RNA polymerase cannot penetrate but complexes of DNA, histones, and acidic proteins form more extended structures in which the DNA is relatively accessible. However, it seems unlikely that regulatory acidic proteins bind to nucleohistone throughout the entire length of a transcriptional unit since this would require that the transcriptional unit would not only encode information for protein synthesis but would also have to encode information for protein binding. It is perhaps more likely that regulatory acidic proteins bind to specific binding sites here designated address sites (Fig. 3). The effect of binding acidic protein to an address site is to cause localized unwinding of the supercoiled DNA, thus allowing polymerase to attach to a nearby promoter site. If the polymerase transcribes as far as the initiator sites at the beginning of the structural gene without being hindered by regulatory molecules which might be bound to the regulator loci, then further unwinding of adjacent nucleohistone can occur due to the increased net negative charge of the accumulated RNA. This RNA might also interact with RNA-binding acidic proteins to form informomer-like structures (Levy and Simpson, 1973; Krichevskaya and Georgiev, 1969), thereby increasing the negative change. In this way the structural gene which was hitherto embedded in the tightly coiled band region is now capable of being transcribed.

This basic model does not explain why the transcriptional unit in eukaryotes is so large nor is it clear how the binding of a single destabilizing molecule to an address site could produce such a drastic unwinding effect. Consequently, Paul postulates that a need for more address sites

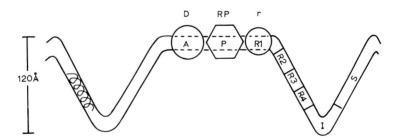

Fig. 3. Model of the transcriptional unit in chromatin. Nucleohistone is depicted as a supercoiled structure with a diameter of about 120 Å. A transcriptional unit is postulated to contain an address site (A) closely linked to a promoter site (P), regulatory sites (R_1, R_2, etc.), and an initiator site (I). A destablizing molecule (D) binds to A loosening the supercoiling, thus permitting a molecule of RNA polymerase (RP) to attach to the promoter. If no repressor molecules (r) are bound to regulator sites, transcription of the structural gene (S) is initiated.

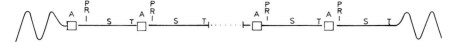

Fig. 4. The configuration of a transcriptional unit after simple gene reduplication. Address loci (A) with destabilizer molecules alternate with cistrons containing promoter (P), regulator (R) and initiator (I) loci, structural information (S), and a terminator locus (T).

was met by gene duplication. In this scheme (Fig. 4) multiple structural genes are separated by control regions to give a structure resembling that proposed for the ribosomal cistron. Such a structure would be maintained if multiple copies of the structural gene were required. Where only a single gene suffices the remainder of the structural genes are free to diverge.

The address sites on the other hand would not be permitted to diverge since a critical number of them would be required for destabilization. The fully evolved transcriptional unit could assume several forms. If the diverged structural DNA were lost then a single initiator, repressor, and structural gene would accompany a large number of address sites. If it proved vital to maintain the size of the transcriptional unit for the unwinding mechanism then the diverged structural DNA would be retained. A consequence of this last model would be a large RNA molecule only part of which would be used for translation and a processing mechanism to remove unwanted RNA before transport to the cytoplasm. The existence of such a processing step is now recognized.

In Paul's model the participation of acidic proteins could occur at two levels of control. The address sites need not be particularly specific and indeed if they comprised repetitive DNA then groups of structural genes might possess address sites of related sequence. The destabilizing protein which interacts with these sites than might operate a coarse control whereby groups of metabolically related genes might be activated. The possibility of a further fine control which determines which genes are to be transcribed (or perhaps their temporal relationship) could be mediated by repressor proteins bound to one or more of the repressor sites adjacent to the structural gene. Hence the acidic proteins could exercise both a positive and negative control.

In the model proposed by Crick, the structural gene is present in the interband region while the control elements are contained in the condensed band regions. The initial activation event is the localization of specific control elements within the condensed DNA and subsequent unwinding of the DNA to permit the binding of polymerase. Transcription then occurs through the unwound band region into the interband region. The idea of gene derepression involving an initial recognition or address site and subsequent unwinding of condensed DNA is common to both models. Whereas Paul suggests that address sites are located in the interband regions and consists of double-stranded DNA sequences accessible to a binding protein, Crick favors the idea that unpaired single-stranded DNA loops in the band regions provide the recognition. Although it is not discussed it is possible that activation simply involves the binding of RNA polymerase to these regions followed by unwinding and that regulation if required is achieved by the binding of inhibitory molecules in a manner analogous to bacterial control systems.

Crick's argument for activation by recognition of single rather than double-stranded DNA is that the tertiary structure of the active sites of proteins are generally concave and as such would not easily fit into the DNA grooves into which the recognizable groups on the bases protrude. Owing to the helical nature of the DNA it would be difficult to construct a protein capable of recognizing more than just a few base pairs. The experiments of Levy and Simpson (1973), however, do not support the concept of a single-stranded DNA regions in chromatin. These workers titrated rabbit liver chromatin with antiserum specific for single-stranded DNA and estimated that a maximum of only about 0.01% of the DNA in chromatin is recognized as single-stranded by this antibody. For an animal genome of 6×10^9 base pairs containing 3×10^4 structural genes (Ohta and Kimura, 1971) each with single-stranded control regions of 100 base pairs as suggested by Crick, then 0.05% of the DNA would react. The studies also suggest in fact that the low figure of 0.01% is probably confined to single-stranded regions at the end of chromatin molecules formed by breakage.

Clearly the question of DNA recognition is the most important and most conjectural aspect of these models. Analogies can be drawn from prokaryotic systems where a number of regulatory molecules, most of which are proteins (Müller-Hill *et al.*, 1971; Zubay *et al.*, 1970; Chadwick *et al.*, 1970). Besides negative control by repressors, other proteins can exert a positive control over DNA transcription. Both RNA polymerase and *lac* repressor bind preferentially to A–T rich regions of DNA (Jones and Berg, 1966; Lin and Riggs, 1970); since the pairing energy of A–T base pairs is less than G–C pairs, this might indicate that a localized melting of DNA is required for complex formation to occur. Alternatively, this might be a characteristic of a special conformation of DNA to which a protein might bind (Braun, 1971). Genetic mapping indicates that the first 50 N-terminal amino acids of each *lac* repressor subunit are involved in DNA binding (Adler *et al.*, 1972). The N-terminus is rich in apolar residues, especially tyrosine, and it is suggested that by forming an α-helix, this region of the protein could bind to the wide groove of DNA. In the case of the *lac* system, despite recent knowledge of the nucleotide sequence of the operator DNA and primary structure of the repressor protein it is still difficult to explain how specific interaction occurs. The situation in eukaryotes is even less complete. Specific interaction between acidic chromosomal proteins and DNA is inferred from chromatin reconstitution studies. Kleinsmith *et al.* (1970) have shown by DNA–cellulose chromatography that a small fraction of acidic phosphoprotein will bind to DNA in a specific fashion. However, the functional significance of the complex was not determined.

Evidence for specific interactions has also come from studies on the binding of steroid–receptor complexes to DNA. Spelsberg *et al.* (1971, 1972) and O'Malley *et al.* (1972) indicated that an acidic chromosomal protein is the acceptor in the progesterone/chick oviduct system, whereas King and Gordon (1972) and Clemens and Kleinsmith (1972) proposed on the basis of their receptor–DNA binding studies in the estradiol/rat uterus system that DNA is the acceptor and that the acidic proteins merely control which regions of the DNA are accessible.

Clearly the role of acidic proteins in eukaryotic gene regulation is complex. The proposed structure for the transcriptional unit raises the possibility that acidic proteins may act at a number of different levels, e.g., RNA recognition at the address site, supercoil destabilization and unwinding, negative regulation at regulator sites, and informofer formation. Thus the idea that gene activation is mediated by a single DNA–acidic protein interaction may be illusory. The eukaryotic field requires a definitive system similar to the *lac* operon in bacteria where the genetic material can be isolated, manipulated *in vitro,* and the product of the structural gene easily recognized.

ACKNOWLEDGMENT

I would like to express thanks to Drs. A. J. MacGillivray and D. Rickwood for extensive help with the literature research and to Dr. J. Paul for his useful comments on the manuscript.

REFERENCES

Adler, K., Beyreuther, K., Fannig, E., Geisler, N., Gronenborn, B., Klein, A., Müller-Hill, B., Ptashne, M., and Schmitz, A. (1972). *Nature (London)* **237,** 322.

Ahmed, K., and Ishida, H. (1971). *Mol. Pharmacol.* **7,** 323.

Allfrey, V. G., Littau, V. C., and Mirsky, A. E. (1963). *Proc. Nat. Acad. Sci. U.S.* **49,** 414.

Andress, D., Mousseron-Canet, M., Borgna, J. C., and Beiziat, Y. (1972). *C. R. Acad. Sci., Ser. D.* **274**, 2606.
Arnold, E. A., and Young K. E. (1972). *Biochim. Biophys. Acta* **257**, 482.
Artman, M., and Roth, J. S. (1971). *J. Mol. Biol.* **60**, 291.
Axel, R., Cedar, H., and Felsenfeld, G. (1973). *Proc. Nat. Acad. Sci. U.S.* **70**, 2029.
Becker, H., and Stanners, C. P. (1972). *J. Cell. Physiol.* **80**, 51.
Bekhor, I., Kung, G. M., and Bonner, J. (1969). *J. Mol. Biol.* **39**, 351.
Benjamin, W., and Gellhorn, A. (1968). *Proc. Nat. Acad. Sci. U.S.* **59**, 262.
Berendes, H. D., and Beermann, W. (1969). In "Handbook of Molecular Biology" (A. Lima-de-Faria, ed.), p. 501. North-Holland Publ., Amsterdam.
Bhorjee, J. S., and Pederson, T. (1972). *Proc. Nat. Acad. Sci. U.S.* **69**, 3345.
Bishop, J. O., and Rosbash, M. (1973). *Nature (London) New Biol.* **241**, 204.
Bishop, J. O., Pemberton, R., and Baglioni, G. (1972). *Nature (London)* **235**, 231.
Borun, T. W., and Stein, G. S. (1972). *J. Cell Biol.* **52**, 308.
Braun, S. (1971). *Nature (London), New Biol.* **232**, 174.
Britten, R. J., and Davidson, E. H. (1969). *Science* **165**, 349.
Britten, R. J., and Kohne, D. E. (1968). *Science* **161**, 529.
Carlsson, S. A., Moore, G. M., and Ringertz, N. (1973). *Exp. Cell Res.* **76**, 234.
Carroll, D., and Botchan, M. (1972). *Biochem. Biophys. Res. Commun.* **46**, 1681.
Caspersson, T., Zech, I., Johansson, C., and Modest, E. (1970). *Chromosoma* **30**, 215.
Chadwick, P., Pirrotta, V., Steinberg, R., Hopkins, N., and Ptashne, M. (1970). *Cold Spring Harbor Symp. Quant. Biol.* **35**, 283.
Chaudhuri, S., Stern, G., and Baserga, R. (1972). *Proc. Soc. Exp. Biol. Med.* **139**, 1363.
Chytil, F., and Spelsberg, T. C. (1971). *Nature (London), New Biol.* **229**, 101.
Clark, R. J., and Felsenfeld, G. (1971). *Nature (London), New Biol.* **229**, 101.
Clemens, L. E., and Kleinsmith, L. J. (1972). *Nature (London), New Biol.* **237**, 206.
Cognetti, G., Settineri, D., and Spinelli, G. (1972). *Exp. Cell Res.* **71**, 465.
Connor, B. J., and Patel, G. L. (1972). *J. Cell Biol.* **55**, 499.
Crick, F. H. C. (1971). *Nature (London)* **234**, 25.
Cross, M. E. (1972). *Biochem. J.* **128**, 1213.
Darnell, J. E. (1968). *Bacteriol. Rev.* **32**, 262.
Drews, J., Brawerman, G., and Morris, H. P. (1968). *Eur. J. Biochem.* **3**, 284.
Dupraw, E. J., and Rae, P. M. M. (1965). *Nature (London)* **212**, 598.
Elgin, S. C. R., and Bonner, J. (1970). *Biochemistry* **9**, 4440.
Elgin, S. C. R., and Bonner, J. (1972). *Biochemistry* **11**, 772.
Epstein, C. J., and Motulsky, A. G. (1966). *Progr. Med. Genet.* **4**, 85.
Flamm, W. G., Bond, H. E., and Burr, H. E. (1966). *Biochim. Biophys. Acta* **129**, 310.
Frenster, J. H. (1965). *Nature (London)* **206**, 680.
Georgiev, G. P. (1969). *J. Theor. Biol.* **25**, 473.
Gershey, E. L., and Kleinsmith, L. J. (1969). *Biochim. Biophys. Acta* **194**, 519.
Gilmour, R. S., and Paul, J. (1969). *J. Mol. Biol.* **34**, 305.
Gilmour, R. S., and Paul, J. (1970). *FEBS Lett.* **9**, 242.
Gilmour, R. S., and Paul, J. (1973). *Proc. Nat. Acad. Sci. U.S.* **70**, 3440.

10. Role of Acidic Proteins in Gene Regulation 315

Gilmour, R. S., Humphries, S. E., Hale, E. C., and Paul, J. (1973). *Inserm. Int. Coloq. Syn. Normal Pathol. Proteins Higher Organisms, Paris 1973*, p. 89.
Goodwin, G. H., and Johns, E. W. (1972). *FEBS Lett.* **21**, 103.
Graziano, S. L., and Huang, R. C. (1971). *Biochemistry* **10**, 4770.
Gronow, M. (1972). *Biochem. J.* **130**, 11P.
Grouse, L., Chilton, M., and McCarthy, B. J. (1972). *Biochemistry* **11**, 798.
Harlow, R., Tolstoshev, P., and Wells, J. R. E. (1972). *Cell Differentiation* **2**, 341.
Harrison, P. R., Hell, A., Birnie, G. D., and Paul, J. (1972). *Nature (London)* **239**, 219.
Helmsing, P. J. (1972). *Cell Differentiation* **1**, 19.
Helmsing, P. J., and Berendes, H. D. (1971). *J. Cell Biol.* **50**, 893.
Heyden, H. W. U., and Zachau, W. G. (1971). *Biochim. Biophys. Acta* **232**, 651.
Hill, R. J., Poccia, D. C., and Doty, P. (1971). *J. Mol. Biol.* **61**, 445.
Huang, R. C., and Bonner, J. (1962). *Proc. Nat. Acad. Sci. U.S.* **48**, 1216.
Huang, R. C., and Huang, P. C. (1969). *J. Mol. Biol.* **39**, 356.
Itzhaki, R. (1971a). *Biochem. J.* **122**, 583.
Itzhaki, R. (1971b). *Biochem. J.* **125**, 221.
Johns, E. W., and Forrester, S. (1969). *Eur. J. Biochem.* **8**, 547.
Jones, O. W., and Berg, P. (1966). *J. Mol. Biol.* **22**, 199.
Judd, B. H., Shen, M. W., and Kaufman, T. W. (1972). *Genetics* **71**, 139.
Jungmann, R. A., and Schweppe, J. S. (1972). *J. Biol. Chem.* **247**, 5535.
Kacian, D. L., Spiegelman, S., Bank, A., Terada, M., Metafora, S., Dow, L., and Marks, P. A. (1972). *Nature (London), New Biol.* **235**, 167.
Kamiyama, M., and Wang, T. Y. (1971). *Biochim. Biophys. Acta* **228**, 563.
King, R. J. B., and Gordon, J. (1972). *Nature (London), New Biol.* **240**, 185.
Kleinsmith, L. J., and Allfrey, V. G. (1969). *Biochim. Biophys. Acta* **175**, 123.
Kleinsmith, L. J., Allfrey, V. G., and Mirsky, A. E. (1966). *Science* **154**, 780.
Kleinsmith, L. J., Heidema, J., and Carroll, A. (1970). *Nature (London)* **226**, 1025.
Kostraba, N. C., and Wang, T. Y. (1970). *Int. J. Biochem.* **1**, 327.
Krichevskaya, A. A., and Georgiev, G. P. (1969). *Biochim. Biophys. Acta* **194**, 619.
Kruh, J., Tichonicky, L., and Wajcmax, H. (1969). *Biochim. Biophys. Acta* **195**, 549.
Kruh, J., Tichonicky, L., and Dastugue, B. (1970). *Bull. Soc. Chim. Biol.* **52**, 1287.
Langan, T. A. (1967). *In* "Regulation of Nucleic Acid and Protein Biosynthesis" (V. V. Koningsberger and L. Bosch, eds.), B.B.A. Library, Vol. 10, p. 233. Elsevier, Amsterdam.
LeStourgeon, W. M., and Rusch, H. P. (1971). *Science* **174**, 1233.
Levy, S., and Simpson, R. T. (1973). *Nature (London), New Biol.* **241**, 139.
Levy, S., Simpson, R. T., and Sober, H. A. (1972). *Biochemistry* **11**, 1547.
Lin, S. Y., and Riggs, A. D. (1970). *Nature (London)* **228**, 1184.
Lurguin, P. F., Saligy, V. L., and Neelin, J. M. (1972). *Arch. Int. Physiol. Biochim.* **80**, 202.
McClure, M. E., and Hnilica, L. S. (1970). *J. Cell Biol.* **47**, 1339.
MacGillivray, A. J., and Rickwood, D. (1973a). *In* "Biochemistry of Differentiation and Development" (J. Paul, ed.), Biochem. Ser., Vol. 9. Medical and Technical Publishing Co. Ltd., Oxford (in press).
MacGillivray, A. J., and Rickwood, D. (1973b). *Trans. Biochem. Soc.* **1**, 685.

MacGillivray, A. J., Cameron, A., Krauze, R. J., Rickwood, D., and Paul, J. (1972). *Biochim. Biophys. Acta* **277**, 384.
Means, A. R., Spelsberg, T. C., and O'Malley, B. W. (1974). *In* "Methods in Enzymology" (in press).
Mirsky, A. E. (1971). *Proc. Nat. Acad. Sci. U.S.* **68**, 2945.
Miura, A., and Ohba, Y. (1967). *Biochim. Biophys. Acta* **145**, 436.
Müller-Hill, B., Beyreuther, K., and Gilbert, W. (1971). *In* "Methods in Enzymology" (L. Grossman and K. Moldave, eds.), Vol. 21, p. 483. Academic Press, New York.
Nabholz, M., Miggiano, V., and Bodmer, V. (1969). *Nature (London)* **223**, 358.
Ohta, T., and Kimura, M. (1971). *Nature (London)* **233**, 118.
Olins, D. E. (1969). *J. Mol. Biol.* **43**, 439.
O'Malley, B. W., Spelsberg, T. C., Schrader, W. T., Chytil, F., and Steggles, A. W. (1972). *Nature (London)* **235**, 141.
Paul, J. (1972). *Nature (London)* **238**, 444.
Paul, J., and Gilmour, R. S. (1966). *J. Mol. Biol.* **16**, 241.
Paul, J., and Gilmour, R. S. (1968). *J. Mol. Biol.* **34**, 305.
Platz, R., Kish, V., and Kleinsmith, L. J. (1970). *FEBS Lett.* **12**, 38.
Platz, R., Stein, G. S., and Kleinsmith, L. J. (1973). *Biochem. Biophys. Res. Commun.* **51**, 735.
Richter, K. H., and Sekeris, C. E. (1972). *Arch. Biochem. Biophys.* **148**, 44.
Rickwood, D., Riches, P. G., and MacGillivray, A. J. (1973). *Biochim. Biophys. Acta* **299**, 162.
Ross, J., Aviv, H., Scolnick, E., and Leder, P. (1972). *Proc. Nat. Acad. Sci. U.S.* **69**, 264.
Schiltz, E., and Sekeris, C. E. (1969). *Hoppe-Seyler's Z. Physiol. Chem.* **350**, 317.
Schiltz, E., and Sekeris, C. E. (1971). *Experientia* **27**, 30.
Shirey, T., and Huang, R. C. (1969). *Biochemistry* **8**, 4138.
Shaw, L. M. J., and Huang, R. C. C. (1970). *Biochemistry* **9**, 4530.
Shearer, R. W., and McCarthy, B. J. (1967). *Biochemistry* **6**, 283.
Shelton, K. R., and Allfrey, V. G. (1970). *Nature (London)* **228**, 132.
Shelton, K. R., and Neelin, J. M. (1971). *Biochemistry* **10**, 2342.
Southern, E. C. (1970). *Nature (London)* **227**, 794.
Spelsberg, T. C., and Hnilica, L. S. (1970). *Biochem. J.* **120**, 435.
Spelsberg, T. C., Steggles, A. W., and O'Malley, B. W. (1971). *J. Biol. Chem.* **246**, 4186.
Spelsberg, T. C., Steggles, A. W., Chytil, F., and O'Malley, B. W. (1972a). *J. Biol. Chem.* **247**, 1368.
Spelsberg, T. C., Mitchell, W. M., Chytil, F., Wilson, E. M., and O'Malley, B. W. (1973). *Biochim. Biophys. Acta* **312**, 465.
Stedman, E., and Stedman, E. (1950). *Nature (London)* **166**, 780.
Stein, G., and Baserga, R. (1970). *Biochem. Biophys. Res. Commun.* **41**, 715.
Stein, G., and Borun, T. W. (1972). *J. Cell Biol.* **52**, 292.
Stein, G. S., and Farber, J. (1972). *Proc. Nat. Acad. Sci. U.S.* **69**, 2918.
Stollar, B. D. (1970). *Biochim. Biophys. Acta* **209**, 541.
Szeszak, F., and Phil, A. (1971). *Biochim. Biophys. Acta* **247**, 363.
Teng, C. S., and Hamilton, T. H. (1970). *Biochem. Biophys. Res. Commun.* **40**, 1231.
Teng, C. S., Teng, C. T., and Allfrey, V. G. (1971). *J. Biol. Chem.* **246**, 3597.
Tsuboi, A., and Baserga, R. (1972). *J. Cell. Physiol.* **80**, 107.

Turkington, R. W., and Riddle, M. (1969). *J. Biol. Chem.* **244**, 6040.
Umansky, S. R., Tokarskaya, V. I., Zotova, R. N., and Migushina, V. C. (1971). *Mol. Biol. (USSR)* **5**, 270.
Verma, I. M., Temple, G. F., Fan, H., and Baltimore, D. (1972). *Nature (London), New Biol.* **235**, 163.
Wang, T. Y. (1967). *J. Biol. Chem.* **242**, 1220.
Wang, T. Y. (1970). *Exp. Cell Res.* **61**, 455.
Wang, T. Y. (1971). *Exp. Cell Res.* **69**, 217.
Wilhelm, J. A., Ansevin, A. T., Johnson, A. W., and Hnilica, L. S. (1972a). *Biochim. Biophys. Acta* **272**, 220.
Wilhelm, J. A., Groves, C. M., and Hnilica, L. S. (1972b). *Experientia* **28**, 514.
Yoshida, M., and Shimura, K. (1972). *Biochim. Biophys. Acta* **263**, 690.
Zubay, G., Schwartz, D., and Beckwith, J. (1970). *Cold Spring Harbor Symp. Quant. Biol.* **35**, 433.

Author Index

Numbers in parentheses are reference numbers and indicate that an author's work is referred to although his name is not cited in the text. Numbers in italics show the page on which the complete reference is listed.

A

Adelman, M. R., *187*
Adhya, S., 9(60, 61), *23*
Adler, K., 5(44), 312, *23*, *313*
Ahmed, I., 138, *158*
Ahmed, K., 14(152, 153), 21(152, 153), 26, 112, 122, 123, 126, *133*, 138, *158*, 301, *313*
Alberga, A., 262, *290*
Alberti, K. G. M., 261, *294*
Alberts, B. M., 11(85, 86, 87, 88), 16(85, 86), *24*
Alfert, M., 195, *209*
Algranati, I. D., 64, 65, *102*
Allan, M. A., 46, *56*
Allfrey, K. M., 14(136), 26, 119, *135*
Allfrey, V. G., 2(2), 3(6, 16, 19, 24), 10(78, 79, 80, 83), 12(102), 13(113, 114), 14(78, 79, 80, 118, 125, 136, 139, 141, 145, 151, 154, 158, 159, 160), 15(24, 79, 80, 83, 125, 154, 172), 16(24, 80, 125, 172), 17(24, 172), 18(172), 19(172), 20(172), 21(78, 79, 80, 83, 125, 145, 151), *21, 22, 24, 25, 26, 27, 30, 36, 37, 42, 45, 50, 53, 54, 55, 56, 57*, 63, 64, 65, 66, 69, 70, 77, 78, 79, 81, 83, 87, 89, 90, *100, 101, 102*, 104, 107, 108, 109, 112, 113, 114, 117, 118, 119, 121, 122, 123, 124, 126, 129, 132, *133, 134, 135*, 138, 143, 147, *158*, 162, 163, 167, 171, 174, 184, 186, 187, *188, 189, 190*, 199, 200, 204, *209, 211, 212*, 214, 215, 224, 230, 234, 235, *242, 243, 244, 245*, 284, 285, *294*, 297, 298, 299, 300, 301, 302, *313, 315, 316*
Alonso, C., 194, *210*

Amer, S. M., 62, *100*
Ames, B. H., 139, *158*
Amodio, F. J., 11(86), 16(86), *24*, 227, *242*
Amos, H., 262, *294*
Anderson, K. M., 223, *242*, 251, 261, 262, *290, 292, 293*
Anderson, N. G., 183, *187*
Anderson, S. L., 115, *134*, 200, *212*
Anderson, W. B., 3(33), 9(64), 10(64, 73, 74), *22, 23, 24*
Andress, D., 304, *314*
Ansevin, A. T., 3(27), 20(27), *22, 33, 35, 37, 39, 47, 48, 49, 53, 57*, 274, 277, *294*, 300, *317*
Appel, S. H., 104, *134*
Arditti, R., 10(75, 77), *24*
Arnold, E. A., 39, 40, 41, 44, *53*, 298, 299, *314*
Aronson, A. I., 14(133), 26, 37, 38, *56*, 83, *101*
Arrighi, F. E., 222, *243*
Artman, M., 307, *314*
Asakura, S., 187, *189*
Ashburner, M., 193, 194, 197, 206, *209, 210*
Atger, M., 251, 252, 253, *293*
Attramadal, A., 261, *295*
Ausiello, D. A., 261, *291*
Aviv, H., 305, *316*
Axel, R., 174, 187, 306, *314*

B

Babcock, K. L., 154, *158*
Baggett, B., 251, *295*
Baglioni, G., 304, *314*
Bahr, G. F., 198, *210*

319

Bajszar, G., 20(194), 27
Bakay, B., 51, 52, 53
Balhorn, R., 3(21, 23), 22
Balls, M., 219, 220, 243
Baltimore, D., 305, 317
Bank, A., 305, 315
Barbiroli, B., 249, 291
Bardin, C. W., 251, 291
Barrett, T., 17(178), 27, 44, 54, 191, 210
Barry, J., 249, 291
Barton, A. D., 50, 54
Barton, R. W., 262, 291, 293
Baserga, R., 3(29), 14(128, 143, 146), 15(171), 20(29), 22, 25, 26, 27, 32, 35, 45, 54, 57, 81, 83, 101, 102, 161, 168, 171, 187, 189, 190, 191, 200, 210, 212, 213, 214, 215, 216, 222, 225, 226, 243, 244, 245, 299, 302, 304, 314, 316
Baudisch, W., 194, 210
Baulieu, E. E., 251, 252, 253, 262, 290, 293, 294
Baxandall, J., 173, 188
Baxter, J. D., 261, 267, 271, 291
Beard, V., 252, 260, 262, 267, 293
Beato, M., 265, 291
Beccari, E., 20(196), 27
Becker, H., 229, 243, 302, 314
Beckers, P. G. A., 194, 211
Beckwith, J., 3(32), 9(65), 10(75, 77), 22, 23, 24, 312, 317
Beerman, W., 309, 314
Beermann, W., 192, 193, 194, 198, 210
Behnke, O., 177, 180, 187, 188
Behrens, M., 33, 54
Beiziat, Y., 304, 314
Bekhor, I., 15(167), 18(167), 27, 39, 54, 274, 291, 314
Bendall, J. R., 183, 187
Bender, M. A., 248, 294
Benjamin, W. B., 14(156, 163), 15(156), 26, 38, 44, 54, 104, 112, 121, 133, 201, 210, 211, 214, 243, 298, 314
Bennett, T. P., 183, 188
Bensen, C. M., 267, 271, 291
Berendes, H. D., 14(142), 26, 65, 101, 174, 188, 193, 194, 195, 197, 198, 199, 201, 202, 203, 207, 208, 210, 211, 214, 227, 243, 301, 309, 314, 315
Berg, P., 312, 315

Bergstrand, A., 13(112), 25
Berlowitz, L., 3(17), 22
Bernard, W., 199, 210, 212
Bernardi, G., 39, 54
Bertram, J. S., 168, 172, 173, 174, 175, 178, 187, 189
Bess, L. G., 264, 265, 292
Beyreuther, K., 5(44), 23, 312, 313, 316
Bhorjee, J. S., 79, 100, 230, 243, 299, 300, 302, 314
Bibring, T., 173, 188
Bieswig, D., 265, 291
Billet, F. S., 219, 220, 243
Billing, R. J., 249, 291
Birnie, G. D., 304, 315
Birnstiel, M. L., 47, 54
Bishop, J. O., 20(192), 27, 304, 314
Blattner, F. R., 7(53), 23
Blobel, G., 264, 291
Block, R., 10(76), 24
Bodmer, V., 307, 316
Boettiger, J. K., 7(53), 23
Boffa, L. C., 14(136), 26, 119, 132, 135, 163, 167, 184, 186, 190
Bogoroch, R., 262, 291, 294
Bond, H. E., 307, 314
Bonner, J., 3(7, 9), 14(121), 15(167, 168), 16(168), 17(168), 18(167), 22, 25, 27, 30, 33, 37, 38, 39, 40, 41, 44, 45, 54, 55, 56, 57, 161, 188, 191, 211, 214, 243, 244, 274, 291, 297, 298, 299, 300, 314, 315
Bordwell, L., 3(23), 22
Borgna, J. C., 304, 314
Borun, T. W., 3(14), 22, 83, 102, 215, 231, 232, 243, 244, 302, 314, 316
Botchan, M., 308, 314
Bourgeois, S., 4(34, 36, 37), 5(34, 42, 45), 6(47), 23
Boyd, J. B., 142, 158
Braendle, W., 265, 291
Bramwell, M. E., 186, 188
Braun, S., 312, 314
Brawerman, G., 262, 291, 308, 314
Brecher, P. I., 251, 252, 253, 260, 261, 291, 292
Bresciani, F., 251, 257, 258, 259, 294
Breuer, M. E., 194, 195, 210, 211
Brinck-Johnson, T., 261, 293

Author Index

Britten, R. J., 17(179, 180, 181), 18(179, 180, 181), 27, 208, 210, 307, 308, 309, 314
Brown, D. F., 177, 188
Bruchovsky, N., 251, 261, 265, 291
Bruskov, V. I., 50, 56
Brutlag, D., 38, 45, 46, 56
Buck, M. D., 14(147), 26
Buller, R. F., 254, 260, 268, 269, 291
Burgess, R. R., 139, 158
Burr, H. E., 307, 314
Busch, H., 13(115), 25, 30, 33, 34, 35, 47, 48, 49, 51, 54, 55, 56, 57, 62, 64, 77, 100, 101, 190, 248, 291
Butler, J. A. V., 30, 54, 224, 243
Butterworth, P. H. W., 12(105), 25
Buttin, G., 9(62), 23
Byers, T. J., 248, 291
Byvoet, P., 126, 133

C

Cameron, A., 39, 42, 55, 167, 168, 189, 298, 299, 300, 306, 316
Cameron, I. L., 215, 216, 217, 218, 219, 235, 236, 240, 243, 244
Carlsson, S. A., 186, 188, 189, 300, 314
Carlsson, U., 15(175), 27
Carroll, A., 15(166), 16(166), 17(166), 27, 36, 45, 55, 123, 134, 313, 315
Carroll, D., 39, 42, 44, 55, 191, 211, 308, 314
Caspersson, T., 309, 314
Castles, T. R., 262, 291
Cedar, H., 174, 187, 306, 314
Chader, G. J., 252, 260, 293
Chadwick, P., 312, 314
Chalkley, G. R., 265, 293
Chalkley, R., 3(21, 23), 22, 104, 135
Chamberlin, M., 12(90), 24
Chambon, P., 12(104, 106), 25
Chamness, G. C., 255, 258, 260, 273, 291, 293
Chanda, S. K., 37, 54
Chaudhuri, S., 3(29), 15(171), 20(29), 22, 27, 35, 45, 54, 57, 225, 226, 245, 304, 314
Chauveau, J., 32, 50, 54, 56, 264, 291
Chen, B., 10(71, 73), 24
Chentsov, J. S., 32, 35, 48, 50, 54

Cherian, M. G., 37, 54
Chesterton, C. J., 12(105), 25
Chet, I., 155, 158
Chilton, M. D., 168, 188, 307, 315
Chipchase, M. I. H., 47, 54
Chiu, J. F., 138, 158
Choi, Y. C., 33, 34, 54
Chu, L. L. H., 14(148), 26, 262, 263, 265, 295
Chung, L. W. K., 223, 243
Church, R. B., 18(182), 20(182, 191, 194), 27, 225, 243, 249, 291
Chytil, F., 14(126, 127, 134), 25, 26, 174, 189, 214, 243, 248, 249, 250, 275, 277, 278, 279, 280, 281, 282, 283, 287, 288, 291, 292, 294, 295, 298, 303, 313, 314, 316
Clark, J. H., 260, 291
Clark, R. J., 308, 314
Claycomb, W. C., 207, 210
Clemens, L. E., 267, 291, 313, 314
Clever, U., 193, 194, 195, 197, 200, 206, 210
Coffey, D. S., 223, 240, 243, 244
Cognetti, G., 301, 314
Cohen, C., 177, 188
Cohen, I., 177, 188
Cohen, S., 186, 186, 189
Cohn, M., 4(34), 5(34, 42), 23
Cohn, P., 224, 243
Cole, R. D., 3(8), 22, 30, 57, 83, 102, 161, 190, 191, 212, 222, 245, 264, 265, 295
Comings, D. E., 221, 243
Connaway, S., 10(75), 24
Conner, B. J., 301, 314
Conner, J., 38, 45, 46, 56
Cooper, S., 172, 188
Corlette, S. L., 194, 212
Corvol, P. L., 251, 252, 294
Couch, R. M., 223, 242
Cowan, D. M., 265, 293
Crabbé, J., 261, 262, 291
Craddock, C., 138, 158
Craig, L. C., 61, 100
Crampton, C. F., 222, 243
Creuzet, C., 14(123), 25, 43, 44, 55, 214, 244
Crick, F. H. C., 309, 314
Crippa, M., 20(196), 27, 53, 54

Crocker, T. T., 40, 55
Crooke, S. T., 51, 56
Cross, M. E., 302, 314
Crouse, H. V., 194, 210
Cukier-Kahn, R., 11(84), 24

D

Dahlberg, J. E., 7(53), 23
Dahmus, M. E., 13(108), 25, 33, 54
Daly, M. M., 13(113, 114), 25, 30, 54
Daneholt, B., 192, 193, 210, 211
Darnell, J. E., 20(193), 27, 308, 314
Daskal, I., 33, 34, 54
Dastugue, B., 20(186), 21(186), 27, 112, 115, 125, 126, 127, 133, 174, 188, 224, 243, 300, 314, 315
Davidson, E. H., 208, 210, 308, 309, 314
Davidson, O. W., 265, 295
Davie, E. W., 131, 134
DeAngelo, A. B., 249, 291
Debault, L. E., 218, 243
DeCrombrugge, B., 9(60, 63), 10(71, 73), 23, 24
Defer, N., 20(186), 21(186), 27, 115, 125, 127, 133
DeLange, R. J., 3(12), 22, 30, 54
Delius, H., 11(87), 24
Derksen, J., 194, 198, 199, 208, 210, 211
DeRosier, D. J., 187, 189
DeSella, C., 253, 293
Desjardins, P. R., 47, 55, 115, 133
DeSombre, E. R., 251, 252, 253, 255, 257, 258, 259, 260, 261, 262, 271, 291, 292
DeWeer, P., 261, 262, 291
DiMauro, E., 12(107), 25
Dina, D., 20(196), 27
Dingman, C. W., 13(117), 25, 89, 100, 161, 162, 188, 189, 214, 243, 37, 54
Dixon, F. H., 214, 244
Dixon, G. H., 5(43), 23, 162, 189
Dmitrieva, N. P., 32, 35, 48, 57
Dolbeare, F., 14(120), 25
Doty, P., 14(132), 25, 33, 39, 40, 41, 44, 55, 57
Dounce, A. L., 32, 33, 34, 38, 52, 54, 56
Dow, L., 305, 315
Drews, J., 262, 291, 308, 314

Dubin, D. T., 139, 158
Duntze, W., 151, 158
Dupraw, E. J., 309, 314

E

Echols, H., 8(54), 9(61), 23
Edelman, I. S., 14(148), 26, 251, 261, 262, 263, 265, 266, 291, 292, 294, 295
Edström, J.-E., 192, 193, 210, 211
Egyházi, E., 192, 193, 210, 211
Elgin, S. C. R., 3(9), 22, 30, 38, 39, 40, 41, 44, 54, 142, 158, 161, 167, 168, 188, 190, 191, 211, 214, 243, 298, 299, 300, 314
Eliasson, N. A., 13(112), 25
Ellgaard, E. G., 200, 210, 211
Emmer, M., 3(33), 9(63, 64), 10(64, 71), 22, 23, 24
Emmerich, H., 207, 211
Enea, V., 14(141), 26, 83, 101, 171, 188, 214, 243
Epstein, C. J., 307, 314
Erdos, T., 255, 294
Eron, L., 10(75, 76), 24
Evans, A., 265, 295
Evans, A. K., 223, 242
Evans, J. H., 224, 243
Evans, J. N., 224, 243

F

Faiferman, I., 50, 51, 54
Fairbanks, G., 84, 101
Falk, R. J., 251, 291
Fambrough, D. M., 31, 33, 54
Fan, H., 305, 317
Fanestil, D. D., 262, 266, 291
Fang, S., 251, 261, 262, 292, 293
Fannig, E., 5(44), 23, 312, 313
Fanshier, L., 40, 55
Farber, J., 171, 189, 233, 245, 304, 311
Farr, A. L., 97, 101
Feigelson, M., 83, 101, 262, 292
Feigelson, P., 83, 101, 102, 262, 292
Felsenfeld, G., 174, 187, 306, 308, 314
Ferris, F. L., 11(86), 16(86), 24, 227, 242
Fiandt, M., 7(53), 23
Fievez, M., 81, 102

Author Index

Files, J. G., 5(41), 23
Fimognari, G. M., 251, 261, 291, 292
Finch, J. T., 187, 190
Firszt, D. C., 62, 100
Flamm, W. G., 47, 54, 307, 314
Forer, A., 168, 172, 173, 174, 175, 178, 187, 189
Forrester, S., 31, 33, 55, 299, 315
Forte, L., 262, 292
Fox, M., 216, 217, 218, 244
Fox, T. O., 228, 229, 243
Frankel, J., 218, 243
Frenster, J. H., 14(118, 119), 25, 37, 50, 54, 162, 187, 188, 189, 214, 243, 300, 314
Frey, L., 11(85), 16(85), 24
Froehner, S. C., 3(9), 22, 30, 54, 161, 188

G

Gabrusewycz-Garcia, N., 194, 211
Ganem, D., 5(39, 40), 23
Garcea, R. L., 267, 271, 291
Garren, L. D., 262, 292
Gefter, M. L., 11(87), 24
Geisinger, F., 12(106), 25
Geisler, N., 5(44), 23, 312, 313
Gelehrter, T. D., 262, 292
Gellhorn, A., 14(163), 26, 38, 44, 54, 104, 133, 214, 243, 298, 314
Georgiev, G. P., 20(194, 195), 27, 32, 35, 48, 50, 51, 54, 55, 56, 57, 174, 188, 209, 211, 308, 310, 314, 315
Gerner, E. W., 230, 232, 243
Gershey, E. L., 14(138), 26, 106, 122, 129, 133, 138, 158, 162, 188, 199, 200, 209, 211, 212, 299, 302, 314
Geschwind, I. I., 195, 209
Getz, S., 138, 158
Ghosh, S., 8(54), 23
Giannopolous, G., 252, 292
Gilbert, W., 3(31), 4(31, 35, 38), 6(38), 22, 23, 142, 158, 312, 316
Gilmour, R. S., 2(3), 3(28), 15(174), 20(28, 174, 184), 22, 27, 33, 35, 39, 41, 54, 55, 56, 89, 101, 174, 188, 214, 243, 274, 292, 297, 303, 304, 307, 314, 315, 316
Glascock, R. F., 250, 292

Glasser, S. R., 249, 250, 272, 273, 292, 295
Gniazdowski, M., 12(106), 25
Goff, C., 12(94), 24
Goldstein, L., 50, 51, 55, 248, 291
Goodman, E. M., 66, 70, 92, 96, 101, 188
Goodman, R. H., 201, 210, 211
Goodman, R. M., 14(156), 15(156), 26, 112, 121, 133
Goodwin, G. H., 299, 315
Gordon, J., 252, 255, 260, 262, 263, 264, 265, 266, 267, 293, 313, 315
Gornall, A. G., 115, 133
Gorovski, M. A., 195, 211
Gorovsky, M. A., 3(18), 22
Gorski, J., 249, 251, 252, 253, 255, 257, 258, 259, 260, 268, 291, 292, 293, 294, 295
Gotestan, M., 9(60), 10(71, 72, 73, 74), 15(72), 23, 24
Gould, H. J., 17(178), 27, 44, 54, 191, 210
Grace, D. M., 186, 188
Granner, D., 3(21, 23), 22
Graziano, S. L., 39, 40, 41, 55, 298, 299, 315
Green, H., 227, 228, 229, 244
Greenblatt, J., 8(55), 23
Greenleaf, A. L., 12(100), 25
Greenman, D. L., 262, 292
Griffin, E. E., 215, 235, 236, 240, 243
Griffiths, G., 55
Grodzicker, T., 10(77), 24
Grogan, D. E., 47, 49, 55
Gronenborn, B., 5(44), 23, 312, 313
Gronow, M., 55, 300, 315
Gros, F., 11(84), 24
Gross, P. R., 83, 101
Grossbach, U., 194, 211
Grouse, L., 168, 188, 307, 315
Groves, C. M., 300, 317
Grunicke, H., 162, 188
Gutmann, E. D., 227, 242
Gurdon, J., 2(1), 21
Gurley, L. R., 3(22), 22
Gutman, E., 11(86), 16(86), 24
Guttes, E., 183, 188
Guttes, S., 183, 188

H

Hagen, D. C., 6(49), 23
Hale, E. C., 307, 315
Hall, B. D., 12(107), 25
Hall, R. H., 39, 40, 41, 56
Hamilton, M. G., 50, 51, 54
Hamilton, T. H., 14(140), 26, 83, 102, 171, 174, 190, 204, 212, 214, 223, 245, 249, 292, 293, 295, 301, 316
Hammarsten, E., 13(112), 25
Hancock, R., 3(15), 22
Hanoune, J., 224, 243
Hardin, J. M., 104, 135
Harlow, R., 299, 315
Harrad, K. R., 14(161), 26, 121, 122, 134
Harris, G. S., 266, 292
Harris, H., 186, 188, 240, 243
Harrison, P. R., 304, 315
Harrow, R., 53, 55
Hatano, S., 177, 188, 190
Hatzel, I., 252, 292
Hausen, F., 13(109), 25
Heidema, J., 15(166), 16(166), 17(166), 27, 36, 45, 55, 123, 134, 313, 315
Heilmeyer, L. M. G., Jr., 104, 134
Hell, A., 304, 315
Helmsing, P., 14(142), 26
Helmsing, P. J., 65, 101, 174, 188, 193, 194, 203, 210, 211, 214, 227, 243, 301, 315
Helmstetter, C. E., 172, 188, 216, 217, 243
Henderson, L. E., 15(175), 27
Herman, T. S., 251, 261, 292
Herrick, G., 11(88), 24
Hershey, J., 7(50), 23
Herzog, R. D., 62, 101
Heyden, H. W. U., 307, 315
Higashinakagawa, T., 13(111), 25, 34, 48, 55
Higashi, S., 187, 189
Highland, E., 261, 295
Hilgartner, C. A., 38, 56
Hill, R. J., 14(132), 25, 39, 40, 41, 44, 55

Hnilica, L. S., 3(10, 26, 27), 14(150), 20(27), 21(150), 22, 26, 30, 31, 32, 33, 35, 37, 39, 40, 46, 47, 48, 49, 53, 55, 57, 89, 101, 109, 122, 133, 134, 138, 158, 161, 189, 214, 222, 233, 243, 244, 248, 264, 265, 274, 277, 292, 294, 295, 300, 302, 303, 315, 316, 317
Hoekstra, W. G., 250, 292
Hogeboom, G. H., 32, 55
Holbrook, D. J., 224, 243
Hollenberg, C. P., 12(107), 25
Holoubek, V., 40, 55
Holt, T. K. H., 195, 197, 198, 200, 201, 202, 211
Holzer, H., 151, 158
Hood, L. E., 142, 158, 167, 168, 190
Hopkins, N., 312, 314
Hord, G., 46, 57
Hosick, H., 193, 210
Hotta, Y., 220, 243
Howell, R. R., 262, 292
Howk, R., 43, 55
Hoyer, B. H., 2(4), 22
Huang, P. C., 15(173), 27, 39, 55, 303, 315
Huang, R. C. C., 3(7), 15(173), 22, 27, 33, 39, 40, 41, 54, 55, 56, 191, 212, 214, 243, 297, 298, 299, 300, 303, 315, 316
Huff, K., 273, 293
Humphries, S. E., 307, 315
Humphrey, R. M., 230, 232, 243
Huppert, J., 62, 101
Hurlbert, R. B., 47, 55
Hurst, J. D., 251, 292
Huxley, H. E., 177, 182, 187, 188, 189, 190
Hyman, R. W., 12(91), 24

I

Ickowicz, R., 52
Inagaki, A., 33, 34, 54
Inman, D. R., 255, 265, 293
Inoue, A., 3(24), 15(24, 172), 16(24, 172), 17(24, 172), 18(172), 19(172), 20(172), 22, 27
Irvin, J. L., 224, 243
Ishida, H., 14(153), 21(153), 26, 122, 123, 133, 138, 158, 301, 313

Ito, J., 267, 271, *291*
Ito, M., 220, *243*
Itzhaki, R., 308, *315*

J

Jacob, F., 3(30), 4(30), 7(51), *22, 23*
Jacob, J., 248, *293*
Jacobson, H. I., 250, 251, *292*
Jacquet, M., 11(84), *24*
James, T. W., 216, *243*
Jeffrery, W. R., 218, *243*
Jelinek, W., 20(193), *27*
Jenkins, L. M., 218, *243*
Jenkins, M., 11(86), 16(86), *24,* 227, *242*
Jennings, A. W., 260, *291*
Jensen, E. V., 250, 251, 252, 253, 255, 257, 258, 259, 260, 261, 262, 271, *291, 292*
Jeter, J. R., Jr., 92, *101,* 215, 217, 218, 224, 235, *243, 244*
Jobe, A., 4(37), 6(47), *23*
Jockusch, B. M., 177, 180, *188*
Johansson, C., 309, *314*
Johmann, C. A., 3(18), *22*
Johns, E. W., 30, 31, 33, 41, 42, *54, 55,* 57, 299, *315*
Johnson, A. W., 33, 47, 49, 53, 57, 300, *317*
Johnson, E. M., 3(24), 10(79, 83), 14(79, 145, 154), 15(24, 79, 83, 154), 16(24), 17(24), 21(79, 83, 145), *22, 24, 26,* 78, 79, 81, *100, 101,* 126, 129, *133,* 138, 143, 147, *158,* 230, 234, 235, *244*
Johnson, G., 104, *135*
Jokela, H. A., 51, *55*
Jolinek, W., 51, *55*
Jones, O. W., 312, *315*
Judd, B. H., 309, *315*
Jung, I., 262, *290*
Jungleblut, P. W., 251, 252, 253, 257, 260, *292*
Jungmann, R. A., 14(165), 21(165), *27,* 123, *133,* 138, *158,* 171, *188,* 302, *315*

K

Kabat, D., 104, *133*
Kacian, D. L., 305, *315*

Kaiser, A. D., 7(51), *23*
Kajiwara, K., 3(13), *22*
Kamiyama, K., 214, *244*
Kamiyama, M., 20(185, 186), 21(186), 27, 36, 42, 55, 112, 115, 125, 126, *133,* 174, *188,* 304, *315*
Kaplowitz, P. B., 126, 128, 129, *133*
Karlson, P., 194, *210*
Karn, J., 3(24), 10(79), 14(79, 145, 154), 15(24, 79, 145, 154), 16(24), 17(24), 21(79), *22, 24, 26,* 78, 79, 81, *100, 101,* 129, *133,* 143, *158,* 230, 234, 235, *244*
Kasai, M., 187, *189*
Katzman, P. A., 249, *293*
Kaufman, T. W., 309, *315*
Kawamura, H., 177, *189*
Kawasaki, T., 39, *54*
Kawashima, T., 252, 253, 257, 260, *292*
Kay, E. R. M., 34, *56*
Kedinger, C., 12(104, 106), *25*
Keevert, J. B., 3(18), *22*
Kellar, K. L., 45, *56*
Kenney, F. T., 262, *292*
Kessler, D., 177, *189*
Keyl, H. G., 194, *210*
Kidson, C., 263, *293*
Kimura, M., 312, *316*
King, R. J. B., 223, *244,* 252, 255, 260, 262, 263, 264, 265, 266, 267, *293,* 313, *315*
Kirby, K. S., 61, 63, *101,* 263, *293*
Kirkham, W. R., 50, *55*
Kish, V. M., 14(124), *25,* 106, 110, 115, 117, 119, 126, *133, 134,* 138, *158,* 302, *316*
Klein, A., 312, *313*
Klein, P., 184, *189*
Kleinsmith, L. J., 10(81), 14(81, 124, 138, 151, 155, 158, 159, 160), 15(155, 166, 169), 16(81, 166, 169), 17(166), 21(81, 151, 169), *24, 25, 26, 27,* 36, 42, 45, *55, 56,* 104, 106, 107, 109, 110, 112, 113, 114, 115, 117, 118, 119, 121, 122, 123, 124, 125, 126, 128, 129, *133, 134,* 138, *158,* 162, 174, *188, 189,* 215, 234, 235, *244,* 267, *291,* 299, 302, 313, *314, 315, 316*
Klemm, A., 5(44), *23*
Knecht, M., 49, *55*

Knowler, J. T., 249, *293*
Kobayashi, S., 187, *189*
Koenig, H., 14(120), *25*
Kohen, E., 62, *100*
Kohler, P. O., 251, 253, *294*
Kohne, D. E., 17(179, 181), 18(179, 181), 27, 307, *314*
Koide, S. S., 263, *295*
Korenman, S. G., 251, 253, *293*, *294*
Korn, E. D., 177, *189*
Kornberg, T., 11(87), *24*
Kostraba, N. C., 20(187, 188), 27, 35, 36, 42, 44, 55, 108, 125, *134*, 174, *188*, 214, 224, *244*, 299, 300, *315*
Kraemer, R. J., 240, *244*
Krahn, E., 62, *101*
Krauze, R. J., 39, 42, 55, 167, 168, *189*, 298, 299, 300, 306, *316*
Krebs, E. G., 104, *134*, 151, *158*
Kress, H., 195, *211*
Krichevskaya, A. A., 50, 51, 55, 56, 310, *315*
Kriegstein, H., 183, *188*
Kristensen, B. I., 177, *187*
Kroeger, H., 193, *211*, 248, *293*
Kruh, J., 20(186), 21(186), 27, 115, 125, 127, *133*, 224, *243*, 300, *314*, *315*
Kung, G. M., 15(167), 18(167), 27, 39, 54, 274, *291*, *314*
Kuzmich, M. J., 217, *244*
Kyser, K. A., 255, *293*

L

Laemmli, U. K., 98, *101*
Landon, E. J., 262, *292*
LaFond, R. E., 207, *210*
Lambert, B., 192, 193, *210*, *211*
Lang, N., 262, 263, *294*
Langan, T. A., 14(157), *26*, 42, 55, 104, 113, 118, 119, 125, 127, *134*, 200, *211*, 299, *315*
Larson, D. A., 77, *101*, 167, *189*
Leder, P., 305, *316*
Ledinko, N., 171, *188*
Lee, S. C., 13(108), *25*
Leenders, H. J., 194, 198, 201, 202, 207, 208, *210*, *211*
Lemborska, S. A., 174, *188*
LeStourgeon, W. M., 14(135), *26*, 52, 55, 66, 69, 70, 73, 75, 77, 78, 79, 83, 89, 90, 91, 92, 93, 96, *101*, 119, 122, *134*, 138, 139, 140, 142, 154, *158*, 162, 163, 167, 168, 171, 172, 173, 174, 175, 178, 180, 181, 184, 185, 187, *188*, *189*, 214, 227, 233, *244*, 298, 301, 302, *315*
Leung, K. M. T., 255, 258, 259, *295*
Levina, L., 222, *244*
Levinthal, C., 84, *101*
Levy, R., 14(144), *26*, 83, *101*, 171, *189*, 223, *244*
Levy, S., 14(144), 17(177), *26*, 27, 39, 40, 41, 44, 55, 83, *101*, 171, *189*, 223, *244*, 298, 299, 310, 312, *315*
Lezzi, M., 193, *211*
Liang, T., 261, *293*
Liau, M. C., 47, 48, 55
Liao, S., 251, 261, 262, *290*, *291*, *292*, *293*, *294*,
Liew, C. C., 115, *133*
Lin, A. H., 262, *293*
Lin, J. C., 14(128), *25*
Lin, S. Y., 6(46), 6(48), *23*, 312, *315*
Lindskog, S., 15(175), *27*
Linn, T. G., 12(100), *25*
Lipsey, A., 62, *100*
Litt, M., 34, *56*
Littau, U. C., 50, *56*
Littau, V. C., 3(6), 12(102), 14(136), 15(172), 16(172), 17(172), 18(172), 19(172), 22, 25, *26*, 27, 78, 81, *101*, 119, *135*, 162, 163, 167, 184, 186, *189*, *190*, 200, *209*, 214, *242*, 297, *313*
Litwack, G., 263, 264, *293*
Loeb, J. E., 14(123), *25*, 43, 44, 55, 214, *244*
Loeb, P. M., 262, *295*
Losick, R., 12(98, 99, 100), *24*, *25*
Love, R., 118, *135*
Lowry, O. H., 97, *101*
Lue, P. F., 115, *133*
Lukanidin, E. M., 50, 51, *56*
Lurguin, P. F., 301, *315*

M

Maas, P. M. J. M., 194, 198, 208, *211*
McCarthy, B. J., 2(4), 18(182), 20(182, 191), *22*, 27, 168, *188*, 225, *243*, 249, *291*, 307, 308, *315*, *316*

Author Index

McCarty, K. S., 50, 51, 56
McClure, M. E., 109, *133*, 161, *189*, 214, 222, *244*, 302, *315*
MacGillivray, A. J., 10(82), 14(161, 162), 16(82), *24*, *26*, 32, 36, 39, 42, 44, *55*, *56*, 110, 111, 112, 119, 121, 122, 126, *134*, 167, 168, *189*, 191, 200, *211*, 214, *244*, 298, 299, 300, 301, 302, 306, *315*, *316*
McGrath, J., 12(90), *24*
McGuire, J. L., 253, *293*
McGuire, W. L., 251, 253, 255, 258, 260, 273, *291*, *293*, *294*
MacKay, M., 38, *56*
McParland, R., 51, *56*
Magasanik, B., 6(49), *23*
Magun, B. E., 139, *158*, 174, 175, 180, 184, *189*, 214, 227, *244*
Maizel, J. V., 141, *158*
Maizels, N., 4(38), 6(38), *23*
Mainwaring, W. I. P., 261, 273, 274, *293*
Makman, R. S., 10(66), *23*
Malpoix, P. J., 81, *102*, 223, *244*
Mandel, J. L., 12(106), *25*
Mangan, F. R., 261, *293*
Maniatas, T., 7(52), *23*
Maramatsu, M., 34, 47, 48, 55, *56*
Marks, P. A., 305, *315*
Martelo, O. J., 131, *134*
Martin, D. W., Jr., 126, *135*
Martin, L., 223, *244*, 265, *293*
Martin, T. E., 209, *211*
Marushige, K., 14(121, 131), *25*, 33, 38, 45, 46, *54*, *56*, 162, *189*, 214, *244*
Marx, J., 252, 260, 262, 267, *293*
Massol, N., 262, *290*
Mateyko, G. M., 196, *211*
Matthews, D. E., 215, 226, 232, *245*
Maurer, H. R., 265, *293*
Maxam, A., 4(38), 6(38), *23*
Means, A. R., 249, *293*, 300, *316*
Meihlac, M., 12(106), *25*
Meisler, M. H., 118, *134*
Melli, M., 20(192), *27*
Metafora, S., 305, *315*
Miggiano, V., 307, *316*
Migushina, V. C., 298, *317*
Migushina, V. L., 39, 40, 41, 46, *57*
Milgrom, E., 251, 252, 253, *293*
Miller, J. H., 5(39, 40), *23*

Mirsky, A. E., 2(2), 3(6), 12(102), 13(113, 114, 116), 14(118, 151, 158), 21(151), *21*, *22*, *25*, *26*, 30, 35, 37, 38, 50, *54*, *56*, 63, *101*, 104, 107, 112, 113, 119, 122, 129, *134*, 138, *158*, 161, 162, 174, *188*, *189*, 199, 200, *209*, *212*, 214, 215, *242*, *244*, 297, 302, 308, *313*, *315*, *316*
Musky, A. E., 224, *244*
Mitchell, W. M., 14(134), *26*, 279, *295*, 298, 303, *316*
Mitchison, J. M., 213, 216, 220, *244*
Miura, A., 308, *316*
Miyagi, M., 20(189), *27*
Modest, E., 309, *314*
Mohberg, J., 67, *101*, 140, 142, *158*, 172, *189*
Mohla, S., 253, 259, 271, *291*
Molnar, J., 50, 51, *56*
Molnar, Y., 50, *56*
Monahan, J. J., 39, 40, 41, *56*
Monod, J., 3(30), 4(30), 10(69), *22*, *23*
Monty, K. J., 34, *56*
Moore, G. P. M., 186, *188*, 300, *314*
Moore, P. B., 187, *189*
Moore, S., 222, *243*
Morey, K. S., 263, 264, *293*
Morita, T., 48, *57*
Morris, H. P., 162, *188*, 308, *314*
Moss, B., 111, *134*
Motulsky, A. G., 307, *314*
Moulé, Y., 32, 50, *54*, *56*, 264, *291*
Mousseron-Canet, M., 304, *314*
Mueller, G. C., 3(13), *22*, 213, 215, 230, *244*, 249, *293*
Müller-Hill, B., 3(31), 4(31, 35), 5(44), *22*, *23*, 142, *158*, 312, *313*, *316*
Munck, A., 261, *293*, *296*
Munns, T. W., 249, *293*
Muramatsu, M., 13(111), *25*
Murray, J. M., 177, 182, *190*
Musliner, T. A., 252, 260, *293*

N

Nabholz, M., 307, *316*
Nachmias, V. T., 177, *189*
Nakano, E., 187, *189*
Nakao, K., 3(25), *22*

Nations, C., 66, 70, 83, 91, 92, 96, *101*, 167, 168, 171, 172, 173, 174, 175, 180, 184, 185, *188*, *189*, 214, 227, *244*
Neelin, J. M., 64, 66, 70, 73, 77, 78, *101*, 298, 299, 300, 301, *315*, *316*
Newby, R. F., 4(34), 5(34), 23
Nias, A. H. W., 216, 217, 218, *244*
Nicolette, J. A., 249, *292*, *293*
Nielsen, L. E., 177, *187*
Nissley, P., 10(73, 74), *24*
Nobis, P., 37, 53, *56*
Nola, E., 251, 258, 259, *294*
Noland, B. J., 104, *135*
Nooden, L. D., 15(168), 16(168), 17(168), *27*, 39, 40, 41, 45, *57*
Norberg, B., 13(112), *25*
Noteboom, W. D., 249, 255, *293*
Notides, A., 252, 253, *292*
Numata, M., 251, 252, 253, 260, 261, *291*, *292*
Nuret, P., 12(104), *25*
Nyhan, W. L., 62, *100*

O

Ochoa, S., 64, 65, *102*
Oda, T., 8(59), *23*
Oelschlager, W., 34, *56*
Ogata, K., 48, *57*
Ohba, Y., 308, *316*
Ohga, Y., 115, *135*
Ohnishi, T., 177, *187*, *189*
Ohnuma, J., 177, *188*
Ohta, T., 312, *316*
Okamoto, T., 8(57), *23*
Olins, A. L., 184, *189*
Olins, D. E., 184, *189*, 308, *316*
Oliver, D., 3(21), *22*
Olson, M. O. J., 33, 34, 48, 49, *54*, *56*
O'Malley, B. W., 14(127, 134), *25*, *26*, 174, *189*, 248, 249, 251, 252, 253, 254, 255, 259, 260, 261, 262, 267, 268, 269, 270, 271, 272, 273, 274, 275, 277, 278, 279, 280, 287, 288, *291*, *293*, *294*, *295*, 298, 300, 303, 313, *316*
Onishi, T., 13(111), *25*
Ono, T., 3(25), *22*
Oosawa, F., 177, *187*, *188*, *189*
Ord, M. G., 3(20), *22*, 104, *134*

Orrick, L. R., 48, 49, *56*
Ozaki, H., 14(131), *25*, 214, *244*

P

Padilla, G. M., 216, 217, 219, *243*
Pallotta, D., 3(17), *22*
Panitz, R., 193, 194, *210*, *211*
Pardee, A. B., 228, 229, *243*
Parks, J., 10(72), 15(72), *24*
Parsons, J. T., 50, 51, *56*
Pastan, I., 3(33), 9(60, 63, 64), 10(64, 67, 68, 70, 71, 72, 73, 74), 15(72), *22*, *23*, *24*
Patel, G. L., 15(170), 16(170), *27*, 35, 36, 39, 42, 43, 45, 46, 51, 52, *56*, *57*, 301, *314*
Patel, V., 35, 49, *56*
Paul, J. S., 2(3), 3(28), 10(82), 15 (174), 16(82), 20(28, 174, 184), *21*, *22*, *24*, *27*, 32, 33, 35, 36, 39, 41, 42, 44, *54*, *55*, *56*, 89, 90, *101*, 126, *134*, 167, 168, 174, *187*, *188*, *189*, 191, 196, 200, *211*, 214, 235, 240, *243*, *244*, 274, 292, 297, 298, 299, 300, 303, 304, 306, 307, 309, *314*, *315*, *316*
Pavan, C., 194, 195, *210*, *211*
Pavlat, W. A., 235, *244*
Pederson, T., 79, *100*, 184, *189*, 230, *243*, 299, 300, 302, *314*
Pelling, C., 192, 193, 194, *211*
Pelmont, J., 62, *101*
Pemberton, R., 304, *314*
Penit-Soria, J., *314*
Penman, S., 33, *56*
Perlman, R. L., 3(33), 9(63, 64), 10(64, 67, 68, 70, 71, 72, 73, 74), 15(72), *23*, *24*
Perris, A. D., 129, *135*
Perry, R. P., 48, *55*
Perry, S. V., 177, 182, *189*
Peterkin, B. M., 273, 274, *293*
Pettit, B. J., 197, *211*
Pfahl, M., 5(44), *23*
Phil, A., 307, *316*
Phillips, D. M. P., 3(11), *22*, 30, *54*, *56*, 214, *244*
Philpot, J. St. L., 184, *189*
Piha, R. S., 51, *55*
Pipkin, J. L., 77, *101*

Author Index

Pipkin, J. L., Jr., 167, *189*
Pirrotta, V., 312, *314*
Platt, D. B., 248, *291*
Platt, T., 5(39, 40, 41), *23*
Platz, R., 138, *158*, 302, *316*
Platz, R. D., 14(124, 150, 155), 15(155), 21(150), 25, 26, 119, 122, 126, 128, 129, 133, *134*, 138, *158*, 234, 235, *244*
Pleger, G. L., 3(18), *22*
Poccia, D. L., 14(132), 25, 39, 40, 41, 44, *55*
Poels, C. L. M., 194, 197, 200, *212*
Pogo, A. O., 12(102), 25, 50, 51, *54*, 56, 199, *212*, 224, *244*
Pogo, B. G. T., 50, *56*, 199, 200, *209*, *212*, 224, *244*
Polikapova, S. T., 222, *244*
Pollard, T. D., 177, *189*
Polister, A. W., 30, 35, 38, *56*, 63, *101*
Pooley, A. S., 252, 260, 262, 267, *293*
Poort, C., 50, 52, *56*
Porter, G. A., 262, *291*, *294*
Potter, V. R., 162, *188*, 264, *291*
Pratje, E., 104, *134*
Prescott, D. M., 51, 55, 248, *294*
Ptashne, M., 7(52), *23*, 312, *313*, *314*
Puca, G. A., 251, 257, 258, 259, *291*, *294*
Pusztai, A., 15(176), 27, 61, 62, *101*

R

Rae, P. M. M., 309, *314*
Raina, P. N., 262, *294*
Rainey, C. H., 14(149), 26, 171, *189*
Randall, R. J., 97, *101*
Rao, B. R., 251, 252, 253, 255, *294*, *295*
Rasch, R. W., 197, *211*
Raspé, G., 249, 257, *294*
Raynaud, J. P., 262, *290*
Raynaud-Jammet, C., 262, *290*
Reddy, R., 33, 34, *54*
Reeck, G. R., 14(122), 25, 167, *189*
Reeder, R. H., 20(190), 21(190), 27, 84, *101*
Reichard, P., 13(112), *25*
Reif-Lehrer, L., 262, *294*
Reimann, E. M., 131, *134*
Remichek, J., 9(61), *23*

Reti, I., 255, *294*
Riches, P. G., 10(82), 14(161, 162), 16(82), *24*, *26*, 110, 112, 119, 121, 122, *134*, 301, 302, *316*
Richter, K. H., 14(130), 25, 39, 40, 41, 45, *56*, 108, *134*, 298, 299, *316*
Rickwood, D., 10(82), 14(161, 162), 16(82), *24*, *26*, 36, 39, 42, 55, *56*, 110, 111, 112, 119, 121, 122, 126, *134*, 298, 299, 300, 301, 302, 306, *315*, *316*
Rickwood, R. J., 167, 168, *189*
Riddle, M., 123, *135*, 138, *158*, 200, *212*, 302, *317*
Riggs, A. D., 4(34, 36, 37), 5(34, 42, 45), 6(46, 48), *23*, 312, *315*
Ringborg, U., 192, 193, *210*, *211*
Ringertz, N., 300, *314*
Ringertz, N. R., 186, *188*, *189*
Ris, H., 13(116), 25, 30, *56*, 161, *189*
Ritosso, F. M., 194, *212*
Robbins, E., 3(14), *22*, 184, *189*
Roberts, J., 8(56, 58), *23*
Rochefort, H., 253, 262, *290*, *294*
Ro-Choi, T. S., 33, 34, *54*
Roeder, R. G., 12(103), *25*
Rogers, A. W., 255, *294*
Romball, C. G., 197, *210*
Romen, W., 34, *56*
Rosbash, M., 304, *314*
Rosebrough, N. J., 97, *101*
Rosen, F., 262, *294*
Rosenberg, S. A., 14(144), 26, 83, *101*, 171, *189*, 223, *244*
Rosenblum, E. N., 111, *134*
Roses, A. D., 104, *134*
Ross, J., 305, *316*
Roth, J. S., 307, *314*
Rouiller, C., 264, *291*
Rouiller, C. H., 32, *54*
Rousseau, C. G., 267, 271, *291*
Rovera, G., 14(146), 26, 83, *101*, 168, 171, *189*, 215, 225, *244*
Ruddon, R. W., 14(149), 26, 115, *134*, 167, 171, *189*, 190, 200, *212*, 223, 224, *245*
Rudick, M. J., 215, 235, 236, 240, *243*, *244*
Rudkin, G. T., 194, 195, *212*
Ruiz-Carrillo, A., 3(19), *22*

Rusch, H. P., 14(135), 26, 52, 55, 66, 67, 69, 70, 73, 75, 77, 78, 79, 83, 90, 91, 92, 93, 96, 101, 119, 122, 134, 138, 139, 140, 142, 154, 155, 158, 162, 163, 167, 168, 171, 172, 173, 174, 175, 177, 178, 180, 181, 184, 185, 187, 188, 189, 214, 227, 233, 244, 299, 301, 302, 315
Rutter, W. J., 12(103), 25
Ryser, U., 177, 180, 188, 189
Ryskov, A. P., 20(194), 27

S

Sadgopal, A., 37, 56
Salas, J., 227, 228, 229, 244
Saligy, V. L., 301, 315
Samarina, O. P., 50, 51, 56, 209, 211
Sar, M., 261, 294
Sarff, M., 253, 294
Sauer, H. W., 154, 158
Savage, C. R., Jr., 186, 189
Savage, R. E., 186, 189
Schachner, M., 12(96), 24
Schauder, P., 14(147), 26
Schiltz, E., 14(164), 27, 112, 113, 134, 301, 316
Schlatteres, B., 34, 56
Schmitz, A., 5(44), 23, 312, 313
Schneider, A. B., 3(33), 9(64), 10(64), 22, 23
Schneider, W. C., 32, 55
Schrader, W. T., 251, 252, 253, 259, 260, 261, 262, 267, 280, 287, 288, 294, 313, 316
Schroeder, W. T., 174, 189
Schwartz, D., 3(32), 9(65), 22, 23, 312, 317
Schweppe, J. S., 14(165), 21(165), 27, 123, 133, 138, 158, 171, 188, 302, 315
Scolnick, E., 305, 316
Seale, R. L., 14(133), 26, 37, 38, 56, 83, 101
Segal, H. L., 104, 134
Segal, S. J., 249, 265, 295
Seifart, K. H., 13(110), 25
Sekeris, C. E., 14(130, 164), 25, 27, 39, 40, 41, 45, 56, 108, 112, 113, 134, 262, 263, 265, 291, 294, 298, 299, 301, 316
Seligy, V. L., 64, 78, 101

Sellers, L., 3(23), 22
Sellwood, S. M., 14(161), 26, 121, 122, 134
Serfling, E., 194, 211
Settineri, D., 301, 314
Sevall, J. S., 39, 40, 41, 45, 57
Sevall, S., 15(168), 16(168), 17(168), 27
Shakoori, A. R., 34, 56
Shapiro, A. L., 141, 158
Shapiro, I. M., 222, 244
Sharp, G. W. G., 261, 291, 294
Shaw, L. M. J., 39, 40, 56, 191, 212, 298, 299, 300, 316
Shea, M., 10(81), 14(81), 16(81), 21(81), 24, 36, 56, 125, 126, 134, 174, 189
Shearer, R. W., 308, 316
Shelton, K. R., 14(139), 26, 63, 64, 65, 66, 70, 73, 77, 78, 83, 101, 171, 174, 189, 204, 212, 214, 244, 284, 285, 294, 298, 299, 300, 301, 316
Shen, M. W., 309, 315
Shepherd, G. R., 104, 135
Sherman, M. R., 251, 252, 253, 254, 255, 259, 260, 261, 262, 269, 294
Sherod, D., 104, 135
Shibata, H., 33, 34, 54
Shimura, K., 39, 40, 41, 57, 298, 317
Shirey, T., 39, 40, 56, 298, 299, 316
Shorenstein, R. G., 12(98), 24
Shyamala, G., 251, 252, 253, 257, 292, 294, 295
Sica, V., 251, 258, 259, 294
Sidebottom, E., 186, 188
Siebert, G., 33, 34, 56
Siegel, R. B., 12(89), 24
Sigal, N., 11(87), 24
Simard, R., 34, 57
Simon, M. N., 12(92), 24
Simpson, R. T., 14(122, 144), 17(177), 25, 26, 27, 39, 40, 41, 44, 55, 83, 101, 167, 171, 189, 223, 244, 298, 299, 310, 312, 315
Sirlin, J. L., 248, 293
Slavik, M., 223, 242
Sluyser, M., 263, 264, 265, 294
Smart, J. E., 3(9), 22, 30, 54, 161, 188
Smellie, R. M. S., 249, 291, 293
Smith, D., 252, 253, 292

Smith, E. L., 3(12), 22, 30, 54
Smith, J., 17(180), 18(180), 27
Smith, J. A., 37, 56, 223, 244, 252, 260, 262, 267, 293
Smith, S., 251, 252, 253, 260, 261, 292
Snyder, L., 12(97), 24
Sober, H. A., 14(122), 17(177), 25, 27, 39, 40, 41, 44, 55, 298, 299, 315
Sonenshein, A. L., 12(98), 24
Sonnenbichler, J., 37, 53, 56
Soriano, R. Z., 118, 135
Sorof, S., 51, 52, 53
Sorsa, M., 198, 212
Southern, E. C., 307, 316
Spalding, J., 3(13), 22
Spelsberg, T. C., 3(10, 26, 27), 14(126, 127, 134), 20(27), 22, 25, 26, 30, 32, 35, 37, 38, 39, 48, 57, 89, 101, 133, 161, 174, 189, 214, 243, 248, 249, 250, 251, 253, 259, 260, 261, 262, 269, 270, 271, 272, 273, 274, 275, 277, 278, 279, 280, 281, 282, 283, 285, 287, 288, 291, 292, 294, 295, 298, 300, 303, 313, 314, 316
Spiegelman, S., 305, 315
Spinelli, G., 301, 314
Sporn, M. B., 13(117), 25, 37, 54, 89, 100, 161, 162, 188, 189, 214, 243
Spudich, J. A., 177, 182, 187, 189, 190
Stancel, G. M., 255, 258, 259, 295
Stanier, J. E., 184, 189
Stanners, C. P., 229, 243, 302, 314
Stedman, E., 3(5), 22, 30, 57, 161, 190, 297, 316
Stedman, E., 3(5), 22, 30, 57, 161, 190, 297, 316
Steele, W. J., 13(115), 25, 35, 51, 57, 62, 64, 77, 101, 190
Steggles, A. W., 14(127), 25, 174, 189, 248, 251, 252, 253, 259, 260, 261, 262, 264, 265, 267, 269, 270, 271, 272, 273, 274, 275, 277, 278, 280, 287, 288, 293, 294, 295, 298, 313, 316
Stein, G. S., 3(29), 14(143, 155), 15(155, 171), 20(29), 22, 26, 27, 32, 35, 45, 54, 57, 81, 83, 101, 102, 122, 134, 138, 158, 161, 171, 187, 190, 191, 200, 210, 212, 213, 214, 215, 216, 225, 226, 231, 232, 233, 234, 235, 243, 244, 245, 302, 304, 314, 316

Stein, H., 13(109), 25
Stein, W. H., 222, 243
Steinberg, R., 312, 314
Stellwagen, R. H., 3(8), 22, 30, 57, 83, 102, 161, 190, 191, 212, 222, 245
Stern, G., 304, 314
Stern, H., 220, 243
Stevens, B. J., 199, 212
Stocken, L. A., 3(20), 22, 37, 56, 104, 134
Stollar, B. D., 308, 316
Stone, G. M., 251, 295
Striebich, M. J., 32, 55
Studier, F. W., 12(92, 93), 24
Stumpf, W. E., 252, 253, 255, 257, 260, 261, 292, 294, 295
Sugano, H., 34, 48, 55
Sugiura, M., 8(57), 23
Summers, W. C., 12(89), 24
Sunaga, K., 263, 295
Sung, M. T., 5(43), 23
Sutherland, E. W., 10(66), 23
Suzuki, H., 5(45), 23
Suzuki, T., 251, 252, 253, 257, 260, 261, 292
Swaneck, G. E., 14(148), 26, 261, 262, 263, 265, 295
Swift, H., 194, 195, 198, 199, 207, 209, 211, 212
Szego, C. M., 250, 295
Szent-Györgyi, A., 182, 190
Szeszak, F., 307, 316
Szybalski, W., 7(53), 23

T

Tack, L. O., 221, 243
Takahashi, K., 177, 188
Takaku, F., 3(25), 22
Takanami, M., 8(57, 59), 23
Takeda, M., 115, 135
Talwar, G. P., 265, 295
Tan, C. H., 20(189), 27
Tanaka, H., 177, 190
Tanaka, Y., 177, 189
Taniguchi, M., 187, 189
Tata, J. R., 12(101), 25, 249, 263, 292, 295
Taylor, E. W., 183, 190
Tazawa, M., 177, 188

Temple, G. F., 305, *317*
Teng, C. S., 10(78, 80), 14(78, 80, 125, 140), 15(80, 125), 16(80, 125), 21(78, 80, 125), *24, 25, 26*, 36, 45, 57, 65, 66, 69, 70, 77, 78, 79, 81, 83 87, 89, 90, *100, 102*, 108, 109, 119, 121, 123, 124, 126, *133, 135*, 138, *158*, 171, 174, 187, *190*, 200, 204, *212*, 214, 215, 223, *245*, 249, *295*, 298, 300, 301, *316*
Teng, C. T., 10(78, 80), 14(78, 80, 125), 15(80, 125), 16(80, 125), 21(78, 80, 125), *24, 25*, 36, 45, 57, 65, 66, 69, 70, 77, 78, 79, 81, 87, 89 90, *100, 102*, 108, 109, 119, 121, 123, 124, 126, *133, 135*, 138, *158*, 174, 187, *190*, 200, *212*, 214, 215, *245*, 298, 300, *316*
Terada, M., 305, *315*
Terayama, H., 3(25), *22*
Terenius, L., 251, *295*
Thayler, M. M., 225, *245*
Thomas, G. H., 255, *294*
Thomas, L. E., 50, 55
Thomas, T. L., 15(170), 16(170), *27*, 36, 43, 45, 46, *56*, 57
Thrall, C. L., 232, *245*
Threlfall, G., 10(82), 16(82), *24*, 32, 36, *55, 56*, 126, *134*, 191, 200, *211*, 214, *244*
Tichonicky, L., 300, *314, 315*
Tobey, R. A., 3(22), *22*
Toft, D. O., 251, 252, 253, 254, 255, 257, 258, 259, 260, 261, 262, 267, 268, 269, 288, *291, 292, 294, 295*
Toft, P. O., 252, 253, 254, 255, 259, 260, 269, *294*
Tokarskaya, V. I., 39, 40, 41, 46, 57, 298, *317*
Tolstoshev, P., 53, *55*, 299, *315*
Tomkins, G. M., 126, *135*, 261, 262, 267, 271, *291, 292*
Trachewsky, D., 249, *295*
Travers, A. A., 12(95), *24*
Truong, H., 262, *290*
Tsai, Y. H., 264, 265, *295*
Tsuboi, A., 168, 171, *190*, 225, 226, *245*, 299, *316*
Tsurugi, K., 48, *57*
Tsvett, M., 61, *102*

Tuan, Y. H., 33, *54*
Turkington, R. W., 123, *135*, 138, *158*, 200, *212*, 302, *317*
Tveter, K. J., 261, *295*
Tymoczko, J. L., 261, *293*

U

Ullman, A., 10(69), *23*
Umanskii, S. R., 39, 40, 41, 46, 57
Umansky, S. R., 298, *317*

V

van den Broek, H. W. J., 15(168), 16(168), 17(168), *27*, 39, 40, 41, 45, 57
van Eupen, O., 193, *211*
Varmus, H. E., 10(70, 71), *24*
Vazquez-Nin, G., 199, *212*
Verma, I. M., 305, *317*
Vertes, M., 223, *244*, 252, 260, 262, 267, *293*
Vidali, G., 3(24), 10(79), 14(79, 136, 145, 154), 15(24, 79, 154), 16(24), 17(24), 21(79, 145), *22, 24, 26*, 78, 79, 81, *100, 101*, 119, 129, 132, *133, 135*, 143, *158*, 163, 167, 184, 186, *190*, 199, *211, 212*, 230, 234, 235, *244*
Vigersky, R., 253, 260, *291*
Villee, C. A., 207, *210*, 225, *245*, 252, 260, *293*
Vincent, W. S., 34, 47, 57
Viñuela, E., 64, 65, *102*, 141, *158*
von Ubisch, H., 13(112), *25*
Vorobyev, V. I., 184, *190*

W

Wagner, T. E., 263, 265, *295*
Wajcmax, H., 300, *315*
Wakabayashi, K., 46, 57
Walker, P. M. B., 18(183), *27*
Walters, R. A., 3(22), *22*
Wang, S., 46, 57
Wang, T. Y., 14(129), 20(185, 187, 188), *25, 27*, 33, 35, 36, 39, 41, 42, 43, 44, 45, 48, 49, 50, 51, 52, *55, 56*, 57, 89, *102*, 108, 125, *134*, 174, *188*, 214, 224, *244, 245*, 298, 299, 300, 304, *315, 317*

Wangh, L. J., 3(19), 22
Waskell, L., 12(90), 24
Watts, S., 177, 182, 189
Weber, A., 177, 182, 190
Weber, K., 5(39, 40, 41), 12(94), 23, 24
Wells, J. R. E., 53, 55, 299, 315
Weisenthal, L. M., 167, 190, 223, 224, 245
Whitfield, J. F., 129, 135
Wicks, W. D., 262, 292
Widnell, C. C., 12(101), 25, 292
Wieslander, L., 193, 211
Wiest, W. G., 251, 252, 253, 255, 294, 295
Wilbur, K. M., 183, 187
Wilhelm, J. A., 3(10), 22, 30, 32, 33, 47, 49, 53, 57, 161, 189, 248, 294, 300, 317
Willart, E., 198, 199, 211
Williams, D., 253, 255, 268, 295
Williamson, H. E., 261, 262, 291, 295
Wilson, E. M., 14(134), 26, 38, 57, 279, 282, 285, 295, 298, 303, 316
Wilson, J. D., 249, 251, 261, 262, 265, 291, 295
Wira, C., 261, 293, 296
Wobus, U., 194, 211
Woo, S. L. C., 131, 134
Woodard, J., 195, 211
Wotiz, H. S., 253, 260, 291
Wotiz, H. S., 253, 260, 291
Wray, W., 66, 70, 75, 83, 89, 90, 92, 96, 101, 142, 158, 163, 168, 175, 188
Wu, A. M., 8(54), 23

Wu, F. C., 142, 158, 167, 168, 190
Wullems, G. J., 201, 211

Y

Yamamoto, H., 198, 212
Yamamoto, T., 177, 189
Yamamura, H., 115, 135
Yang, Y., 168, 172, 173, 174, 175, 178, 187, 189
Yates, K. M., 255, 294
Yeoman, L. C., 33, 34, 54
Yermolayeva, L. P., 32, 35, 48, 57
Yoshida, M., 39, 40, 41, 57, 298, 317
Young, D. A., 262, 296
Young, K. E., 39, 40, 41, 44, 53, 298, 299, 314
Yu, F. L., 83, 102

Z

Zachau, W. G., 307, 315
Zalta, J., 34, 57
Zalta, J. P., 34, 57
Zampeth-Bosseler, F., 81, 102
Zardi, L., 14(128), 25
Zbarskii, I. B., 32, 35, 48, 50, 57
Zech, I., 309, 314
Zeuthen, E., 216, 217, 245
Zillig, W., 12(96), 24
Zimmerman, A. M., 217, 219, 243, 244
Zobel, C. R., 35, 49, 56
Zotova, R. N., 39, 40, 41, 46, 57, 298, 317
Zubay, G., 3(32), 9(65), 10(75), 22, 23, 24, 33, 57, 312, 317

Subject Index

A

Acrylamide gels, preparation of, 98–99
Actin
  adenosine triphosphatase and, 183
  heterochromatin and, 186
  myosin–DNA complex and, 186–187
  nuclear proteins and, 173, 174, 177–180, 184, 185
Actinomycin D, acidic nuclear proteins and, 225, 232
Adenosine diphosphate F-actin filaments and, 187
Adenosine triphosphatase, nuclear, actomyosin and, 177, 178
Adenosine triphosphate
  nuclear protein-bound phosphate metabolism in, 112–113
  γ-^{32}P, labeling of nuclei *in vitro*, 147–151
S-Adenosylmethionine, histone methylation and, 201
Aldosterone
  binding and localization of, 261–262
  nuclear proteins and, 14
Alkylating agents, nuclear proteins and, 62
α-Amanitin, ribonucleic acid polymerases and, 12
Amethopterin, cell synchronization and, 217
Amino acid(s)
  acidic nuclear protein composition, 73, 109
  nuclear phosphoprotein composition, 109
  sequence, *lac* repressor, 5
p-Aminosalicylate, nuclear proteins and, 63
Aniline blue black, gel staining and, 98
Antisera, progesterone-receptor complex acceptor and, 279–281
Azure B, chromosome puffs and, 194

B

Bacillus subtilis, ribonucleic acid polymerases, sporulation and, 12
Bio-Rex 70, phosphoprotein isolation and, 106
Bovine serum albumin (BSA)
  phenol and, 62
  renaturation of, 15
Brain
  chromatin, phosphoprotein and, 121
  embryonic, ribonucleic acid polymerase in, 161
  high molecular weight acidic proteins in, 168
5-Bromodeoxyuridine, *lac* repressor binding and, 6
Bromphenol blue, electrophoresis and, 98
Burkitt lymphoma cells, nuclear acidic proteins of, 167, 223–224

C

Calcium ions
  actomyosin ATPase and, 183
  chromatin dispersion and, 184
Calcium phosphate gel, phosphoprotein isolation and, 106–107
Carbonic anhydrases, renaturation of, 15
Cell(s), proliferation, nuclear acidic protein heterogeneity and, 161–171
Cell cycle
  acidic proteins and
    cell proliferation and, 222–227
    electron microscopy and, 235–240
    general synthesis, 221–222, 302
    highly synchronous populations, 227–233
    phosphorylation of, 234–235
  embryonic, 220
  nuclear phosphoproteins and, 122
  protein phosphorylation and, 78–81
Cell cycle systems, synchronous, comparative analysis, 215–221

Subject Index

Chinese hamster cells
  acidic nuclear protein synthesis in, 222
  synchronized, acidic nuclear proteins in, 233
Chironomus tentans, chromosome puffs in, 192–194
Chironomus thummi, chromosome puffs, histone acetylation and, 200
CHO cells, synchronized, acidic nuclear proteins in, 228–229, 230–231
Chromatin
  acidic proteins
    fractionation and characteristics, 43–46
    general considerations, 35–36
    methods for isolation, 36–43
  condensation, nuclear condensation and, 183–184
  deoxyribonucleic acid in, 308
  isolation of, 33–34
  phosphoprotein localization and quantitation, 121
  steroid hormone binding, 262–263
  structure, histones and, 2–3
  transcriptional units, 310–311
  transcription of, 303
Chromosomes
  banding of, 309–312
  polytene, chemical modification of proteins, 199–201
  puffs
    cytochemistry, 194–199
    gene activity and, 192–194
    protein phosphorylation and, 15
Chymotrypsinogen
  phenol and, 62
  renaturation of, 15
Citrate, nucleolar proteins and, 47
Citric acid, isolation of nuclei and, 32–33
Colchicine, cell synchronization and, 217
Conconavalin A, nonhistone proteins and, 14
Contractile systems, chemistry of, 182–183
Coomassie brilliant blue, gel staining and, 98
Corticosterone, acidic nuclear proteins and, 127
Cortisol
  binding and localization of, 261–262
  gene activation and, 14
  histones and, 264
  nonhistone protein and, 204, 301
  nuclear protein and, 63, 81, 82, 83
  protein phosphorylation and, 123
Cortisone, histones and, 263, 264
Cyclic adenosine monophosphate
  chromosome puffs and, 200–201
  galactose operon and, 9–10
  nuclear proteins and, 15
  protein kinases and, 116–117, 126, 129
Cycloheximide
  acidic nuclear proteins and, 225
  chromosome puff protein and, 197
  nuclear actin and, 180
Cytoplasm
  acidic proteins, as steroid hormone receptors, 250–252
  nuclear acidic protein synthesis and, 83
Cytosine arabinoside, acidic nuclear proteins and, 232

D

Dedifferentiation, nuclear acidic proteins and, 167–171, 173
Deoxycholate
  nuclear residue and, 35
  nucleolar proteins and, 47, 48
Deoxynucleohistone
  reconstitution, acidic proteins and, 42
  solubility of, 41–42
Deoxyribonuclease
  acidic protein isolation and, 38
  bound steroid hormones and, 266–267
Deoxyribonucleic acid
  affinity of "acceptor" molecule for, 281–284
  amplification, chromosome puffs and, 194
  binding
    sequence-specific, 17–20
    specificity in, 15
    species-selective, 16–17
  eukaryotic, nature and amount, 307–308
  globin messenger RNA transcription and, 303–307

homologous, affinity for acidic nuclear proteins, 86–88, 89
inhibition of synthesis, synchronization and, 216, 217, 219
lac repressor binding and, 5–6
lambda repressor and, 7–8
myosin complex, 184, 186–187
phenol extraction and, 96
phosphoprotein-binding specificity, 123–124
proteins associated, transcription and, 2, 9–11, 303–304
repetitive sequences, function, 308–309
requirements for "acceptor" action, 284–286
steroid hormone binding by, 266–267
synthesis
    acidic protein synthesis and, 171
    histone methylation and, 201
    transcription, unique and repetitive sequences, 20–21
    unwinding, proteins and, 11
Deoxyribonucleic acid polymerase, synthesis, refeeding and, 235, 239
Derepression, acidic proteins and, 36
Detergents, isolation of nuclei and, 33
Dexamethasone, binding of, 267, 271
Differentiation
    nuclear nonhistone proteins and, 14, 160
    ribonucleic acid polymerases and, 12
Dihydrotestosterone, binding and localization, 261–262, 272–273
Drosophila
    nuclear proteins, 65–66
    polytene chromosomes of, 174
Drosphila hydei
    chromosome puffs
        histone acetylation and, 200
        nonhistone proteins and, 193, 195, 196, 197, 199, 205
    histone methylation in, 201
Drosphila melanogaster
    chromosome puffs, 195, 196
    histone acetylation and, 200

E

Ecdysone
    chromosome puffs and, 194, 201, 202–203, 301

nonhistone proteins and, 14, 205–206
transport of, 207–208
Electron microscopy, gene activiation and, 235–240
Electrophoresis
    nuclear acidic proteins, 44, 45, 68–72
    nucleolar acidic proteins, 49
    phenol-soluble proteins, 98–99
    phosphoproteins, acidic nuclear, 110–111
Embryo(s), cell synchrony in, 220
Embryonic tissue, nonhistone protein in, 161
Enzymes, residual nuclear proteins and, 173
Equilibrium centrifugation, acidic protein isolation and, 37–38
Erythrocytes
    maturation, phosphoproteins and, 122, 163, 167, 302
    nuclei
        contractile proteins in, 183, 186
        residual protein in, 161
        ribonucleic acid polymerase in, 161–162
Escherichia coli
    ribonucleic acid polymerase
        eukaryotic templates and, 21
        phage infection and, 12
Estradiol-17β
    chromatin binding, 249–250, 272
    deoxyribonucleic acid binding, 267
    gene activation and, 14
    histones and, 263
    localization of, 255
    nonhistone protein and, 204, 301
    uptake of, 251
Estrogen
    acidic nuclear protein and, 304
    cytosol-receptor, transformation of, 253, 257–258
    protein and ribonucleic acid synthesis and, 249–250
Ethionine, histone ethylation and, 201
Eukaryotes
    gene regulation in, 307–313
    transcriptional control in, 13–21

Subject Index

F

Fast green, acidic, chromosome puffs and, 195, 196, 197
Fibroblasts
  acidic nuclear protein synthesis in, 222, 225–227, 302
  density inhibited, nuclei of, 183
  nuclei, contractile proteins in, 175, 181
  oncogenically transformed, contractile proteins in, 187
  refeeding, acidic protein synthesis and, 168–171
  synchronized, acidic nuclear proteins in, 227–228, 232
Ficoll
  isolation of nucleoli and, 34
  selection synchrony and, 219
Formic acid, chromatin dissociation and, 39

G

Galactokinase, transcription and, 9–10
Galactose-1-phosphate-uridyl transferase, transcription and, 9–10
Galactoside, *lac* repressor and, 4
Gel filtration, separation of nucleic acids and proteins by, 40
Gene(s)
  activity
    chromosome puffs and, 192–194
    nuclear acidic proteins and, 13–15, 21, 90–91
    phosphoproteins and, 121–123
    protein changes and, 201–206
    role of protein, 206–209
  regulation
    major acidic nuclear proteins and, 171–174
    phosphoproteins and, 142–143
Globin, messenger RNA, unique DNA sequence and, 304–307
Glucagon
  gene activation and, 14
  nuclear acidic protein synthesis and, 83
Glucocorticoids, chromatin and, 262
Glucose, catabolite repression and, 10
Gonadotropins
  nuclear proteins and, 14
  protein phosphorylation and, 123, 302
Guanidine hydrochloride
  acidic nuclear protein isolation and, 299
  chromatin dissociation and, 39, 40

H

HeLa cells
  acidic nuclear protein synthesis in, 222, 302
  cell cycle, phosphoproteins and, 122
  nuclei, contractile proteins in, 175, 181
  starved, nuclei of, 183
  stress, acidic nuclear proteins and, 163, 166
  synchronized
    acidic nuclear proteins in, 230, 231–233
    protein phosphorylation in, 234–235
Heterochromatin
  contractile proteins and, 185–186
  phosphoproteins and, 121
Hippeastrum belladonna, generative nuclei, nuclear proteins and, 167
Histone(s)
  acetylation of, 199–200
  amino acid composition, 109
  binding of, 308
  chromatin structure and, 2–3
  chromosome puffs and, 195
  deacetylation of, 132
  electrophoresis of, 67
  hormone-receptor complexes and, 263–265
  methylation and ethylation of, 201
  modification of, 3
  nucleoli and, 47
  phosphoprotein interaction, 127–130
  phosphorylation of, 200–201
  protein phosphorylation and, 81
  removal from nuclei, 37–38
Hormones
  acidic nuclear protein phosphorylation and, 122–123
  steroid, binding of, 126–127
Hut operons, repression of, 6–7

Hydrochloric acid
　histone removal and, 37
　nuclear protein extraction by, 93
Hydrocortisone, histones and, 263–265
Hydroxyapatite, chromatin macromolecules and, 39
Hydroxyurea, cell synchronization and, 217, 218, 232–233

I

Informofers, proteins of, 51
Insulin
　acidic protein phosphorylation and, 122–123
　nuclear proteins and, 14
Ion-exchange chromatography
　acidic protein separation and, 41, 43, 44–45
　nucleolar acidic proteins, 49
Isoelectric focusing, acidic proteins, 44, 74, 78
Isopropyl-β-D-thiogalactoside, mutant *lac* repressor and, 5
Isoproterenol
　acidic protein phosphorylation and, 122
　nonhistone proteins and, 14

K

Kidney
　acidic nuclear protein composition, 77
　phosphoprotein content, 108

L

Lac repressor
　binding to DNA, 312–313
　copies per genome, 172
　transcriptional control and, 4–6
Lambda repressor, transcription and, 7–8
Landschutz ascites cells, chromatin, phosphoproteins of, 302
Lanthanum chloride, nucleic acid precipitation and, 40–41
Leucine, incorporation, cell cycle and, 222, 235
Light green-orange G, chromosome puffs and, 195

Lilies, anthers, cell synchrony in, 220
Lipid, extraction, nuclear proteins and, 93–94
Liver
　acidic nuclear proteins, 46, 70
　　amino acid composition, 77
　chromatin, phosphoprotein in, 121
　dispersed chromatin, proteins of, 167–168
　fetal, acidic protein of, 306–307
　nuclei
　　contractile proteins in, 184
　　residual protein in, 161
　phosphoprotein content, 108
　regenerating, nuclear synthesis in, 224–225
　stimulated, nonhistone protein in, 162
Liver cells
　acidic nuclear protein synthesis in, 223
　cooled, nuclei of, 183
Lymphocytes
　acidic nuclear protein synthesis in, 222, 223
　gene activity, phosphoproteins and, 122, 302
Lymphocytic leukemia cells, acidic nuclear proteins of, 167

M

Magnesium ions, chromatin dispersion and, 184
Mammary gland, nuclear acidic protein synthesis in, 223
Mammary tumors, estradiol localization in, 255
Mercuric bromphenol blue, chromosome puffs and, 197
Mercury orange, chromosome puffs and, 197
Meromyosin(s), nuclear proteins and, 178
Methionine, histone methylation and, 201
Methyl green-pyronin, chromosome puffs and, 194–195
N^7-Methyl histidine, nuclear proteins and, 177
Microtubules, protein, nuclear proteins and, 173

Subject Index

Mitotic cycle, nuclear phosphoproteins and, 122
Mitotic index, synchrony and, 220
Morris 5123C hepatoma, acidic chromatin proteins of, 162
Myelocytic leukemia cells, nuclear proteins, 224
Myosin
  adenosine triphosphate and, 183
  DNA complex, 184, 186–187
  nuclear proteins and, 173, 174, 177–180

N

Naphthol yellow S, binding, chromosome puffs and, 195, 197, 198
N-(1-Naphthyl)ethylenediamine dihydrochloride, chromosome puffs and, 197
Ninhydrin-Schiff reagent, chromosome puffs and, 197
Novikoff hepatoma cells, residual nuclear protein in, 162
Nuclease, chromatin and, 308
Nuclei
  condensation, chromatin condensation and, 183–184
  contamination by cytoplasmic contractile proteins, 179–180
  electron microscopy, refeeding and, 237–240
  function, phosphoproteins and, 129–132
  isolation of, 32–33, 106
  labeling *in vitro* with [γ-^{32}P]ATP, 147–151
  protein-bound phosphate metabolism in, 112–113
  steroid-binding proteins in, 255–260
Nucleic acids, hydrolysis, acidic protein isolation and, 37
Nucleoli
  acidic proteins
    fractionation and characteristics, 48–49
    isolation, 47–48
  isolation of, 34–35
  proteins, isolation of, 100

Nucleoplasm acidic proteins
  fractionation and characteristics, 52
  isolation, 50–52
Nucleotides, *lac* operator region sequence, 4

O

Oviduct
  chromatin acceptor binding progesterone–receptor complex
    affinity for DNA, 281–284
    component responsible for binding, 274–276
    DNA or protein as acceptor, 287–288
    in vitro binding, 268–274
    purification and properties of acceptor, 277–281
    requirements for acceptor activity, 284–286
    sequence of progesterone action, 289–290
  progesterone binding by, 256, 260

P

pH, histone removal and, 37
Phenobarbital, nuclear proteins and, 14
Phenol
  nuclear acidic protein solubilization and, 62–66, 94–95, 108–109
  as protein solvent, 61–62
  purification of, 95
Phosphatase, nuclear acidic phosphoproteins and, 117–118
Phosphate
  incorporation, starvation and, 154–157
  isotopic, pulse labeling in, 145–147
  protein bound nature, 145
  turnover, kinetics of, 151–154
Phospholipids, removal, nuclear acidic proteins and, 65
Phosphoprotein(s)
  acidic nuclear, 104–105
    amino acid composition, 109
    electrophoretic fractionation, 110–111
    other extraction techniques, 108–110
    salt extraction, 105–108

functional properties,
  cell, tissue and species specificity, 119–120
  cyclic AMP and, 126
  gene activity and, 121–123
  histone interaction, 127–129
  localization and quantitation, 121
  RNA synthesis and, 125–126
  specificity of DNA binding, 123–124
  steroid hormone binding, 126–127
phosphate content
  calculation of copies per nucleus, 142–145
  calculation of phosphates per polypeptide, 141–142
  correspondence of double labels, 139–141
  demonstration of protein-bound nature, 145
phosphate metabolism, 111–112
  dephosphorylation, 117–118
  phosphorylation reaction, 113–117
  protein bound, 112–113
  summarizing model, 118
nuclear function and, 129–132
Phosphorylation, acidic nuclear proteins, 78–81, 82
Physarum
  dedifferentiation, nuclear proteins and, 167–169
  nuclei, contractile proteins and, 175–178, 181, 184–185
  synchrony in, 220
Physarum polycephalum,
  differentiation, 138
    protein phosphorylation and, 155, 157, 163–166
    mitotic cycle, phosphoproteins and, 122
    nuclear proteins of, 66
      composition, 67–75
      chemical characteristics, 75–77
      synthesis, 83–86
Phytohemagglutinin
  nonhistone proteins and, 14, 167, 302
  nuclear phosphoproteins and, 122
Pi factor, ribonucleic acid polymerases and, 12–13
Pituitary, estradiol localization in, 255

Pollen mother cell, acidic nuclear protein composition, 77
Polyethylene glycol, protein solution concentration and, 97
Polylysine, deoxyribonucleic acid and, 308
Progesterone
  action on oviduct cells, sequence of events, 289–290
  histones and, 263
  localization of, 255–256
  receptors
    binding of, 288
    properties of, 258–259
  uptake of, 251
Progesterone–receptor complex,
  chromatin "acceptor" molecule,
    affinity for DNA, 281–284
    component responsible for binding, 274–276
    DNA or protein as acceptor, 287–288
    in vitro binding, 268–274
    purification and properties, 277–281
    requirements for acceptor activity, 284–286
    sequence of progesterone action, 289–290
Prokaryotes, transcription control in, 3–11
Prolactin, acidic protein phosphorylation and, 122–123, 302
Proliferation, nuclear proteins and, 14
Pronase, chromosome puff staining and, 197
Prostate
  acidic nuclear protein synthesis in, 223
  dihydrotestosterone binding by, 273
Proteases, bound steroid hormones and, 262, 263, 265
Protein(s)
  acidic
    fractionation and characteristics, 43–46
    general considerations, 35–36
    methods for isolation, 36–43
  acidic cytoplasmic, as steroid hormone receptors, 250–252
  acidic nuclear,
    biological assessment, 303–307
    cells stimulated to proliferate and, 222–227

Subject Index

distribution and specificity, 300–301
general synthesis, 221–222
heterogeneity and cell proliferation, 161–171
hormone binding to, 265–266
isolation and characterization, 298–300
phosphorylation, 301–302
progesterone–receptor complex and, 274–276
synchronized cell populations and, 227–233
contractile
  function aspects, 181–187
  as major components of nuclei, 174–178
  preliminary evidence for true nuclear components, 178–181
deoxyribonucleic acid-associated, transcription and, 2
deoxyribonucleic acid-binding, positive transcriptional control by, 9–11
deoxyribonucleic acid-unwinding, 11
major nuclear acidic, specific gene regulation and, 171–174
nonhistone
  functions of, 191–192.
  gene activity and, 13–15
  historical considerations, 30–32
  phosphorylation and gene activation, 21
  sequence-specific DNA binding, 17–20
  species-selective DNA binding, 16–17
  specificity in DNA binding, 15
  transcription of unique and repetitive DNA sequences, 20–21
nuclear, 91–92
  acid extraction, 93
  lipid extraction, 93–94
  notes, 95–96
  roles of, 160
  saline extraction, 92–93
  solubilizing residual acidic proteins, 94–95
phenol-soluble nuclear acidic, 66–67
  chemical characteristics, 75–78
  composition, 67–75
  DNA affinity, 86–88

intranuclear concentration, 88–91
phosphorylation, 78–81
preparation, 62–66
synthesis, 81–86
phenol solvents and, 61–62
phosphorylation, 103–104
  gene activation and, 21
polytene chromosomes,
  acetylation, 199–200
  gene activation and, 201–206
  methylation and ethylation, 201
  phosphorylation, 200–201
  role of, 206–209
preparation for electrophoretic separation, 96–97
ribonucleic acid initiation and, 11
staining, chromosome puffs and, 195–197
synthesis
  estrogens and, 249–250
  refeeding and, 235
  synchronized cells and, 217
Protein kinase, nuclear acidic proteins and, 113–117
Puromycin, chromosome puff protein and, 197

R

Rho factor, geometry of functional oligomers, 8–9
Rhynchosciara, chromosome puffs of, 194
Ribonuclease
  chromosome puff staining and, 197
  phenol and, 62
Ribonucleic acid
  chromosome puffs and, 193–194, 195
  initiation proteins, low molecular weight, 11
  processing of, 308, 311
  synthesis,
    acidic nuclear proteins and, 83, 214, 303
    estrogens and, 249–250
    histones and, 3
    nonhistone proteins and, 15
    phosphoproteins and, 125–126
    refeeding and, 235, 240
    synchronized cells and, 217
  transcription, termination of, 9

Ribonucleic acid polymerase
  bacterial, eukaryotic templates and, 21
  binding to DNA, 312
  chromosome puffs and, 207, 208
  embryonic brain chromatin and, 161
  estrogen and, 249–250
  lambda repressor and, 7–8
  nuclear acidic proteins and, 90–91, 173
  phosphoproteins and, 131
  transcription control and, 12–13
Ribonucleoprotein, particles, chromosome puffs and, 198–199, 208–209
Ribosome(s)
  nuclear, proteins of, 50
  starvation and refeeding and, 237–238, 240
Rifamycin AF-013, ribonucleic acid polymerases and, 13

S

Saline, nuclear protein extraction by, 92–93
Salivary gland, acidic nuclear protein
  phosphorylation of, 122
  synthesis, 225
Salmonella typhimurium, hut operon repression, 6–7
Sciara, chromosome puffs of, 194, 201
Sciara coprophila, histone methylation in, 201
Sea urchin, development
  acidic proteins and, 301
  phosphoproteins and, 122
Sephadex G-100, protein solution concentration and, 97
Sigma factors
  phosphorylation of, 131
  transcription and, 10
Sodium chloride
  acidic protein chromatography and, 44–45
  chromatin dissociation and, 39, 40–41
  hormone receptor binding and, 271–272
  nucleolar proteins and, 47–48
  phosphoprotein extraction by, 106–108
Sodium dodecyl sulfate
  acidic nuclear proteins and, 38, 44, 45, 299
  chromatin dissociation and, 39, 40
  nucleolar proteins and, 47, 48
  nucleoplasmic proteins and, 52
  phenol-extracted proteins and, 64, 65
  phosphoproteins and, 110–111
  protein electrophoresis and, 97
  residual nuclear protein and, 95
Species, specificity of DNA binding and, 16–17
Sperm
  chromatin, acidic protein in, 161, 162
  deoxyribonucleic acid, liver acidic proteins and, 46
Starvation
  cell synchronization and, 218
  phosphorylation and, 154–157
Steroid hormones
  chromatin binding, 262–263
  cytoplasmic acidic proteins and, 250–252
  intracellular distribution
    cell nucleus and, 255–257
    cytonuclear transport, 252–255
    nuclear acidic proteins and, 257–260
    tissue specificity of binding, 260–261
    universality of nuclear localization, 261–262
  nuclear and cytoplasmic receptors, 257–260
  primary site of action, 249–250
  tissue specificity of nuclear binding, 260–261
Steroid hormone–receptor complexes, chromatin-binding sites
  acidic proteins, 265–266
  deoxyribonucleic acid, 266–267
  histones, 263–265
Sucrose, isolation of nuclei and, 32
Sulfuric acid, histone removal and, 37
SV40 virus, nuclear nonhistone proteins and, 14
Synchronized cells, acidic nuclear proteins in, 227–233
Synchrony
  degree, determination of, 220–221
  induction, 216–219
  natural, 216, 217, 220
  selection, 216, 217, 219–220

Subject Index

T

Temperature, cell synchronization and, 218
Testosterone
 histones and, 263, 264
 nuclear proteins and, 14
 protein phosphorylation and, 123, 301
Tetrahydrocortisol, histones and, 263
Tetrahydrocortisone, histones and, 263
Tetrahymena, acidic nuclear proteins, 74
Tetrahymena pyriformis
 acidic nuclear protein synthesis in, 224
 starvation and refeeding, electron microscopy and, 235–240
Thymidine, tritiated, cell synchronization and, 217–218, 221
Thymus
 chromatin, phosphoprotein in, 121
 nuclei, contractile proteins in, 177, 178
 phosphoprotein content, 108
Tissues, acidic nuclear phosphoprotein specificity and, 119–120
Toluidine blue
 chromosome puffs and, 194
 deoxyribonucleic acid phosphate groups and, 308
Transcription
 control in prokaryotes, 3–4
  deoxyribonucleic acid binding proteins, 9–11
  deoxyribonucleic acid unwinding protein, 11
  hut operon repression, 6–7
  lac repressor, 4–6
  lambda repressor, 7–8
  rho factor, 8–9
  ribonucleic acid initiation protein, 11
  ribonucleic acid polymerases and, 12–13
 differential, 2
 nonhistone proteins and, 16
 unique and repetitive DNA sequences, 20–21

Trichloroacetic acid, chromosome puffs and, 197
Tropomyosin
 actomyosin ATPase and, 183
 nuclear proteins and, 173, 174, 177–178, 184, 185
Troponin
 actomyosin ATPase and, 183
 nuclear proteins and, 177, 178
Trypsin
 hormone–receptor complex and, 265
 nuclear actomyosin and, 178
Tryptophan, incorporation, cell cycle and, 222, 227–228, 229
Tubulin, nuclear proteins and, 173

U

Ultracentrifugation, separation of nucleic acids and proteins and, 39–40
Urea
 acidic nuclear protein isolation and, 44–45, 299
 chromatin dissociation and, 39, 40–41
 protein renaturation and, 15
Uridine diphosphate-galactose-4-epimerase, transcription and, 9–10
Urocanate, *hut* operon repression and, 6
Uterus
 nuclear acidic protein synthesis in, 223
 steroid hormone uptake by, 251–252, 255, 260

W

Walker rat carcinoma, nuclear acidic protein in, 161, 162

Y

Yoshida ascites sarcoma cells, growth phase, phosphoproteins and, 122

EDITORS

D. E. BUETOW

Department of Physiology
and Biophysics
University of Illinois
Urbana, Illinois

I. L. CAMERON

Department of Anatomy
University of Texas
Medical School at San Antonio
San Antonio, Texas

G. M. PADILLA

Department of Physiology and Pharmacology
Duke University Medical Center
Durham, North Carolina

G. M. Padilla, G. L. Whitson, and I. L. Cameron (editors). THE CELL CYCLE: Gene-Enzyme Interactions, 1969

A. M. Zimmerman (editor). HIGH PRESSURE EFFECTS ON CELLULAR PROCESSES, 1970

I. L. Cameron and J. D. Thrasher (editors). CELLULAR AND MOLECULAR RENEWAL IN THE MAMMALIAN BODY, 1971

I. L. Cameron, G. M. Padilla, and A. M. Zimmerman (editors). DEVELOPMENTAL ASPECTS OF THE CELL CYCLE, 1971

P. F. Smith. The BIOLOGY OF MYCOPLASMAS, 1971

Gary L. Whitson (editor). CONCEPTS IN RADIATION CELL BIOLOGY, 1972

Donald L. Hill. THE BIOCHEMISTRY AND PHYSIOLOGY OF *TETRAHYMENA*, 1972

Kwang W. Jeon (editor). THE BIOLOGY OF AMOEBA, 1973

Dean F. Martin and George M. Padilla (editors). MARINE PHARMACOGNOSY: Action of Marine Biotoxins at the Cellular Level, 1973

Joseph A. Erwin (editor). LIPIDS AND BIOMEMBRANES OF EUKARYOTIC MICROORGANISMS, 1973

A. M. Zimmerman, G. M. Padilla, and I. L. Cameron (editors). DRUGS AND THE CELL CYCLE, 1973

Stuart Coward (editor). DEVELOPMENTAL REGULATION: Aspects of Cell Differentiation, 1973

I. L. Cameron and J. R. Jeter, Jr. (editors). ACIDIC PROTEINS OF THE NUCLEUS, 1974

In preparation

Govindjee (editor). BIOENERGETICS OF PHOTOSYNTHESIS